PROGRESS IN HPLC
Volume 4

Progress in HPLC

Series Editors: H. Parvez and S. Parvez

Other Titles

Volume 1. Gel permeation and ion-exchange chromatography
Volume 2. Electrochemical detection in medicine and chemistry
Volume 3. Flow through radioactivity detection in HPLC

PROGRESS IN HPLC
Volume 4

Supercritical fluid chromatography and micro-HPLC

Editors
M. Yoshioka, S. Parvez, T. Miyazaki
and H. Parvez

Utrecht, The Netherlands
1989

VSP BV (formerly VNU Science Press BV)
P.O. Box 346
3700 AH Zeist
The Netherlands

© 1989 VSP BV

First published in 1989

CIP-DATA KONINKLIJKE BIBLIOTHEEK, DEN HAAG

Supercritical fluid chromatography and micro-HPLC / eds.
M. Yoshioka . . . [et al.]. — Utrecht : VSP —
Ill. — (Progress in HPLC ; vol. 4)
ISBN 90-6764-113-8 geb.
SISO 544 UDC 543.544
Subject heading: chromatography.

Printed in Great Britain by J.W. Arrowsmith Ltd., Bristol

Preface

This book is the fourth in the highly successful series Progress in HPLC and is a result of sustained research interest in the rapidly growing field of HPLC.

The contributions are divided into two sections dealing with (1) super-critical fluid chromatography (SFC) and (2) micro-HPLC. SFC is an ideal separation technique and its potential is enormous in analytical chemistry. This book describes the basic principles of SFC and the methodology using newly-invented equipment in the analysis of natural products and synthetic polymers. Specific micro-HPLC methods are described for measurements of adenine nucleotides, catecholamines, peptides and proteins with aqueous eluents.

This book, a collection of reviews by acknowledged international experts, should provide an excellent reference source for all workers in the field of HPLC and further stimulate interest in SFC and micro-HPLC.

Contents

Preface   v

SUPERCRITICAL FLUID CHROMATOGRAPHY

Chromatography with sub- and supercritical eluents: dependence of
chromatographic parameters on temperature, pressure, and eluent
composition
F.P. SCHMITZ, D. LEYENDECKER, D. LEYENDECKER
and E. KLESPER   3
    Introduction   3
    Materials and methods   4
    Results and discussion   6
    Conclusions   20
    Summary   21
    References   21

Development of an intelligent cascade pump which can perform
microdelivery independent of compressibility of fluid
T. SAITO and M. TAKEUCHI   25
    Introduction   25
    Problems with microdelivery of plunger pump   26
    Principle of the new pump system   29
    Materials and methods   32
    Experimental results   35
    Design of equipment   40
    Attained performance   42
    Discussion   49
    Conclusion   50
    References   51

Current use of packed columns in SFC
T. GREIBROKK, J. DOEHL and E. LUNDANES   53
    Introduction   53
    Materials and methods   54
    Column packing materials   56
    Modifiers   62
    Packed fused silica columns   67
    References   70

Oligomer separation by supercritical fluid chromatography using
gradient elution
F.P. SCHMITZ and B. GEMMEL   73
    Introduction   73
    Materials and Methods   74
    Results and discussion   76
    Conclusion   82
    Summary   83
    References   84

Microscale supercritical fluid extraction and coupling of microscale
supercritical fluid extraction with supercritical fluid chromatography
M. SAITO, T. HONDO, M. SENDA and K. SUGIYAMA   87
    Introduction   87
    Microscale supercritical fluid extraction   88
    Directly coupled microscale supercritical fluid extraction-
        supercritical fluid chromatography   99
    Conclusion   107
    References   107

Supercritical fluid chromatography—mass spectrometry (SFC—MS)
R.D. SMITH, B.W. WRIGHT and H.T. KALINOSKI   111
    Introduction   111
    Properties of supercritical fluids   112
    Supercritical fluid chromatography   115
    Direct fluid injection (DFI) mass spectrometry   122
    Applications   139
    Summary   149
    References   151

Computer-assisted separation system in supercritical fluid
chromatography
K. JINNO   157
    Introduction   157
    Construction of the retention prediction system   159
    Identification of PAHs in diesel particulate matter extract   165
    Conclusion   177
    References   177

MICRO-HPLC

Fluorescent-HPLC for adenine nucleosides and nucleotides in life
science
M. YOSHIOKA, K. YAMADA, M.M. ABU-ZEID, H. FUJIMORI,
A. FUKE, K. HIRAI, A. GOTO, M. ISHII, T. SUGIMOTO
and H. PARVEZ   181
    Introduction   181
    History of LC and HPLC for nucleic acid compounds   182

Methods 186
Results 190
Discussion 202
Conclusion 205

Microbore HPLC for biological samples: catecholamines, peptides
and proteins
K. KOJIMA, H. PARVEZ, S. PARVEZ and T. NAGATSU 211
Introduction 211
Analysis of catecholamines 211
Analysis and purification of peptides and proteins 221
An example of the experiment using microbore HPLC 224
Conclusion 226
References 226

Review: High-performance liquid chromatography of metabolites of
catecholamines and serotonin in urine, plasma, cerebrospinal fluid
and brain tissue. I. Analytical methodology
A. YOSHIDA, M. YOSHIOKA and H. PARVEZ 229
Abbreviations 229
Introduction 229
Treatment of the physiological fluids and tissue prior to
     chromatographic separation 231
Detection methods 233
HPLC methods of analysis for the metabolities of CA and
     5-HT 234
Summary 254
References 254

Solvent elimination approach for microcolumn liquid chroma-
tography-infrared spectroscopy:
K. JINNO and C. FUJIMOTO 273
Introduction 273
Interfacing device 274
Reversed-phase LC-FTIR interfacing 277
Normal-phase LC-FTIR-microprobe laser Raman
     spectroscopy 281
SFC—FTIR interfacing 285
Conclusion 294
References 295

Author index 297

Subject index 299

SUPERCRITICAL FLUID CHROMATOGRAPHY

Progress in HPLC, Vol. 4, pp. 3—23.
Yoshioka *et al.* (Eds)
© 1989 VSP.

Chromatography with sub- and supercritical eluents: dependence of chromatographic parameters on temperature, pressure, and eluent composition

FRANZ P. SCHMITZ, DIETGER LEYENDECKER, DAGMAR
LEYENDECKER, and ERNST KLESPER

Lehrstuhl für Makromolekulare Chemie, RWTH Aachen, Worringerweg,
D-5100 Aachen, Federal Republic of Germany

INTRODUCTION

Recently, supercritical fluid chromatography (SFC) has gained increasing
interest as a separation method complementary to gas and liquid chromatography. This interest in SFC arises partly from the fact that, with this type of
chromatography, a variety of parameters can be altered to influence the
separation. As an example, by increasing the pressure, the density of the
mobile phase also increases, which results in enhanced dissolution ability of
the mobile phase. In addition, the influences of temperature and mobile
phase composition are of primary importance.

In gas chromatography, using carrier gases such as helium, hydrogen or
nitrogen, molecular interactions in the mobile phase are found to be very
small, and this manifests itself in the low boiling temperatures of these
gases. Even when compressed to high densities the dissolution ability of
such carrier gases cannot be increased substantially (McLaren *et al.*, 1968;
Giddings *et al.*, 1968, 1969). Contrary to this, molecular interaction is
observed even at relatively low densities with phases having higher boiling
points. As an example, this has been made use of in vapor gas chromatography, applying vapors of steam, ammonia, amines, formic acid, alcohols,
alkanes and other compounds (Rudenko *et al.*, 1975a, b; Parcher, 1983).
The occurrence of molecular interactions between mobile phase and solute
(and, additionally, between mobile and stationary phase) also implies that the
elution properties of the mobile phase can be altered by admixing a second
(and third) component, as it is commonly applied in liquid chromatography.

In this contribution we will report on the influence of pressure, temperature, and mobile phase composition on the properties of eluents at sub- and
supercritical conditions.

MATERIALS AND METHODS

Two types of instrumentation have been used in our studies. One type, which was obtained by modification of a commercial HPLC instrument (1084 B from Hewlett-Packard), was capable of pumping two eluents at variable proportions and at a constant overall flow rate, and was therefore used for oligomer separations by means of gradient elution; this type of instrument will be described in detail in a separate chapter (F. P. Schmitz and B. Gemmel, this book). The other type was assembled from individual, commercially available modules. This apparatus was only used for isocratic experiments; it is shown schematically in Fig. 1. The mobile phase is pressurized by helium and supplied from a pressure-resistant steel storage container (1). A cascade of four stainless-steel frits, mounted between the container and the pump, provides filtration. The membrane pump (2; MF 65, Orlita, Gießen, FRG) is equipped with an integrated pump head cooling device. The flow pulsations are equalized by a membrane type dampener (3; PDM 3.350, Orlita). Column inlet pressure is measured by a manometer. The sample is introduced by a loop injector (4; Rheodyne 7125). In a forced-air circulation oven (5; UT 5042 EK, Heraeus, Hanau, FRG) the separation column (6) can be thermostatted up to 300°C. Column end pressure is controlled by a metering valve (7). After having left the oven, the mobile phase is allowed to cool to ambient temperature and thus passes, as a liquid, the variable wavelength UV detector (8; LC 75, Perkin-Elmer,

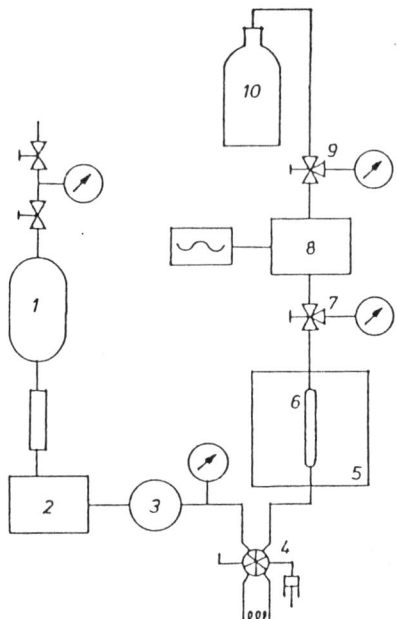

Figure 1. SFC instrument assembled from commercial HPLC parts (schematic). For explanation see text.

Uberlingen, FRG) which is connected to a pen recorder (L 6510, Linseis, Selb, FRG). After having passed another metering valve (9) for controlling the pressure in the detector, the eluate is collected (10). The volume flow rate of the pump was checked at this point and kept for all chromatographic runs at 1 ml/min.

The stainless-steel separation columns (length 25 cm, internal diameter 4.6 mm) were packed with the stationary phase (LiChrosorb Si 60 or Si 100, 10 μm from Merck, Darmstadt, FRG) using a slurry technique (Schmitz *et al.*, 1985). The slurry consisted of the stationary phase (2.5 g) in a mixture of toluene (20 ml) and cyclohexanol (30 ml), which was packed by pumping methanol onto the slurry in a reservoir at the top of the column. Modification of the silica stationary phase was performed as described previously (Schmitz *et al.*, 1987a). At a slow flow rate, dioxane was pumped through the column for 24 h at a temperature of 250°C. Then the column was flushed with the mobile phase under investigation until a constant solute retention was obtained.

Liquid eluents were purified by distillation from sodium. Low-boiling eluents were purchased in sufficient purity as follows: CO_2, 99.995% (Messer-Griesheim, Hürth, FRG); N_2O, 99.5% (Linde, Höllriegelskreuth, FRG); dimethylether, 99% (Merck). The critical data for these eluents are shown in Table 1. The components of the test mixture, naphthalene, anthracene, pyrene and chrysene were purified by repeated crystallization.

Table 1.
Critical temperatures and pressures (according to Reid *et al.*, 1977)

	$T_c/°C$	p_c/bar
Pentane	196.5	33.7
Diethyl ether	193.6	36.4
Dimethyl ether	126.9	53.7
CO_2	31.1	73.8
N_2O	36.5	72.4
1,4-Dioxane	313.9	52.1
Diglyme	356[a]	28[a]

[a] Calculated (cf. Leyendecker *et al.*, 1987c).

Dead times (hold-up times), t_0, were determined by injection of heptane which can serve as an inert solute under the conditions applied here. Capacity ratios, k', selectivities, α, plate numbers, N, and effective plate numbers, N_{eff}, were calculated from retention times, t_r, dead-times, t_0, and peak widths at half height, w', in the usual manner:

$$k_i' = \frac{t_{r,i} - t_0}{t_0} = \frac{t_{r,i}'}{t_0} \qquad (1)$$

$$\alpha_{ij} = \frac{k'_j}{k'_i} \tag{2}$$

$$N_i = 16\left(\frac{t_{r,i}}{w_i}\right)^2 = 8\left(\frac{t_{r,i}}{w'_i}\right)^2 \ln 2 \tag{3}$$

$$N_{eff,i} = 16\left(\frac{t'_{r,i}}{w_i}\right)^2 = 8\left(\frac{t'_{r,i}}{w'_i}\right)^2 \ln 2 \tag{4}$$

Resolution was calculated using eqn (5):

$$R_{ij} = \frac{f_{ij}}{g_{ij}} + \frac{d_{ij}}{w'_i + w'_j}\sqrt{\ln 4} \tag{5}$$

In this equation the f/g ratio is the ratio of average peak height to the depth of the valley, as measured from that peak height, and d_{ij} is the baseline intercept between the inner tangents to two peaks (Schmitz, 1986).

RESULTS AND DISCUSSION

Within the first ten years since the invention of supercritical fluid chromatography by Klesper, Corwin and Turner (1962), some studies were reported on the pressure and temperature dependence of retention at supercritical conditions (Sie et al., 1966; Sie and Rijnders, 1967a–c; Novotny et al., 1971). The dependences were usually found to be nonlinear, which was further evidenced by additional studies (Fuzita et al., 1975; van Wasen and Schneider, 1975; Nieman and Rogers, 1975; van Wasen et al., 1980). When retention data were plotted versus temperature at constant pressure, maxima were observed (Schmitz et al., 1984a, b; Leyendecker et al., 1984, 1985, 1986b, c; Bickmann and Wenclawiak, 1985; Yonker et al., 1985a; Chester and Innis, 1985), as exemplified in Fig. 2. Here the capacity ratios, k', of four aromatic solutes are plotted versus temperature for a pentane mobile phase at a constant column outlet pressure, p_e, of 50 bar. The k' are seen to decrease with temperature, as long as the mobile phase is in the liquid state. At temperatures above critical, the k' first increase and then decrease again, after having passed their maxima. The same applies to other mobile phases, as shown in Figs 3 and 4 for CO_2 and N_2O. This dependence of k' on temperature has been correlated by Yonker et al. to the enthalpy of transfer between the mobile and stationary phase for a given solute, to the heat capacity for that process, and to the variation of the mobile phase density with temperature at constant pressure (Yonker et al., 1985a). The theoretical concept was shown to be in reasonable agreement with experimental observations. At lower pressures, the dependence of k' on temperature is more pronounced, as evidenced by comparing Fig. 5 to Fig. 2.

The dependence of k' on both temperature and pressure can be seen from three-dimensional network plots (Leyendecker et al., 1986b), as shown in Fig. 6 for the capacity ratio of chrysene, k'_c. The height of the k' elevation in the sub- and supercritical gaseous region is seen to decrease with increasing pressure, and simultaneously the position of the maximum is shifted towards

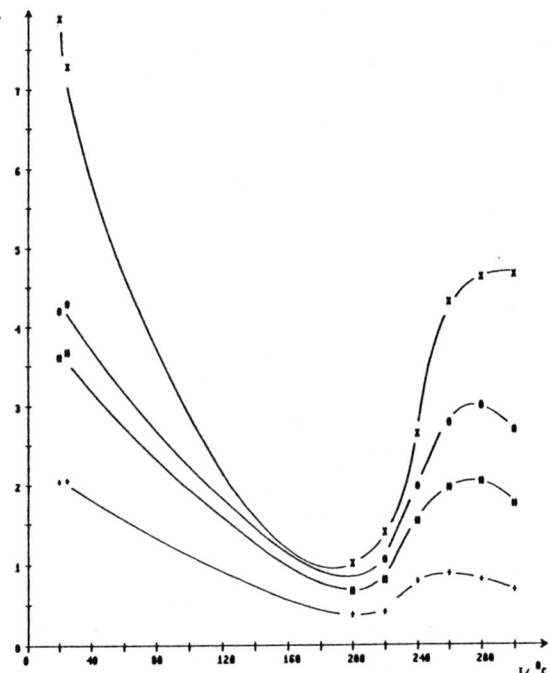

Figure 2. Variation of the capacity ratios of four aromatic test compounds with temperature. Stationary phase: LiChrosorb Si 100, 10 μm, dioxane-modified; mobile phase: pentane; column outlet pressure, p_e: 50 bar; symbols: (+), naphthalene (#), anthracene, (O), pyrene, (×), chrysene.

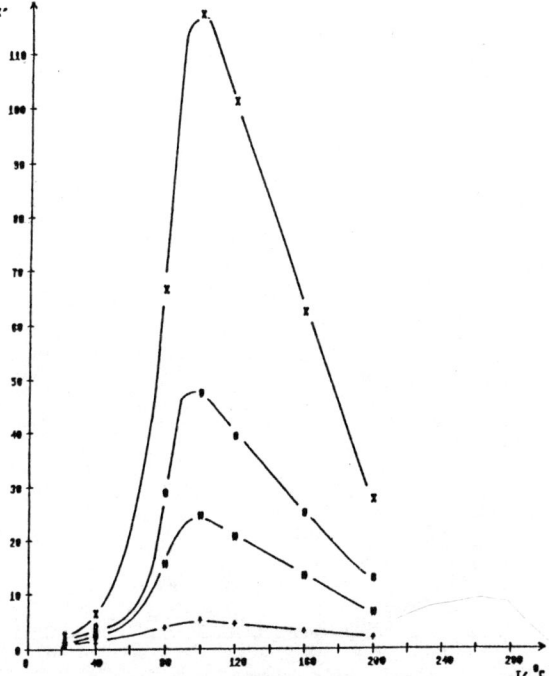

Figure 3. Variation of the capacity ratios of four aromatic test compounds with temperature. Mobile phase: CO_2; p_e: 150 bar; stationary phase and symbols as in Fig. 2.

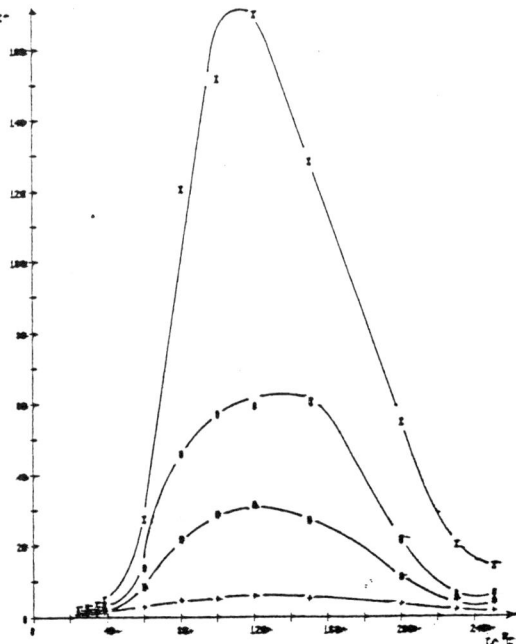

Figure 4. Variation of the capacity ratios of four aromatic test compounds with temperature. Stationary phase: LiChrosorb Si 100, 10 μm; mobile phase: N_2O; p_e: 120 bar; symbols as in Fig. 2.

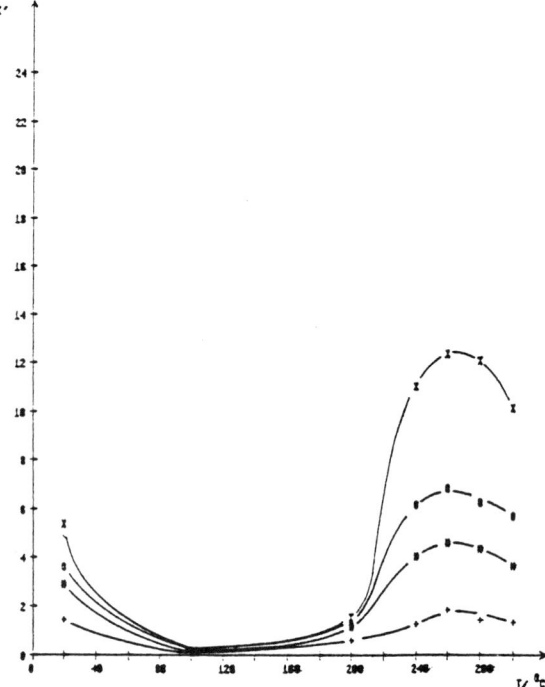

Figure 5. Variation of the capacity ratios of four aromatic test compounds with temperature. Mobile phase: pentane; p_e: 36 bar; stationary phase and symbols as in Fig. 2.

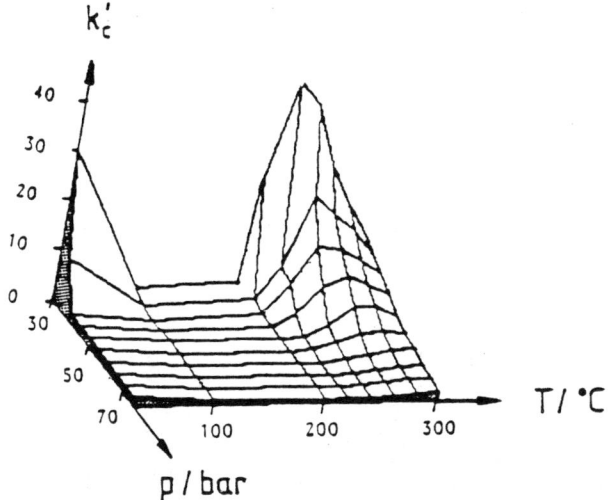

Figure 6. Three-dimensional network plot showing the variation of the capacity ratio for chrysene, k'_c, with temperature and pressure. Stationary phase: LiChrosorb Si 100, 10 μm; mobile phase: pentane.

higher temperatures. A second elevation of k' is observed at low temperatures, i.e. at liquid conditions. When the column outlet pressure is decreased below the limit of 30 bar in Fig. 6, the capacity ratios decrease again at liquid conditions around ambient temperature (Fig. 7a). Thus, similar k' values were obtained at column outlet pressures of 6 and 70 bar. At higher temperatures, in the region of T_c, a steady increase of k' with decreasing pressure is observed (Fig. 7b). At temperatures above T_c, k' maxima were found to occur at pressures somewhat below p_c (Semonian and Rogers, 1978). A thermodynamic description of the dependence of k' on pressure at constant temperature relates changes of k' to the difference of the partial molar volume of the solute in the mobile and in the stationary phase, and to changes of density with pressure (van Wasen and Schneider, 1975; van Wasen *et al.*, 1980; Yonker *et al.*, 1985b). Although the numerical values of k' differ with the type of the stationary phase, the general changes of retention with temperature and pressure do not depend on the type of the stationary phase. This is not only predicted by the thermodynamic models but is also found experimentally by different research groups using different stationary phases (cf. Leyendecker *et al.*, 1986b).

Comparing the results obtained with different mobile phases one finds that in a homologous series of eluents, at equal reduced pressures, $p_r = p/p_c$, the lower members of the series lead to stronger dependence of retention on temperature. This has been demonstrated for alkane eluents (Leyendecker *et al.*, 1985, 1986c), and is evidenced again by Fig. 8 showing the variation of the chrysene capacity ratio with temperature for the eluents diethyl ether and dimethyl ether. This is in agreement with the results of extraction experiments in the supercritical state, where higher extraction

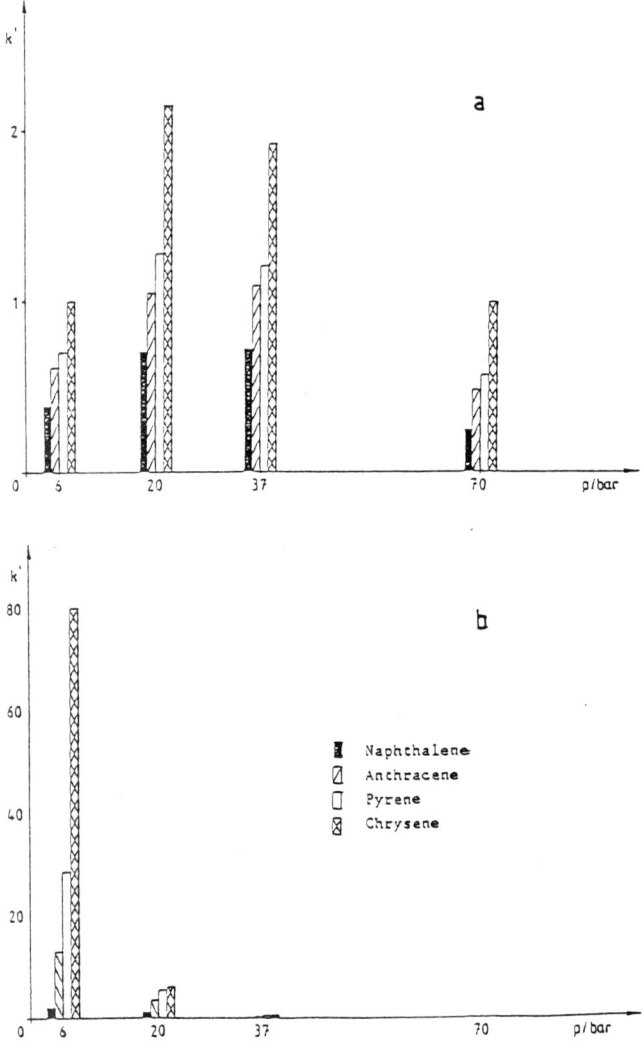

Figure 7. Variation of the capacity ratios of four aromatic test compounds with pressure at two different temperatures of 50°C (**a**) and 192°C (**b**). Stationary phase: LiChrosorb Si 60, 10 μm; mobile phase: pentane.

yields were obtained using eluents with higher T_c. A comparison of eluents having similar critical temperatures, but being of different chemical nature, is reported elsewhere (Leyendecker *et al.*, 1987a).

Similarly to retention with singulary mobile phases, binary eluents show maxima when k' is plotted versus temperature at constant pressure. Owing to the increase of the critical temperature of the mixture, the position of the maximum is shifted towards higher temperatures when a cofluid of higher boiling temperature is added, as shown in Figs 9 and 10 for the addition of 5% (Fig. 9) and 10% (Fig. 10) diethyleneglycol dimethyl ether (diglyme) to

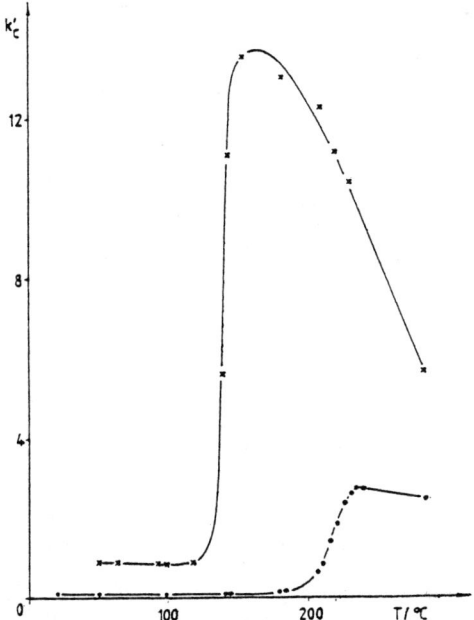

Figure 8. Variation of k'_c with temperature at a reduced pressure of $p_r = 1.06$ with the mobile phases dimethyl ether ($\times$) and diethyl ether ($\bullet$). Stationary phase: LiChrosorb Si 60, 10 µm; $p_e = 38$ bar for diethyl ether and $p_e = 56$ bar for dimethyl ether.

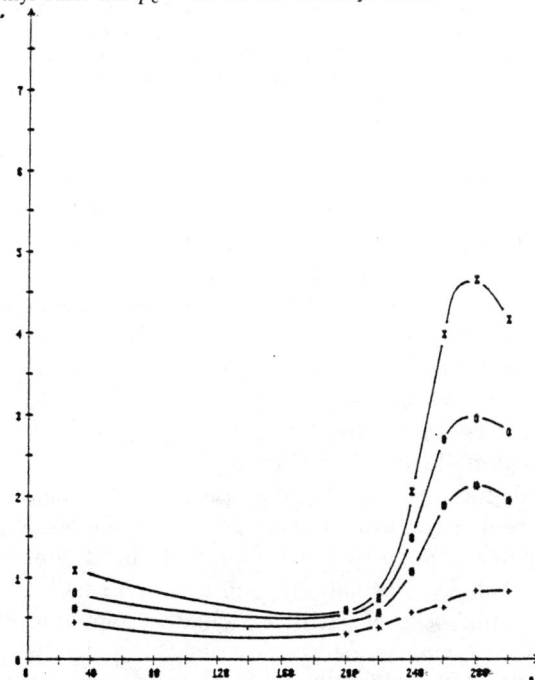

Figure 9. Variation of the capacity ratios of four aromatic test compounds with temperature. Mobile phase: pentane−diglyme, 95:5 (vol/vol); stationary phase, column outlet pressure, and symbols as in Fig. 2.

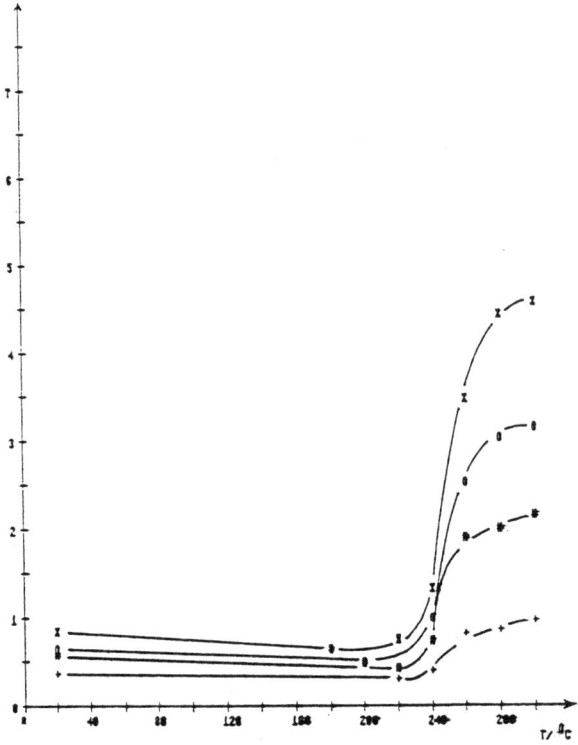

Figure 10. Variation of the capacity ratios of four aromatic test compounds with temperature. Mobile phase: pentane–diglyme, 90:10 (vol/vol); stationary phase, column outlet pressure, and symbols as in Fig. 2.

pentane. The same applies to mixtures of pentane and dioxane (Leyendecker *et al.*, 1987b). By comparison of Figs 9 and 10 to Fig. 2 it is seen that due to the higher solubility of the aromatic test compounds in diglyme, retention is distinctly reduced as long as the mobile phase is in the liquid state. At higher temperatures k'-values can increase intermediately. This is exemplified in Fig. 11 showing the variation of k' for pyrene with the dioxane content at a constant pressure of $p_e = 36$ bar and three different temperatures. This retention behavior can be discussed in terms of preferential absorption of the cofluid at the stationary phase (Leyendecker *et al.*, 1987b).

To be distinguished from the binary mobile phases mentioned here, where the cofluid influences the chromatographic process mainly by altering the mobile phase properties, are eluents where the cofluid affects the chromatographic parameters mainly by interaction with the stationary phase. For the latter case the cofluid is usually applied in much smaller amounts than for the first case. Influences of such low 'modifier' amounts on retention and peak shape have been frequently described, e.g. by Board *et al.* (1983), Gere (1983), Randall (1983, 1984), Blilie and Greibrokk (1985), Levy and Ritchey (1985, 1986), Wright and Smith (1986) as well as by Yonker and Smith (1986). With few exceptions these binary mobile phases have been used with packed columns, where the peak symmetry was often considerably

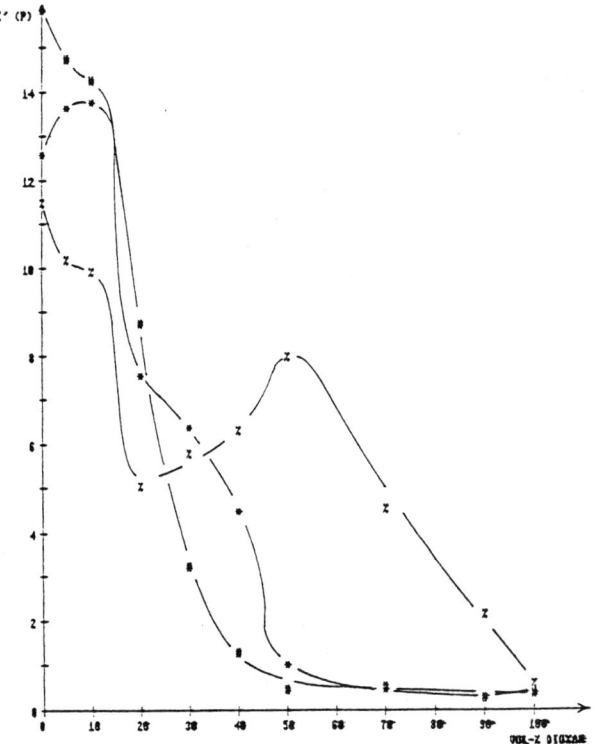

Figure 11. Variation of the capacity ratio of pyrene with the content of dioxane in pentane: p_c: 36 bar; temperatures: (#), 240°C; (*), 260°C; (%) 300°C. Stationary phase as in Fig. 2.

improved upon addition of a modifier. Typical modifier concentrations are in the range of 1%.

Whereas the modification of the stationary phase mentioned above was assumed to be of a reversible type, permanent modification was also reported. Hirata (1984) postulated the formation of ethyl silyl ether linkages upon treatment of silica stationary phases with supercritical binary hexane—ethanol eluents. In our work with supercritical binary eluents containing dioxane we observed, similar to the findings of Hirata, a permanent modification of the silica stationary phase (Schmitz *et al.*, 1987a). Thus, by treating silica with dioxane at elevated temperatures (see Materials and Methods), a modified stationary phase is obtained which is especially useful for separations with a carbon dioxide mobile phase: using unmodified silica, peaks are broad and highly dissymmetric, but peak widths and peak shapes are distinctly improved on the modified stationary phase. Furthermore, retention times are smaller with the dioxane-modified silica for all mobile phases studied, compared to the unmodified material.

Turning now to the dependences of selectivity α, Fig. 12 shows a very distinct maximum for the selectivity between the homologous solutes naphthalene and anthracene, using nitrous oxide as the eluent at a column outlet pressure of 120 bar. For the non-homologous pair anthracene and pyrene

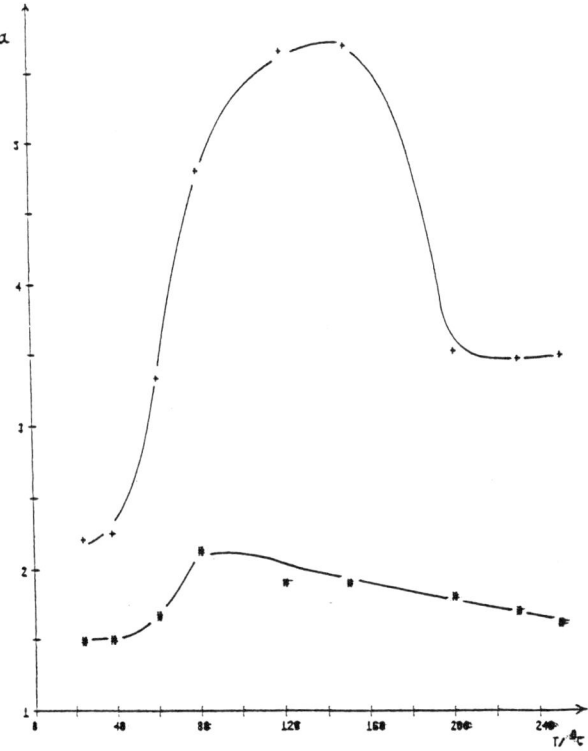

Figure 12. Variation of the selectivity with temperature for two peak pairs: (+), selectivity between naphthalene and anthracene; (#), selectivity between anthracene and pyrene; stationary phase: LiChrosorb Si 100, 10 μm; mobile phase: N_2O; p_e: 120 bar.

the maximum is only small, and the α values are found to be generally lower than for the homologous pair. The selectivity for the pair pyrene−chrysene and its dependence on temperature closely resembles that of anthracene− pyrene, whereas α for anthracene−chrysene behaves similarly to that for naphthalene−anthracene. Thus in this particular case, the more homologous a solute pair the higher is the selectivity, and the stronger is its dependence on temperature. The α maxima are very sensitive to pressure, as a comparison of Figs 12 and 13 shows. Upon roughly doubling the pressure, the maxima disappear almost completely. At $p_e = 250$ bar there is nearly a steady increase of α for naphthalene−anthracene, whereas the selectivity for anthracene−pyrene tends to decrease continuously. Variations of α with temperature and eluent composition are shown in Fig. 14 for pentane− dioxane mobile phases. Again the temperature effects for the homologous peak pair NA exceed those for the AP pair, and so do the concentration effects: the differences between α(NA) and α(AP) increase with increasing dioxane content.

As a third chromatographic parameter, plate numbers and their dependence on the physical state of the mobile phase were studied. As for k' and

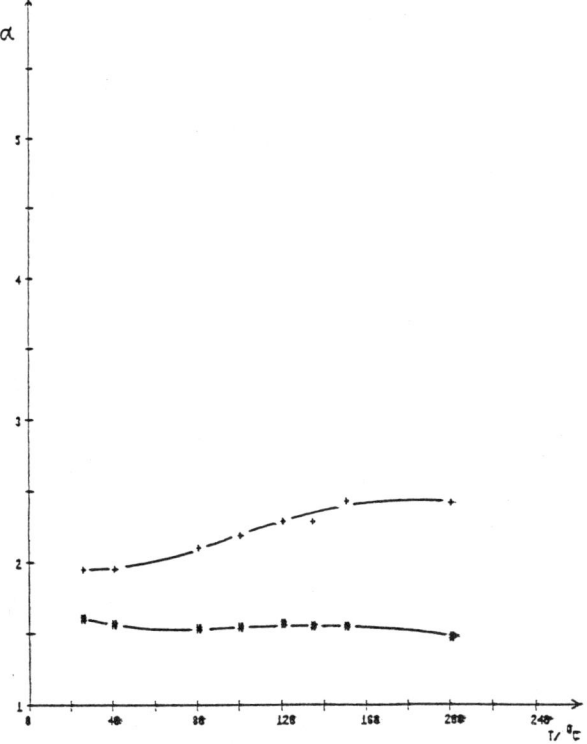

Figure 13. Variation of the selectivity with temperature for the two peak pairs of Fig. 12; p_e: 250 bar; other conditions as in Fig. 12.

for α, maxima were also found to occur for plate numbers, N (Leyendecker *et al.*, 1987b−d). Shape and position of the N maxima resemble those for k', although three-dimensional plots of N_{eff} versus temperature and pressure in a pentane−dioxane mobile phase show a clefted surface rather than the smooth one observed for k' (Fig. 15). Furthermore, the maxima are rather sharp, i.e. the decrease of N towards both higher and lower temperatures is steeper than for k'.

Frequently, the quality of a chromatographic separation is expressed in terms of resolution, R (cf. eqn (5)), which contains information from retention, selectivity and plate number. For closely neighbouring peaks, R can be related to k', α, and N by eqn (6):

$$R = \frac{1}{4} \frac{k'_j}{1 + k'_j} \frac{\alpha - 1}{\alpha} \sqrt{N_j} \tag{6}$$

The dependence of the average resolution R_m, averaged over the three neighbouring peak pairs of the aromatic test mixture is shown in Fig. 16 for carbon dioxide at three different pressures. Like the k' maxima, the height of the R_m maxima decreases with increasing pressure, and the position of

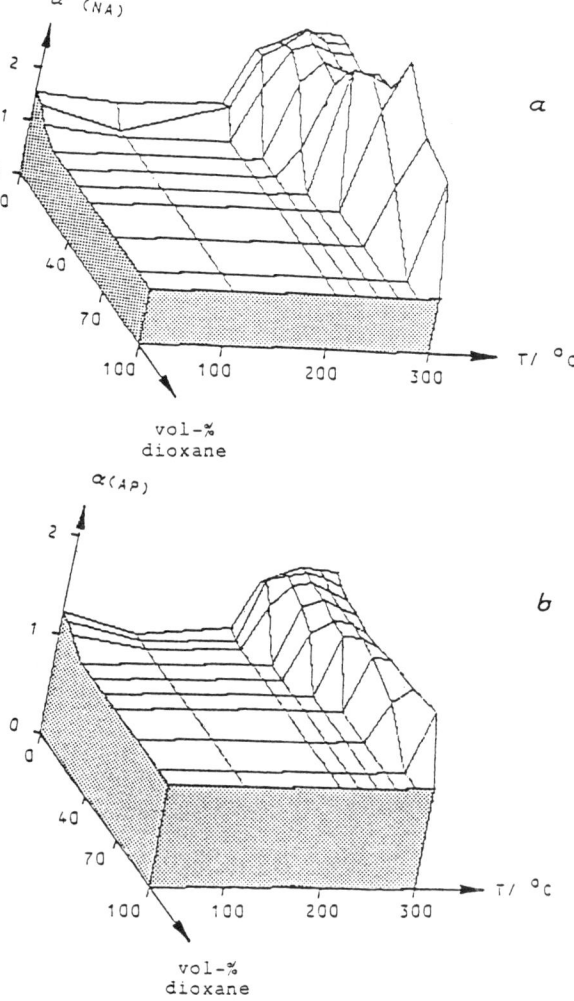

Figure 14. Three-dimensional network plots showing the variation of the selectivity between naphthalene and anthracene (**a**) as well as between anthracene and pyrene (**b**) with temperature and eluent composition (dioxane in pentane, vol/vol). Pressure and stationary phase as in Fig. 12.

the maximum is shifted towards higher temperatures. This can also be derived from the three-dimensional R_m-T-p plot for pentane in Fig. 17. The same behavior also holds true for binary eluent mixtures (Leyendecker *et al.*, 1987b−d). Furthermore, with binary eluents the maxima decrease in height and are shifted towards higher temperatures when the amount of the component with the higher T_c and higher solvating ability in the eluent mixture is raised. This is shown in the three-dimensional R_m-T-x plots of Fig. 18. Comparing the three-dimensional plots of R_m to those of α (Fig. 14)

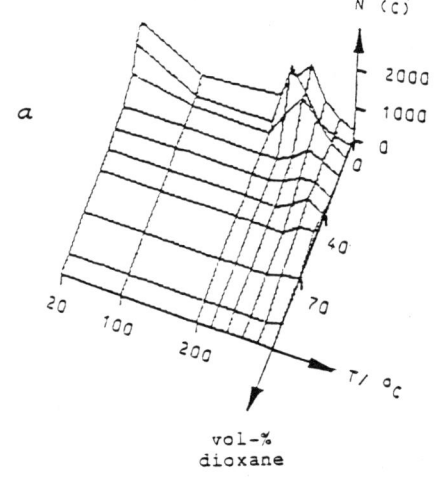

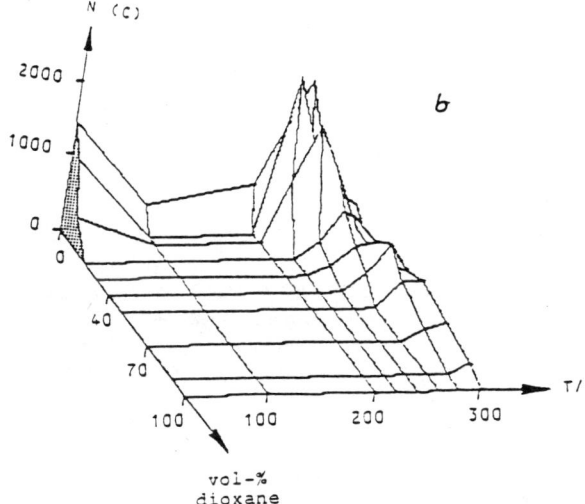

Figure 15. Three-dimensional network plots showing the variation of the effective plate number for chrysene with temperature and eluent composition. Conditions as for Fig. 12. The graph is shown in two perspectives (a, b).

and N_{eff} (Fig. 15), the similarity seems to be most pronounced between R_m and N_{eff}. This corresponds to previous results where an almost linear correlation was found between R_m and N (Leyendecker *et al.*, 1985). The shift of R_m towards higher temperatures with increasing content of the cofluid in the mobile phase suggests that an increase of the temperature during a pressure-programmed or a composition-programmed separation should lead to increased resolution, at least for homologous compounds which are, e.g., encountered with oligomeric samples. Actually this has been confirmed in some preliminary studies (Leyendecker *et al.*, 1986a; Schmitz *et al.*, 1986, 1987b).

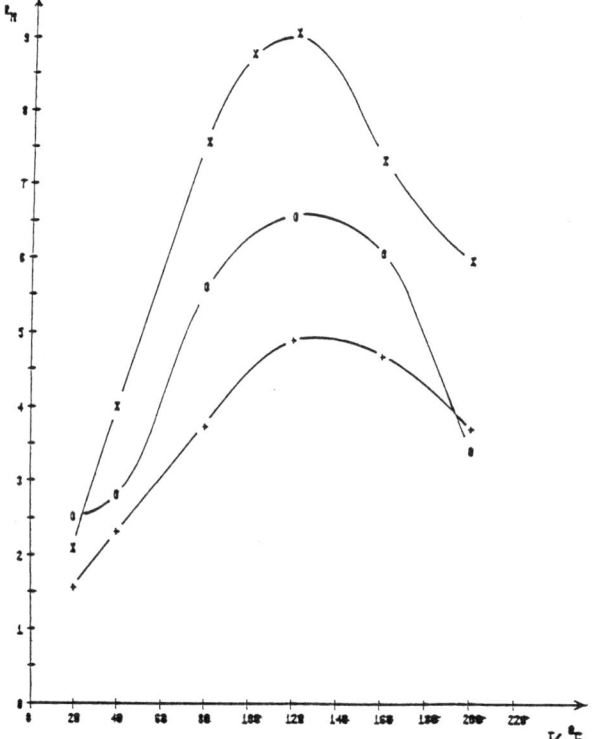

Figure 16. Variation of the average resolution, R_m, between the compounds of the test mixture (naphthalene−anthracene, anthracene−pyrene, pyrene−chrysene) with temperature. Stationary phase: LiChrosorb Si 100, 10 µm, dioxane-modified; mobile phase: CO_2; column outlet pressures, p_e: 150 bar (×), 200 bar (○), 250 bar (+).

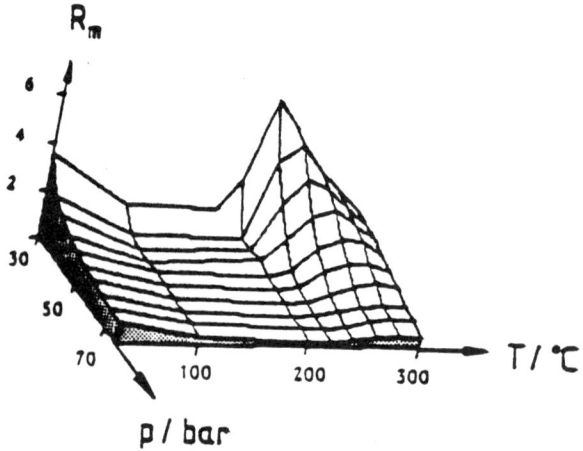

Figure 17. Three-dimensional network plot showing the variation of R_m with temperature and pressure for pentane as the mobile phase. Conditions as for Fig. 6.

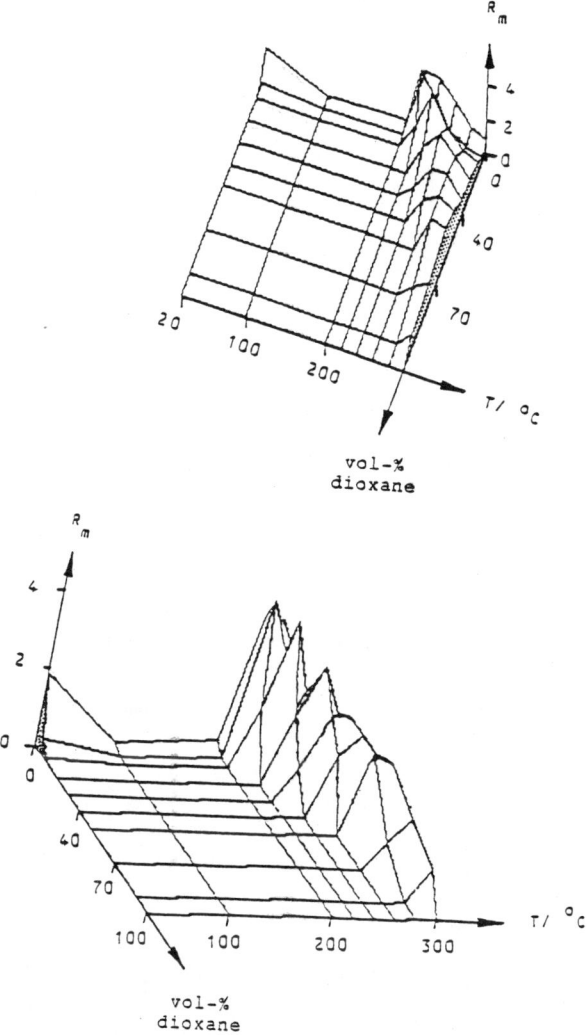

Figure 18. Three-dimensional network plots showing the variation of R_m with temperature and eluent composition (dioxane in pentane, vol/vol). Pressure and stationary phase as for Fig. 12. The graph is shown in two perspectives (a, b).

By extending the studies on R down to low pressures it can be shown that resolution maxima also occur in plots of the average resolution versus pressure. This is demonstrated in Fig. 19a, where the average resolution is plotted as a function of column outlet pressure at four temperatures for pentane as the mobile phase. At low temperatures (with liquid pentane), as well as at temperatures well above T_c, maximum resolution is observed around a pressure of 20 bar, but at a near-critical temperature, resolution decreases with pressure without exhibiting a maximum. At sub- and near-critical temperatures this dependence of resolution corresponds to that of

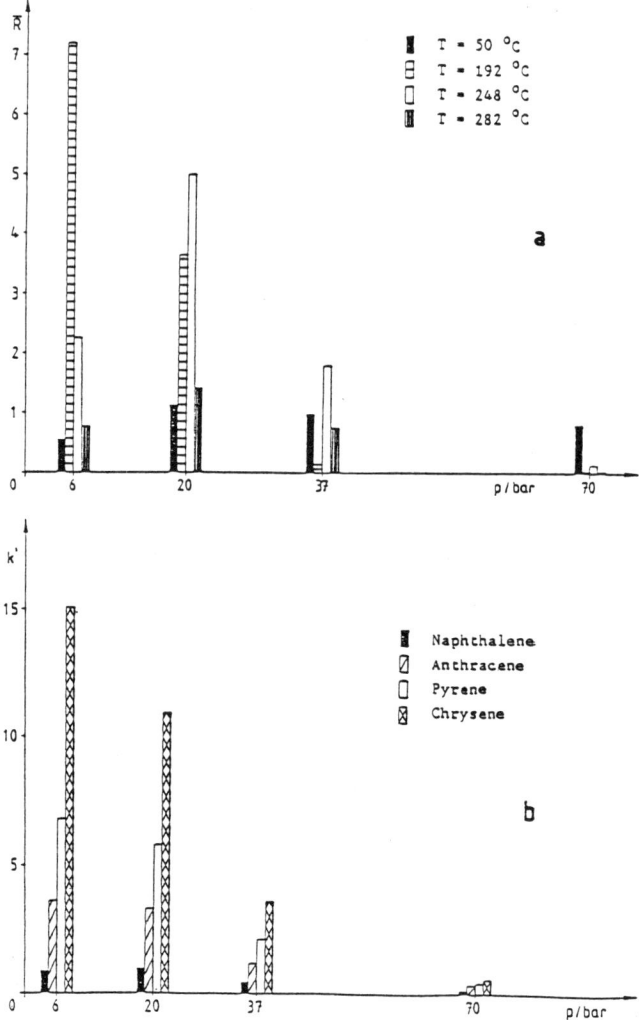

Figure 19. Variation of the average resolution with pressure at four different temperatures (a) and variation of the capacity ratios of the four aromatic test compounds at 248°C (b). Mobile phase: pentane; other conditions as for Fig. 7.

the capacity ratios (cf. Figs. 7 and 19a). At temperatures above T_c, R_m and k' differ in their behavior in as much as a resolution maximum is observed while the capacity ratios only decrease steadily with increasing pressure (Fig. 19b). By comparison of the k' and the R_m plots a pressure may be selected for a given separation problem yielding a needed resolution at a minimum analysis time.

CONCLUSIONS

The results presented here demonstrate the manifold dependences of chromatographic parameters on the physical and chemical conditions of the mobile phase. This is particularly valid for mobile phases in their pressurized gaseous states.

The results can be used for performing separations at optimum conditions, both for non-programmed and for gradient separations. As an example, information is provided for several mobile phases indicating the appropriate temperature ranges where negative temperature programming may be applied successfully, i.e. in what temperature ranges capacity ratios decrease with decreasing temperature. Furthermore, knowledge about the occurrence and the position of the maxima for selectivity, plate numbers and resolution can be of advantage when high-performance chromatograms are needed. Finally, possibilities are opened for multiple gradient separations, which will be demonstrated in another chapter of this book.

SUMMARY

The dependence of retention and separation on temperature, pressure and chemical nature of the mobile phase was studied using different single and binary eluents and a test mixture of four aromatic hydrocarbons. Plots of capacity ratio, k', selectivity, α, effective plate number, N_{eff}, as well as resolution versus temperature, show the occurrence of maxima at temperatures where the eluent is in the sub- or supercritical gaseous state. At certain temperatures, maxima are also observed in plots of resolution versus pressure, even when the capacity ratios decrease monotonously with increasing pressure. These results open possibilities for optimizing chromatographic separations.

ACKNOWLEDGEMENT

Financial support from the Arbeitsgemeinschaft Industrieller Forschungs-vereinigungen (AIF) is gratefully acknowledged.

REFERENCES

Bickmann, F. and Wenclawiak, B. (1985). Überkritische Flüssig-Chromatographie verschiedener Metallchelate von 8-Hydroxychinolin, Diethyldithiocarbamat und Acetylaceton. *Fresenius' Z. Anal. Chem. 320*, 261–264

Blilie, A. L. and Greibrokk, T. (1985). Modifier effects on retention and peak shape in supercritical fluid chromatography. *Anal. Chem. 57*, 2239–2242.

Board, R., McManigill, D., Weaver, H., and Gere D. (1983). The use of modifiers in supercritical fluid chromatography with carbon dioxide. *CHEMSA*, June, 12–24.

Chester, T. L. and Innis, D. P. (1985). Retention in capillary supercritical fluid chromatography. *J. High Res. Chromatogr. Chromatogr. Comm. 8*, 561–566.

Fuzita, K., Shimokobe, I., and Nakazima F. (1975). Supercritical fluid chromatography of polynuclear aromatic hydrocarbon (in Japanese). *Nippon Kagaku Kaishi*, 1348–1351.

Gere D. (1983). Separation of ubiquinones in bacterial cell extracts by supercritical fluid chromatography. Appl. Note 800–2, Publ. No. 43-5953-1691, Hewlett Packard Co., Avondale, PA.

Giddings, J. C., Myers, M. N., McLaren, L., and Keller, R. A. (1968). High pressure gas chromatography of nonvolatile species. *Science 162*, 67–73.

Giddings, J. C., Myers, M. N., and King, J. W. (1969). Dense gas chromatography at pressures to 2000 atmospheres. *J. Chromatogr. Sci. 7*, 276–283.

Hirata, Y. (1984). Comparison of chromatographic behaviour of silica and chemically bonded phases in supercritical fluid chromatography. *J. Chromatogr. 315*, 31–37.

Klesper, E., Corwin, A. H., and Turner, D. A. (1962). High pressure gas chromatography above critical temperatures. *J. Org. Chem. 27*, 700–701.

Levy, J. M. and Ritchey, W. M. (1985). The effects of modifiers in supercritical fluid chromatography. *J. High Res. Chromatogr. Chromatogr. Comm. 8*, 503–509.

Levy, J. M. and Ritchey, W. M. (1986). Investigations of the uses of modifiers in supercritical fluid chromatography. *J. Chromatogr. Sci. 24*, 242–248.

Leyendecker, D., Schmitz, F. P., and Klesper, E. (1984). Chromatography with sub- and supercritical eluents. The influence of temperature, pressure and flow-rate on the behaviour of dimethyl and diethyl ether. *J. Chromatogr. 315*, 19–30.

Leyendecker, D., Schmitz, F. P., Leyendecker, D., and Klesper, E. (1985). Chromatography with sub- and supercritical eluents: The influence of temperature, pressure and flow-rate on the behaviour of lower alkanes. *J. Chromatogr. 321*, 273–286.

Leyendecker, D., Leyendecker, D., Schmitz, F. P., and Klesper, E. (1986a). Multiple gradients in supercritical fluid chromatography (SFC). Ternary gradients by programming eluent composition and temperature at varying pressure. *J. High Res. Chromatogr. Chromatogr. Comm. 9*, 525–527.

Leyendecker, D., Leyendecker, D., Schmitz, F. P., and Klesper, E. (1986b). Isocratic networks in supercritical fluid chromatography. *J. High Res. Chromatogr. Chromatogr. Comm. 9*, 566–571.

Leyendecker, D., Leyendecker, D., Schmitz, F. P., and Klesper, E. (1986c). Chromatographic behaviour of various eluents and eluent mixtures in the liquid and in the supercritical state. *J. Chromatogr. 371*, 93–107.

Leyendecker, D., Leyendecker, D., Schmitz, F. P., and Klesper, E. (1987a). Comparison of eluents in supercritical fluid chromatography. *J. Liq. Chromatogr. 10*, 1917–1947.

Leyendecker, D., Leyendecker, D., Schmitz, F. P., and Klesper, E. (1987b). Influence of temperature, pressure and eluent composition on the behaviour of binary eluents under sub- and supercritical conditions. Pentane and 1,4-dioxane. *J. Chromatogr. 392*, 101–122.

Leyendecker, D., Schmitz, F. P., and Klesper, E. (1987c). Influences of temperature, pressure and eluent composition on the behavior of binary eluents at sub- and supercritical conditions. Pentane and diethyleneglycol dimethyl ether. *Chromatographia 23*, 171–178.

Leyendecker, D., Leyendecker, D., Schmitz, F. P., Lorenschat, B., and Klesper, E. (1987d). Supercritical fluid chromatography using mixtures of carbon dioxide or ethane with 1,4-dioxane as eluents. *J. Chromatogr. 398*, 105–123.

McLaren, L., Myers, M. N., and Giddings, J. C. (1968). Dense gas chromatography of nonvolatile substances of high molecular weight. *Science 159*, 197–199.

Nieman, J. A. and Rogers, L. B. (1975). Supercritical fluid chromatography applied to the characterization of a siloxane-based gas chromatographic stationary phase. *Sep. Sci. 10*, 517–545.

Novotny, M., Bertsch, W., and Zlatkis, A. (1971). Temperature and pressure effects in supercritical-fluid chromatography. *J. Chromatogr. 61*, 17–28.

Parcher, J. F. (1983). A review of vapor phase chromatography: gas chromatography with vapor carrier gases. *J. Chromatogr. Sci. 21*, 346–351.

Randall, L. G. (1983). Choosing a modifier in carbon-dioxide based chromatography: results of a preliminary modifier selectivity survey. Techn. Paper No. 102, Publ. No. 43-5953-1722, Hewlett-Packard Co. Avondale, PA.

Randall, L. G. (1984). Carbon dioxide based supercritical fluid chromatography. In: Ahuja, S. (Ed.), *Ultrahigh Resolution Chromatography. ACS Symp. Ser. 250*, 135–169.

Reid, R. C., Prausnitz, J. M., and Sherwood, T. K. (1977). *The Properties of Gases and Liquids*. McGraw-Hill, New York, 3rd edn.

Rudenko, B. A., Baydarovtseva, M. A., and Kusovkin, V. A. (1975a). Gas chromatography with vapours of low-boiling substances as mobile phases. *J. Chromatogr. 112*, 373–376.

Rudenko, B. A., Baidarovtseva, M. A., and Agaeva, M. A. (1975b). Elution chromatography with vaporous mobile phases (Review). *J. Anal. Chem. USSR 30*, 1003–1027 (*Zh. Anal. Khim. 30*, 1191–1222).

Schmitz, F. P. (1986). Chromatography with sub- and supercritical eluents. Influence of the separation conditions on selectivity, plate number and resolution. *J. Chromatogr. 356*, 261–269.

Schmitz, F. P., Hilgers, H., Leyendecker, D., Lorenschat, B., Setzer, U., and Klesper, E. (1984a). Pressure and temperature dependent chromatographic behavior of *n*-pentane. *J. High Res. Chromatogr. Chromatogr. Comm. 7*, 590–592.

Schmitz, F. P., Leyendecker, D., and Klesper, E. (1984b). Chromatography with mobile phases in the liquid and the supercritical state. *Ber. Bunsenges. Phys. Chem. 88*, 912–915.

Schmitz, F. P., Hilgers, H., Lorenschat, B., and Klesper, E. (1985). Separation of oligomers with UV-absorbing side groups by supercritical fluid chromatography using eluent gradients. *J. Chromatogr. 346*, 69–79.

Schmitz, F. P., Hilgers, H., and Gemmel, B. (1986). High-performance liquid and supercritical fluid chromatographic separation of vinyl oligomers by gradient elution. *J. Chromatogr. 371*, 135–144.

Schmitz, F. P., Leyendecker, D., and Leyendecker, D. (1987a). Permanent 'on column' modification of a silica stationary phase with 1,4-dioxane. *J. Chromatogr. 389*, 245–250.

Schmitz, F. P., Leyendecker, D., Leyendecker, D., and Gemmel, B. (1987b). Possibilities for optimization of oligomer separation with supercritical fluid chromatography. *J. Chromatogr. 395*, 111–123.

Semonian, B. P. and Rogers, L. B. (1978). Unusual gas chromatographic behaviors of naphthalene, pyrene and phenanthrene using pressurized *n*-pentane as a carrier gas. *J. Chromatogr. Sci. 16*, 49–60.

Sie, S. T. and Rijnders, G. W. A. (1967a). High-pressure gas chromatography and chromatography with supercritical fluids. II. Permeability and efficiency of packed columns with high-pressure gases as mobile fluids under conditions of incipient turbulence. *Sep. Sci. 2*, 699–727.

Sie, S. T. and Rijnders, G. W. A. (1967b). High-pressure gas chromatography and chromatography with supercritical fluids. III. Fluid–liquid chromatography. *Sep. Sci. 2*, 729–753.

Sie, S. T. and Rijnders, G. W. A. (1967c). High-pressure gas chromatography and chromatography with supercritical fluids. IV. Fluid–solid chromatography. *Sep. Sci. 2*, 755–777.

Sie, S. T., van Beersum, W., and Rijnders, G. W. A. (1966). High-pressure gas chromatography and chromatography with supercritical fluids. I. The effect of pressure on partition coefficients in gas–liquid chromatography with carbon dioxide as a carrier gas. *Sep. Sci. 1*, 459–490.

van Wasen, U. and Schneider, G. M. (1975). Pressure and density dependence of capacity ratios in supercritical fluid chromatography (SFC) with carbon dioxide as mobile phase. *Chromatographia 8*, 274–276.

van Wasen, U., Swaid, I., and Schneider, G. M. (1980). Physikalisch–chemische Grundlagen und Anwendungen der Fluidchromatographie (SFC). *Angew. Chem. 92*, 585–598; *Angew. Chem., Int. Ed. Engl. 19*, 575–588.

Wright, B. W. and Smith, R. D. (1986). Investigation of polar modifiers in carbon dioxide mobile phases for capillary supercritical fluid chromatography. *J. Chromatogr. 355*, 367–373.

Yonker, C. R. and Smith, R. D. (1986). Study of retention processes in capillary supercritical fluid chromatography with binary fluid mobile phases. *J. Chromatogr. 361*, 25–32.

Yonker, C. R., Wright, B. W., Petersen, R. C., and Smith, R. D. (1985a). Temperature dependence of retention in supercritical fluid chromatography. *J. Phys. Chem. 89*, 5526–5530.

Yonker, C. R., Frye, S. L., Udseth, H. R., Wright, B. W., and Smith, R. D. (1985b). Mechanism of solute retention in supercritical fluid chromatography. *Prepr. Pap. — Am. Chem. Soc., Div. Fuel Chem. 30/3*, 183–188.

Progress in HPLC, Vol. 4, pp. 25—51.
Yoshioka *et al.* (Eds)

Development of an intelligent cascade pump which can perform microdelivery independent of compressibility of fluid

TOSHINORI SAITO and MAKOTO TAKEUCHI

New Project Development Office, JEOL Ltd,
1418 Nakagami, Akishima, Tokyo 196, Japan

INTRODUCTION

Ordinary dual (or single) plunger pumps are usually used in HPLC. When used for eluent delivery of micro or semi-micro column experiments at high pressure, they frequently perform badly due to pulsating flow and degradation of the flow rate. To minimize these unexpected problems certain improvements have been introduced. They are: a feedback from the outlet pressure to the plunger reciprocation cycle to keep the flow rate constant (typical examples are BIP-I HPLC pump (Jasco) and L6000 (Hitachi)); introduction of an active dumping system to suppress pulsation (Patterson, 1982) and simple addition of a pump to back up the inlet pressure (typical example: PLC-20, Eyela, Tokyo Rika Kiki Co. Ltd), because the smaller the difference between the outlet and inlet pressures the smaller is the effect.

However, such pulsating flow and degradation of the flow rate are largely dependent on working pressure, column volume size and properties of the used eluent and its temperature; therefore complete compensation cannot be attained.

We studied quantitatively the causes of the problems and found a relationship with such factors as working pressure, compressibility of the fluid and seal material, dead volume and plunger volume of the pump, and so on.

Then we discovered that in all of the commonly used pumps when the pump plunger is discharging the fluid in the cylinder against the loaded pressure, there always exist two processes of 'compression' and 'delivery of fluid' which are mixed, and the 'compression' process makes the role of 'metering' which is needed in a quantitative delivery pump, uncertain.

We then discovered a way to resolve the problem by dividing the two roles, confining the role of 'compression' to one pump and the role of 'metering delivery' to the other (Saito and Takeuchi, 1986).

PROBLEMS WITH MICRODELIVERY OF PLUNGER PUMP

Waveform analysis of outlet pressure of pump

Figure 1 shows a typical waveform of the outlet pressure of an ordinary reciprocating dual plunger pump. The inner pressure of the pump is zero at the beginning of discharging process and, immediately afterwards it increases along with the increase of pressure until it is equal to the outer pressure. After it exceeds the outer pressure, actual transfer of liquid begins. Consequently, the time between the beginning of the discharge process and the actual beginning of liquid transfer corresponds to the lost volume in a stroke, and the generation of the lost volume results in the outlet pressure drop, which is the cause of the actual pressure ripple, as shown in Fig. 2. Comparing pressure profiles in Figs 1 and 2, the pressure rise after point X in Fig. 2 shows a more gentle slope than that in Fig. 1, because of the lost volume due to the liquid in the outer vessels described below, during the actual discharging process.

The waveform of the outlet pressure after the point X may be expressed in the following equation, with αV_0 being the compression volume of all liquid contained in the pump and outer vessels being connected to the outlet of the pump, including the column, detector cell and so on. The pressure is expressed as

$$P = (P_x - P_e) \times \exp(-t/K\alpha V_0) + KU \tag{1}$$

where U is flow rate (ml/s), P_e is equilibrium pressure at the flow rate U, α is compression coefficient of liquid, V_0 is total liquid volume contained in the pump and outer vessels, K is the ratio of P_e to U, and P_x is the pressure at the point X.

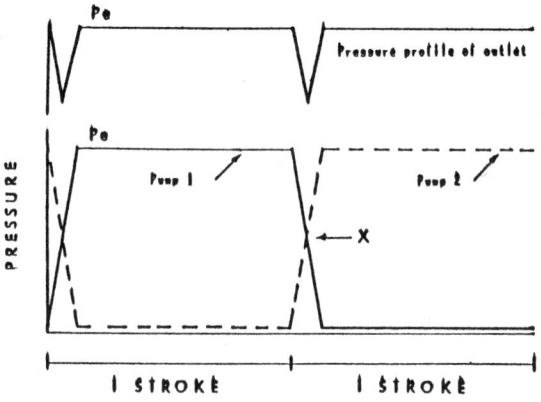

Figure 1. Typical predicted waveform of the outlet pressure of an ordinary reciprocating double-plunger pump.

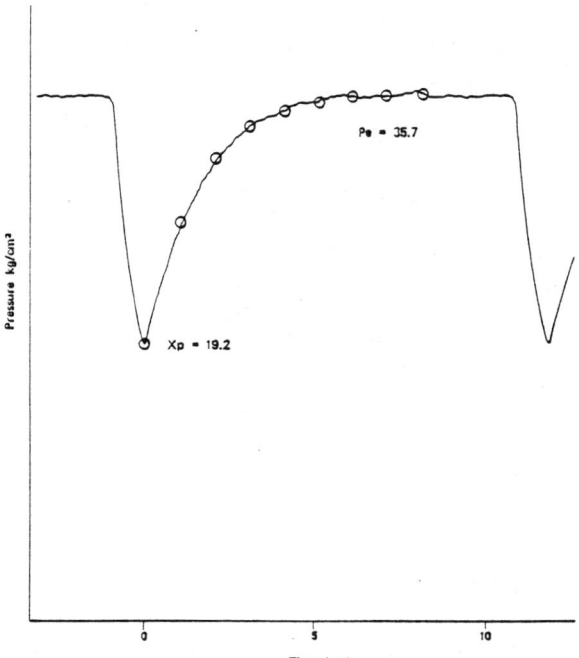

Figure 2. Actual pressure profile of the ordinary plunger pump observed without damping device (full line), and plotting of calculated values (○) by eqn (1), using the following parameters: flow rate $U = 0.4$ ml/min, compression coefficient $\alpha = 0.96 \times 10^{-4}$, ratio of equilibrium pressure P_e to U, $K = 5355$, total liquid volume contained in the pump and the outer vessels $V_0 = 2.89$ ml, time constant $t = 3$ s. The eluent used in ethanol.

The calculation results at several points from eqn (1) are plotted in Fig. 2 (○), which shows very good agreement with the observed result. This means that it takes longer than predicted by Fig. 1 to reach equilibrium pressure. This delay depends largely on the properties of fluid and the volume of the outer vessels.

Consequently, it is preferable to evaluate the pressure ripple not by the amplitude of pressure change, the method generally used, but by the time integral of the pressure change.

The plunger volume and lost volume

The lost volume in the pump comes mainly from two sources: one is the volume construction of liquid corresponding to the dead volume and the plunger volume, which is defined as the volume change of a given amount of plunger movement; the other is that of solid, such as plastic material, used for the seal.

Figure 3 is a general scheme of the pump head. In this figure the lost volume ΔV at pressure P may be described by the following equation:

$$\Delta V = P(\alpha_1 V_d + \alpha_1 V_p + \alpha_2 V_m) \qquad (2)$$

where V_p, V_d and V_m are the plunger volume, dead volume and volume of sealing plastics respectively, and α_1 and α_2 are the compression coefficients of liquid and plastics, respectively.

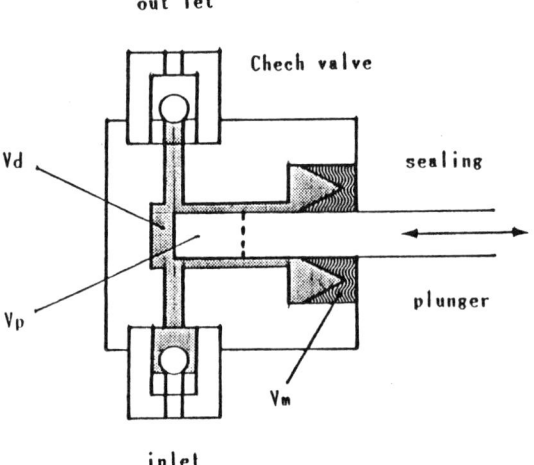

Figure 3. Structural scheme of the pump head; V_p, V_d and V_m are plunger volume, dead volume and volume of sealing material, respectively.

Discrepancy between the actual flow rate and simple predicted value

The lost volume due to compressibility at high pressure depends on the pressure and compression coefficient of the liquid. We assume here the down rate, R, may be defined as follows:

$$R = (V_p - V_e)/V_p \times 100, \tag{3}$$

$$V_e = [V_p - P\alpha(V_p + V_d)](1 + P\alpha) \tag{4}$$

and $\quad R = [P\alpha(V_d/V_p) + (P\alpha)^2(1 + V_d/V_p)] \times 100 \tag{5}$

Here, V_p is the predicted value, V_e is the actual flow volume, V_d is total dead volume and α is the average coefficient of compressibility of liquid and solid.

α is the order of 10^{-4}, so that the second term may be neglected; then we have:

$$R = P\alpha(V_d/V_p) \times 100 \tag{6}$$

As seen from eqn (6), the down rate of flow volume depends on the ratio of dead volume V_d to plunger volume V_p. Figure 4 shows a graphical expression using eqn (6). When the plunger volume is ten times smaller, a ten times larger down rate of the flow volume results; so that if $V_p = 1$ μl, no actual flow is observed at a pressure over 100 kg/cm^2.

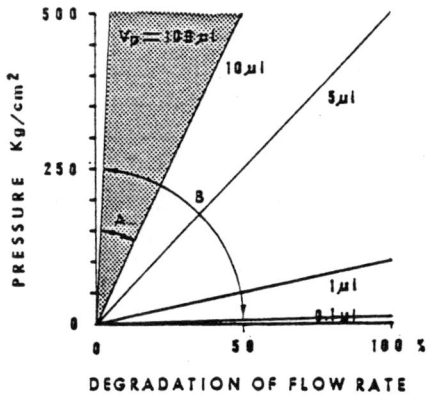

Figure 4. Graphical expression of eqn (6). Degradation of flow rate versus pressure, assuming $V_d = 100 \mu l$, $\alpha = 10^{-4}$

PRINCIPLE OF THE NEW PUMP SYSTEM

Basic concept of the new pump system

From the concept described above, degradations of flow rate are caused by the lost volume due to the compressibility of liquid and solid in the pump. It turns out that the smaller the plunger volume, the larger the effect of pulsating and degradation of the flow rate are.

Consequently it can be said that all problems discussed above can be eliminated only by compensating the lost volume due to the compressibility of all types of material in the pump.

The essential roles required on a pump system suitable for chromatographic experiment can be reduced to two distinct roles. They are:

(1) compressing a given liquid up to the pressure required, and keeping it, and
(2) delivering the compressed liquid continuously at a precise rate.

Simple scheme of the new pump system

The new pump system shown in Fig. 5 is a serial combination of two independent pumps. One of them takes role (1) and the other takes role (2). The first pump, called the 'compression pump' compensates the lost volume mentioned above completely, and the second pump, the 'metering pump', delivers the compressed liquid at a precise rate.

The stroke length of the plunger of the compression pump is designed so as to be variable for compensating the lost volume at any pressure and for liquids with different compressibilities. The stroke length of the plunger of the metering pump is also changeable to restrict the reciprocation time.

At both sides, the inlet and outlet, of the metering pump, the pressure is almost the same; in other words the differential pressure is zero, so that the lost volume in the metering pump is zero. This may be easily understood from Fig. 4. If the pressure is zero in Fig. 4, the lost volume is zero at any plunger volume.

Consequently, the accurate flow rate can be determined from the plunger volume for a unit of time, independent of the column pressure and the property of liquid.

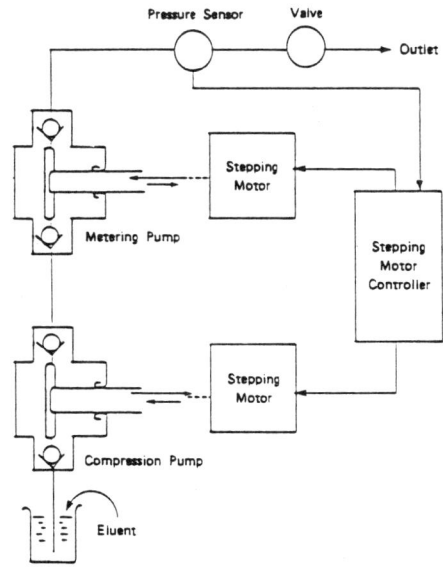

Figure 5. Simple scheme of the cascade pump.

Operation timing of the cascade pump

The principal operation of the compression pump and metering pump may be as follows. The compression pump acts slowly in the suction process and quickly in the discharging process, while the metering pump acts slowly in discharging and quickly in suction, and they operate synchronously. This operation is preferable to avoid the bubble due to cavitation in the compression pump and to give a longer time for measuring the pressure gradient during the time of the discharging process of the metering pump. Figure 6 shows a typical pressure profile of the cascade pump. Since the metering pump sucks the liquid compressed by the compression pump, no pressure drop and no cavitation appear. The outlet pressure of the pump system may be the same as the pressure in the metering pump.

Figure 7 shows more details of operation of the cascade pump. Operation of the metering pump simply consists of a long discharging process and a quick suction process. On the other hand, operation of the compression pump consists of a long suction process and fairly quick discharging processes of compression and real transfer of liquid. The suction process of the metering pump is synchronized with the real transfer of liquid, and the required time for these processes is 0.2 s constantly over all conditions.

The compression process of the compression pump changes in stroke length depending on compressibility of the liquid and pressure used, and therefore the suction process of the compression pump also changes to the same extent as stroke length.

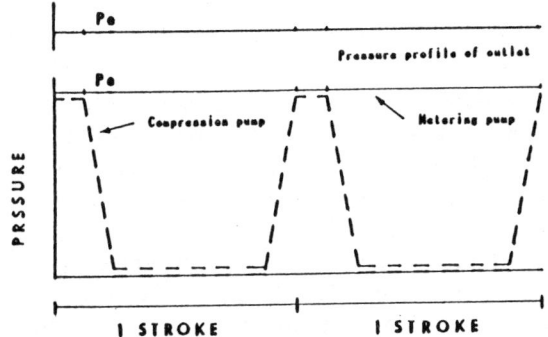

Figure 6. Predicted waveform of the outlet pressure of the cascade pump.

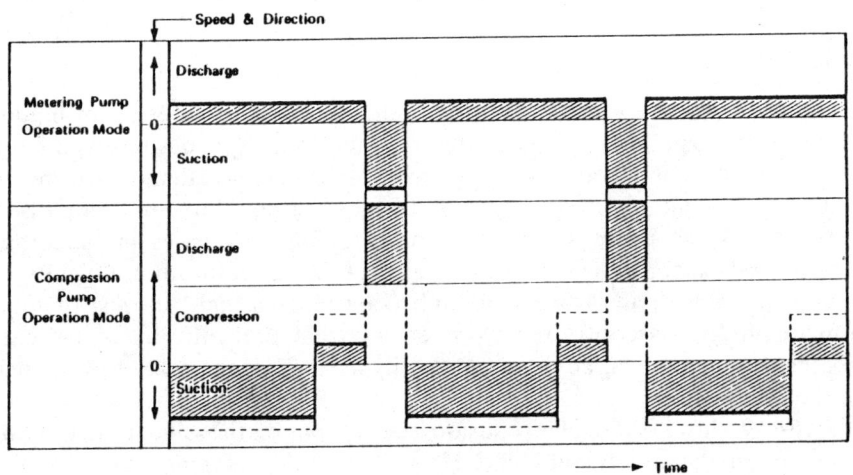

Figure 7. Operation mode and timing of the intelligent cascade pump. The dotted line indicates the change of plunger stroke depending on the compressibility of the liquid used.

MATERIALS AND METHODS

Trial manufacture of cascade pump

We have made several types of cascade pump system so far, by modifying ordinary double plunger pumps, with motors for each plunger being replaced by two independent stepping motors.

In a typical case for liquids, the diameters of the two plungers used are 3.2 mm, but for liquefied CO_2 a plunger diameter of the compression pump of 4.6 mm was used.

To control two stepping motors (Model M092-FC09), at the start of the experiment, we used a universal stepping motor control system, 'Modulynx' (Superior Electric Co. Ltd) with a personal computer PC-9801 F (NEC).

The check valves used in a reciprocating plunger pump are usually one of the most important parts of it. Especially for microdelivery at high pressure, good performance of the check valve is important. At the start of the experiment we used a check valve of an ordinary pump, but its performance was insufficient for use in microdelivery, even for ordinary liquid, and it was almost useless for liquefied gas. Then we developed two types of check valve, one suitable for liquid, and the other for liquefied gas. Both are essentially of the same design, and not so much different from the ordinary one, but their dead volumes are minimized to the extreme. Special care was taken as regards accurate manufacture, especially for the surface finishing of the valve for liquefied gas.

The attained performance of flow isolation as the ratio of the back to the forward is about −60 db, in both liquid and liquefied gas.

Materials

Figures 8 and 9 are graphical representations of some properties of liquids used in this experiment. The compressibility, density and viscosity of the eluents, such as liquefied CO_2, *n*-pentane, *n*-hexane and ethanol, are shown in the same scale as used in Fig. 8, to make straightforward comparison possible. As the figure shows, the compressibility increases in the order ethanol, *n*-hexane, *n*-pentane and liquefied CO_2; therefore the difficulty of microdelivery with an ordinary pump becomes increasingly serious. In addition to this the viscosity of CO_2 is so low that the difficulty of delivery becomes more serious, because CO_2 easily leaks from the seal of parts such as the check valve.

Figure 9 shows thermal expansion coefficients of the eluent such as *n*-pentane, methanol, ethanol and H_2O. The order of magnitudes of the thermal expansion coefficients is similar to that of their compressibilities, and the thermal expansion of stainless steel is about ten time less than those of the organic materials listed here.

All organic solvents (Wako Pur Chem.) used were filtered with a 0.22 μm Millipore filter and degassed with sonication in advance. The liquefied carbon dioxide (Tomoe Shokai), with purity of the order 10^{-4}, was used after filtering with a 0.1 μm Millipore filter. In evaluation of the intelligent cascade pump we used a semi-micro size column, made by packing the Deverosil ODS-5 or ODS-3 (5 μm o.d. or 3 μm o.d. silica gel with ODS Nomura Chem.) into a specially prepared stainless-steel column having 1.7 mm i.d. and 150 mm length (SUS 316).

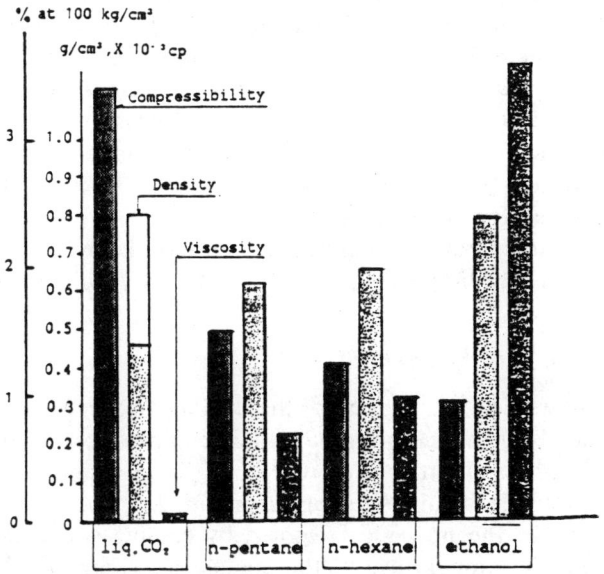

Figure 8. Graphical expression of the compressibility, density and viscosity of each liquid used as eluent in this experiment. The density of liquefied CO_2 changes with pressure and temperature. The blank area indicates the existence of this change.

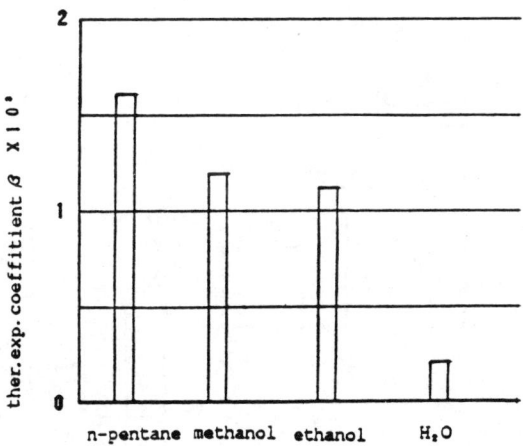

Figure 9. Graphical expression of the thermal expansion coefficient of *n*-pentane, methanol, ethanol and H_2O. These organic materials have ten times greater value than that of stainless steel, of which the pump head is made.

Flow diagram of the experimental equipment for SFC

Figure 10 shows the flow diagram of supercritical fluid chromatography. When using liquefied carbon dioxide as eluent the pump head is cooled by a cooling system which will be described later, but when using *n*-hexane as eluent the pump is operated at environmental temperature. For controlling the back pressure of the column we used the second intelligent cascade pump prepared for liquid. The sampler used here is a micro size sampler, injection volume 0.5 µl (Rheodyne Inc., Model 7520). The UV detector (Souma, Model 320 S) used here is provided with the specially prepared cell whose volume is 1 µl, and tested to withstand pressures of 500 kg/cm^2.

As the eluents for the second pump we used ethanol when the eluent for analysis was the liquefied CO_2 and we used 20% THF in ethanol when *n*-hexane was used for analysis.

The column temperature is controlled by use of the oven for gas chromatography (of GC–MS, JMS-D300, Jeol), temperature regulation within ±0.5°C.

Cooling system of the pump head

Figure 11 shows a block diagram of the cooling system provided for cooling the pump heads of the metering pump and compression pump. The electric thermomodule, maximum power of 125 W, is provided, and on the cold side of the module a mini-size fan is provided for cooling the head with cooled air, while the hot side is cooled by cooling water circulation. The lowest attainable temperature of the cooling air is about −10°C, the head temperature is about −5°C. To attain this performance thermal insulation between the pump bodies and the pump heads is important, Tight construction of the box on which the thermomodule is mounted is also important.

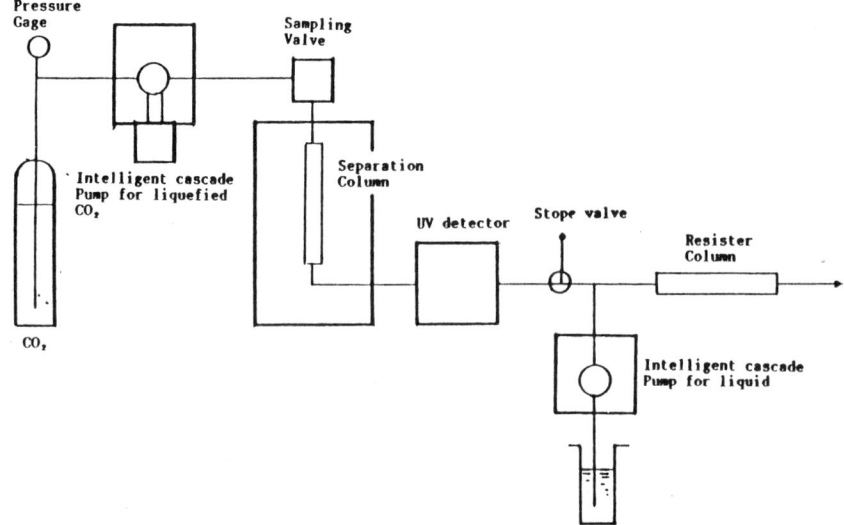

Figure 10. Flow diagram of supercritical fluid chromatography employed in this investigation.

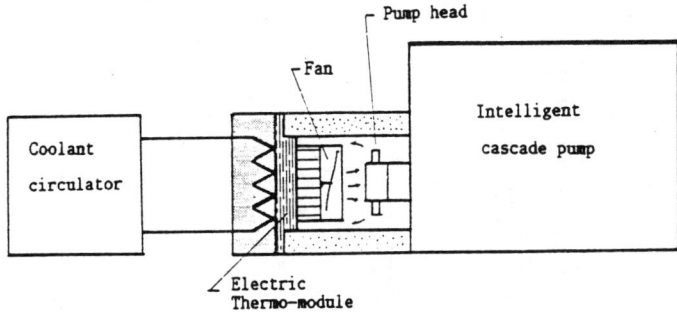

Figure 11. Block diagram of the system used for cooling the pump head.

EXPERIMENTAL RESULTS

Measurement of the lost volume

The lost volume in the pump at a given pressure can be evaluated by examining the relationship between the attained pressure and the plunger volume when the pump is operated, closing its outlet.

Figure 12 shows the results of such an experiment, with a relationship between the plunger volume and the attained pressure, for several different liquids. Linear relationships with different slopes have been observed for H_2O, methanol, *n*-hexane and *n*-pentane, the gradient of the slope decreasing in the order of their compressibility. From linear extrapolation we can estimate that the lost volume at 500 kg/cm^2 may be 11 µl for H_2O and 24 µl for *n*-hexane, for example. For a plunger with 3.2 mm o.d. each value corresponds to a stroke length of 1.4 mm for H_2O and 3 mm for *n*-hexane.

Measurement of the pressure ripple

For straightforward comparison, a pressure ripple measurement had been performed by conventional methods on an ordinary dual plunger pump without damper and the cascade pump.

Figure 13 shows the measurement results of pressure ripple on an ordinary dual plunger pump at two different flow rates; one 400 µl/min, the other 2000 µl/min.

The plunger volume is set at 85 µl and the interval of the ripple is expanded with decrement of the flow rate.

Figure 14 shows the results of the same measurement on the cascade pump. The observed pressure ripple seems to be fairly small.

If a direct comparison is made of the result at 400 µl/min in Fig. 13 and the result at 54 µl/min in Fig. 14, each integral value shows that the pressure ripple of the cascade pump is about 200 times less than that of the ordinary pump.

Since the plunger volume used is as small as 2 µl, the ripple interval of the cascade pump can be made even shorter than 2–10 s, down to a microvolume flow rate of 54–8 µl/min. Moreover, the frequency component of the ripple

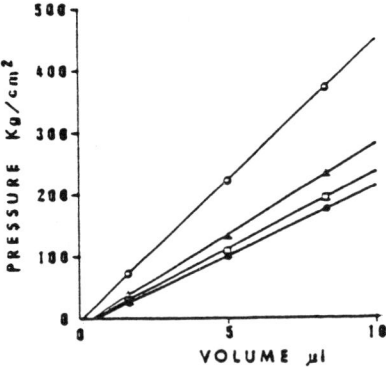

Figure 12. Measurement of lost volume. The relation between attained pressure and plunger volume is plotted here. The results are obtained by closing the outlet of the cascade pump. Open circle: H$_2$O (23.0°C), triangle: methanol (24.6°C), square: n-hexane (24.8°C), solid circle: n-pentane (20.0°C).

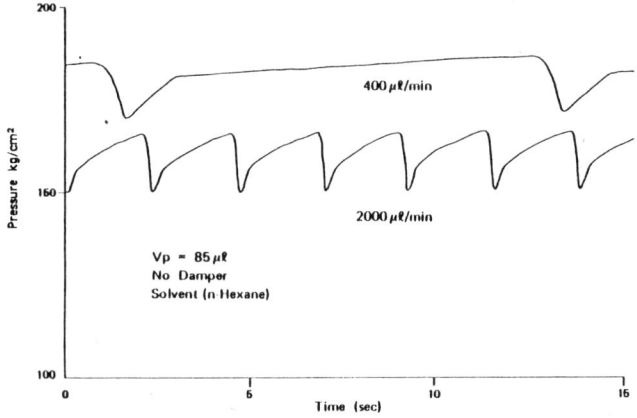

Figure 13. Pressure monitoring of the ordinary double-plunger pump without damping device.

is dominated by high frequency component, which can be easily dampened by use of small capacitance such as the micro column itself.

Confirmation of the constant flow rate independent of property of liquid

From the concept mentioned above, in an ordinary pump system, the flow rate cannot be kept constant among the measurements using different liquids with different compressibility values, even with the same column. On the other hand, on the cascade pump system we can obtain a constant flow rate independent of properties of liquid.

Figure 15 shows an example of reproducibility measurement of the flow rate for several different kinds of liquid, with the proper operation, which will be mentioned later.

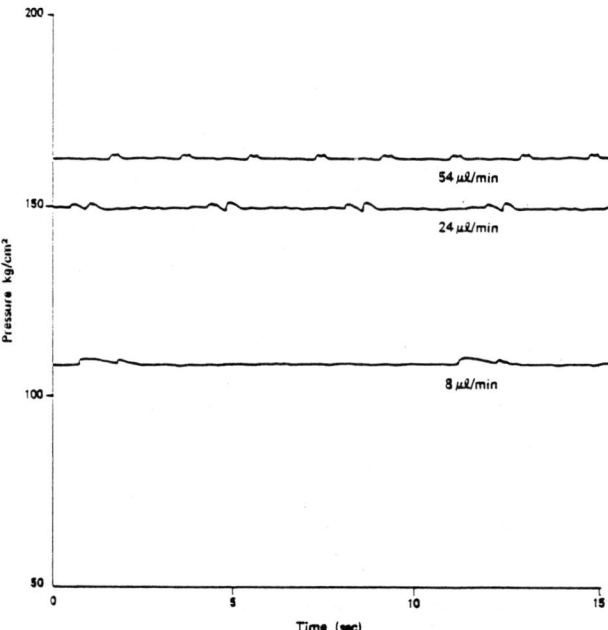

Figure 14. Measurement result of pressure monitoring of the cascade pump without damping device. The plunger volume used is 2 μl, and three flow rates, i.e. 8, 24 and 54 μl/min, are used. The liquid used is ethanol.

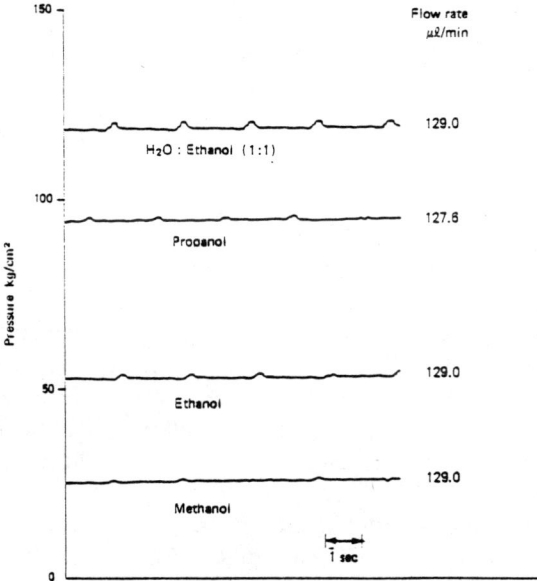

Figure 15. Reproducibility measurement of the flow rate for several different liquids (methanol, ethanol, *n*-propanol and the mixture of even volume of H_2O and ethanol). The column used for loading the pump is 1.5 mm i.d., 25 mm long, packed with 5 μm ODS silica. The predicted value of the flow rate is 129.0 μl/min, and the observed flow rates are shown at the right of each pressure profile.

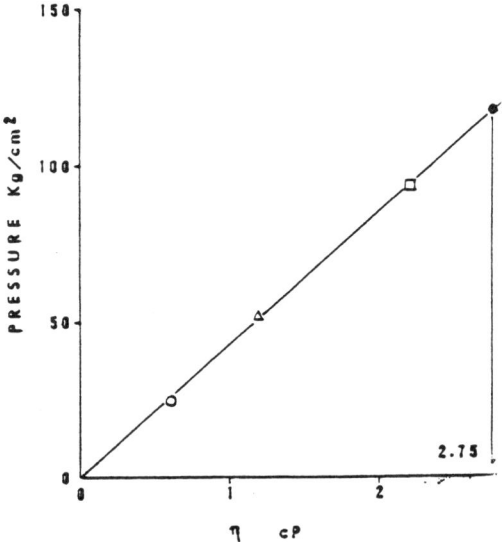

Figure 16. Plotting of the observed pressure in Fig. 15 in terms of the viscosity of each liquid cited from a handbook.

If the same column is used as a load the observed pressure at constant flow rate should correspond to the pressure drop in the column, which represents the viscosity of the liquid.

The actual flow rate is gauged by observing meniscus movement in a pipette provided at the outlet of the column. Figure 16 shows the plotting of the pressure observed in Fig. 15, in terms of the viscosity of each liquid cited from a handbook. Fairly good reproducibility of the flow rate and good linear correlation of the observed pressure to the viscosity are obtained.

Proper operation of the cascade pump

On the operation of the cascade pump system, the compression of liquid should be just enough to compensate the lost volume. However, if it is not enough, or exceeds the lost volume, a discrepancy appears between observed flow rate and the predicted value, which is defined as the plunger volume for a unit time, and the pressure ripples arise at the same time.

Proper operation can be attained by controlling so as to make the pressure gradient at the outlet of the pump, during the discharging process of the metering pump, zero.

Figure 17 shows an example of the pressure monitoring on three different conditions: the compression is (1) too much, $\Delta P < 0$; (2) just enough, $\Delta P = 0$; and (3) not enough, $\Delta P > 0$, to compensate the lost volume.

From the above results it is very clear that by using the pressure gradient ΔP as an error signal, automatic control for keeping the best condition can be performed. For that purpose we have designed a special intelligent control system for a stepping motor using a microprocessor.

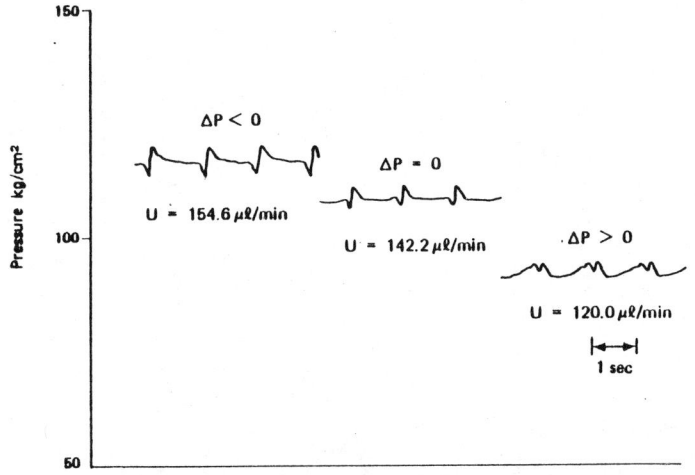

Figure 17. Pressure monitoring of three different conditions. The compression value is more than the lost volume: $\Delta P < 0$, just enough: $\Delta P = 0$, not enough: $\Delta P > 0$. The liquid used is ethanol, ΔP is the pressure gradient of the outlet of the pump during the discharging process of the metering pump.

Delivery of liquefied carbon dioxide

It is difficult to deliver the liquefied carbon dioxide at room temperature using a pump with a small plunger volume, because the liquefied carbon dioxide has so large a compressibility and so small a viscosity. It is necessary to cool it down for delivery.

The upper limit of temperature necessary for delivery may be determined by studying the relation of the pressure arising through the compression for a given plunger volume, and the cooled temperature.

Figure 18 shows such a relation for the cascade pump which delivers the liquefied carbon dioxide, for temperatures below about 10°C; this shows that delivery of the liquefied carbon dioxide is possible below 5°C. It may be preferable to cool down below 0°C, especially for microdelivery at high pressure.

Figure 19 shows pressure monitoring of the same cascade pump at 0°C. The observed pressure ripple seems to be very small, even though no damper was used.

The density of the liquefied CO_2 in the head of the metering pump can be predicted to be 1.04 g/ml from the observed pressure, 196 kg/cm², and temperature, 0°C. Assuming the plunger volume of 8.3 µl, and repetition time of the stroke of 1.84 s, the mass flow rate can be derived. The result obtained corresponds to 74.67 ml/min on the standard condition of gas phase. The actual flow rate observed at the outlet of the column by use of a mass flow meter is 75.92 ml/min on the same standard condition. The predicted value shows good agreement with the observed value within an inevitable error.

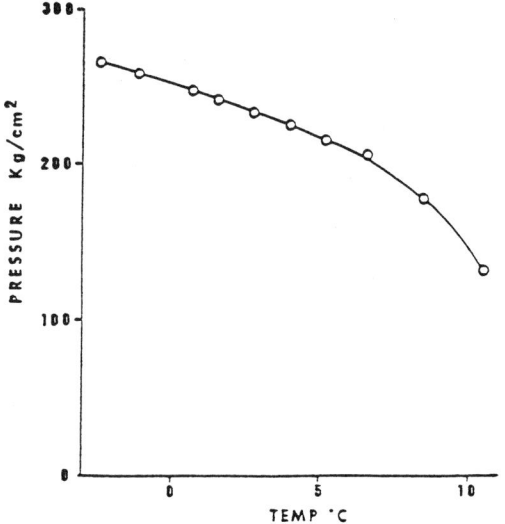

Figure 18. Temperature dependency of the compressibility of liquefied CO_2. The pressure caused by operation of the cascade pump with the plunger volume $V_p = 20$ µl, closing the outlet of the pump, is plotted in terms of the temperature.

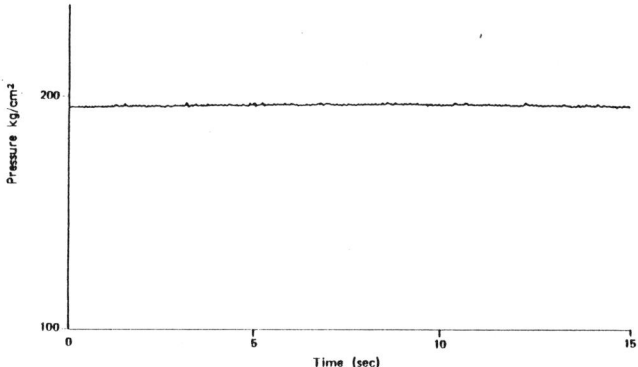

Figure 19. Pressure monitoring of the cascade pump which delivers the liquefied CO_2. The predicted flow rate, on std conditions in gas phase, is 74.67 ml/min. Here, a micro packed column 0.5 mm i.d., 300 mm long, of fused silica capillary packed with 3 µm ODS silica, is used for achieving the appropriate pressure. The plunger value is 8.3 µl, temperature of the head is 0°C. The flow rate observed by mass flow meter is 75.92 std ml/min.

DESIGN OF EQUIPMENT

Control system of the intelligent cascade pump

Figure 20 shows a block diagram of the control system of the intelligent cascade pump. Serially combined, two plunger pumps, the 'compression pump' and the 'metering pump', are designed to work through instructions from a microprocessor, through the respective stepping motor with driver.

A pressure sensor mounted at the outlet of the metering pump feeds the signal to the microprocessor through a 12-bit analog to digital converter with a multiplexer and a gain controller. The multiplexer is used for time sharing service of temperature monitoring and pressure monitoring. The gain controller is provided to adjust the input signal to the full scale of ADC, in three pressure ranges, 0–125, 0–25(and 0–500 kg/cm². Thus the least significant bit of ADC corresponds to 0.1, 0.2 and 0.4 kg/cm² for the ranges of 0–125, 0–250 and 0–500 kg/cm², respectively.

The position of the plunger of each pump is determined by the pulse counting initiated with the signal of position sensor.

The operation and display unit (ODU) consists of 60 × 100-dot liquid crystal display (LCD) and necessary function keys. A graphic LCD displays the pressure profile and gradient programming profile in addition to alpha-numeric displays of pressure, flow rate, temperature, pressure limit and other parameters of pump operation.

The relation between flow rate and plunger volume

Generally, analytical liquid chromatography may be classified into micro, semi-micro and conventional chromatography, depending on the size of column. According to these classifications the flow rate of the pump must be changed.

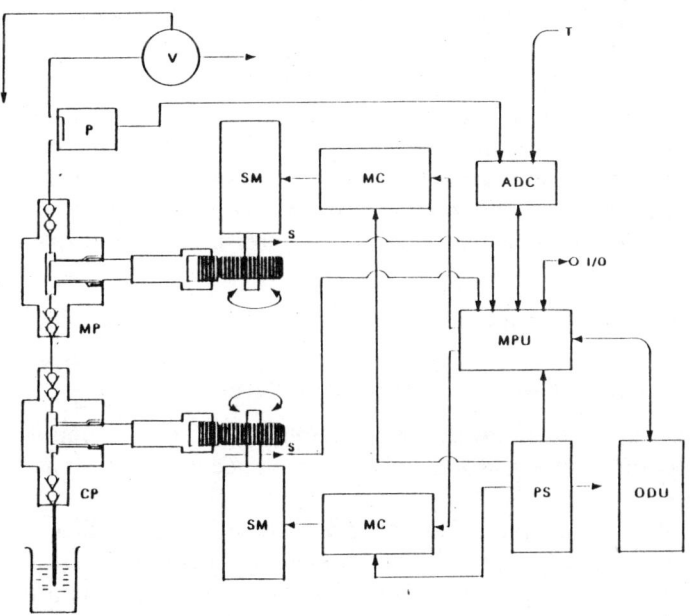

Figure 20. Block diagram of the control system of the intelligent cascade pump. T: temperature sensor, V: drain valve, S: position sensor, P: pressure sensor, ADC: analog to digital converter, MPU: microprocessor, PS: power supply, ODU: operation and display unit, I/O: input and output of signals for external device.

The intelligent cascade pump is designed to cover all three ranges by selecting the plunger volume of the metering pump as listed in Table 1.

As mentioned before, the flow rate can be determined by the plunger volume and repetition time of the metering pump at any pressure and for any liquid.

Table 1.
Plunger volume selection of the metering pump, depending on the flow rate

Range	Flow rate (μl/min)	Plunger volume (μl)	Time of a stroke (s)
Micro	0.1–3.0	0.05	1–30
	3.1–10.0	0.15	3–10
Semi-micro	1–30	0.5	1–30
	31–200	3.0	1–6
Conventional	10–300	5.0	1–30
	310–4000	60.0	1–12

Design of the cam

The cam used to drive the plunger is designed to give three different plunger movements whose magnitude ratios are 1:10:100. Each slope of the cam corresponds to the each flow rate range as listed in Table 1. The plunger stroke is determined by selection of slope and rotating angle of the cam, which corresponds to the number of pulses given to the stepping motor. The operation of the cam is not an ordinary rotation; it turns back and forth within a given angle. The stroke length of the metering pump is always determined by stroke range and slope, once the flow rate is selected. The stroke length of the compression pump changes over the three slopes until complete compensation of the lost volume is attained. The change from one slope to the other is continuous, so that the flow rate gradient operation can be performed continuously, as well as the pressure gradient operation.

ATTAINED PERFORMANCE

Delivery of liquid in semi-micro and micro ranges

Figure 21 shows the measurement results of pressure monitoring in the delivery of methanol using the latest developed intelligent cascade pump with three different flow rates.

The plunger volume used for delivery of the flow rates, 100 and 50 μl/min, is 3 μl in both cases, so that the corresponding repetition times of the stroke are 3.6 and 1.8 s, respectively. When the flow rate is 25 μl/min the plunger volume of 0.5 μl is adopted so that the repetition time required is 1.2 s. The plunger volume is selected by the built-in microprocessor so as to restrict the reciprocation time within 30 s for the selected flow rate.

Figure 22 shows results of pressure monitoring in the delivery of ethanol using the same equipment with three micro range flow rates, 1, 2 and 3 μl/min. The column used as load is a hand-made micro-packing column, packed with 3 μm ODS silica gel in a 0.3 mm i.d., 320 mm long polyimide-coated, fused silica capillary. The same plunger volume, 0.5 μl, is used for three flow rates, and reciprocation times are 30, 15 and 10 s for 1, 2 and 3 μl/min flow rates, respectively.

The pressure ripples observed without damper in the above-mentioned six cases are so small, irrespective of rather high pressure operation, that they can be damped easily by the column itself. Absence of damper operation of a pump gives certain advantages in solvent exchange or cold start to attain the equilibrium condition in a short time.

Application of the intelligent cascade pump for SFC (supercritical fluid chromatography)

Chromatographic analysis of polystyrene oligomer by super critical n-hexane. By means of SFC with *n*-hexane or *n*-pentane as eluent, for the first time, possibility of analysis of the oligomer had been shown by Klesper (1978); in a later, highly efficient analysis using a packed capillary column had been shown by Hirata and Nakata (1984). For analysis of the oligomers with molecular weights spread over a continuous range from low to high, it is important to use density gradient techniques is to attain the efficient analysis of all components in a short time.

We tried SFC with *n*-hexane for polystyrene oligomer analysis, using a semi-micro packed column which we made; this time packed with 5 μm

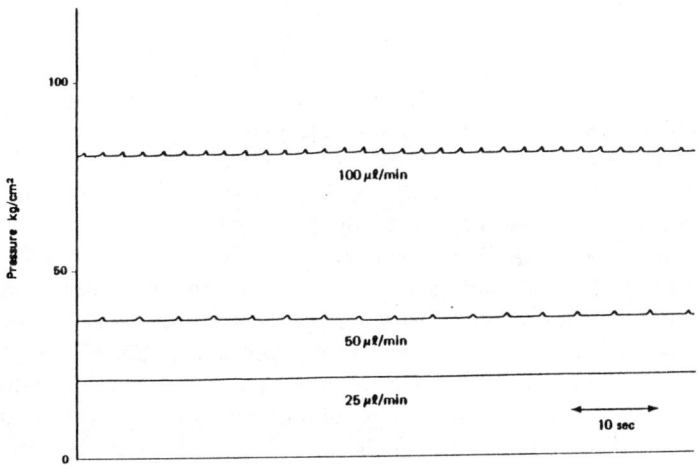

Figure 21. Pressure profile of the intelligent cascade pump at the flow rate of semi-micro range. The liquid used is methanol, the column is 1.7 mm i.d., 150 mm long, of stainless steel, packed with 5 μm ODS silica. The plunger volume used for 25 μl/min is 0.5 μl, and for 50 and 100 μl/min, 3 μl.

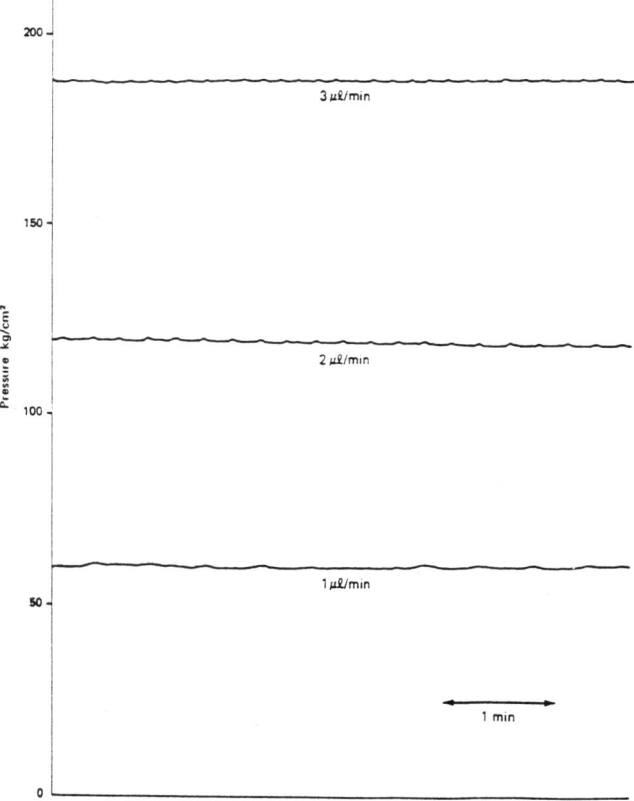

Figure 22. Pressure profile of the intelligent cascade pump at the micro range flow rate. The liquid used is ethanol, and the column is 0.3 mm i.d., 320 mm long, of stainless steel, packed with 3 μm ODS silica. The plunger volume used is 0.5 μl.

ODS silica gel in a 1.7 mm i.d., 150 mm long specially prepared stainless-steel column. For attaining the pressure gradient in the column the back pressure of the outlet of the column was controlled by the pressure programming operation of the second pump as described before. Meanwhile, the eluent delivering pump was kept in a constant flow mode.

Figure 23 shows an example of such experiments on isobaric condition of average pressure as 49 kg/cm^2, using 10% ethanol in n-hexane as eluent. Figure 24 shows the result obtained under the same conditions except that the average pressure is 54 kg/cm^2. As can be seen in Fig. 23, the low molecular weight components show good separation in low pressure (with low density), but the high molecular weight components do not elute smoothly, thus producing a broaden peak. On the contrary, in Fig. 24, obtained by operation at high pressure (with high density), the high molecular weight components show good separation, but the low molecular weight components show poor separation.

The average pressure difference between Fig. 23 and Fig. 24 is only 5 kg/cm^2, and it is clear that the separation is very sensitive to the operating pressure.

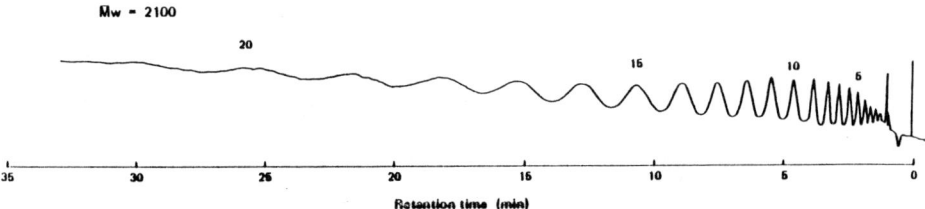

Figure 23. Analysis of polystyrene oligomer by SFC in isobaric condition. Column: 1.7 mm i.d., 150 mm long, of stainless steel, packed with Develosil ODS 3 μm. Column temperature is 250°C. Pressure at the top of column is 60 kg/cm², pressure at the end of column is 38 kg/cm². Eluent is 10% ethanol in *n*-hexane. The average molecular weight of the polystyrene oligomer is 2100.

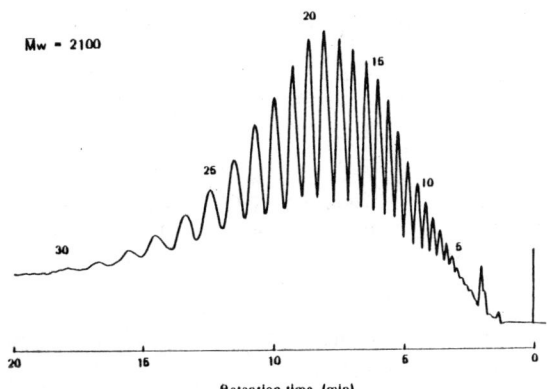

Figure 24. Analysis of polystyrene oligomer by SFC in isobaric condition. The same conditions as in Fig. 23, but the pressure at the top of column is 62 kg/cm², the pressure at the end of column is 44 kg/cm².

Figure 25 shows a similar result obtained with a linear pressure gradient instead of isobaric conditions, using 20% ethanol in *n*-hexane as eluent, on the same column. A shows an experiment where the gradient is 1.07 kg/cm²/min, and B shows a case where the gradient is 0.63 kg/cm²/min. The elution profile with the linear pressure gradient makes for separation of low molecular weight components as well as that of high molecular weight components to the same extent, in contrast to the results on the isobaric condition.

Analysis of polyaromatic compounds using supercritical CO_2. Figure 26 shows a chromatogram of a polyaromatic hydrocarbon mixture obtained by SFC on three different isobaric conditions. Two chromatograms, A and B, are results of an experiment attempting to reduce the analysis time by increasing the flow rate while the temperature is kept constant at 40°C, as usual in HPLC. The elution time is inversely proportional to the flow rate in liquid phase; however the results showed that in SFC the decrease of elution time exceeds the value predicted in liquid phase. This phenomenon can be explained as follows: the elution time is shortened due to the high solubility

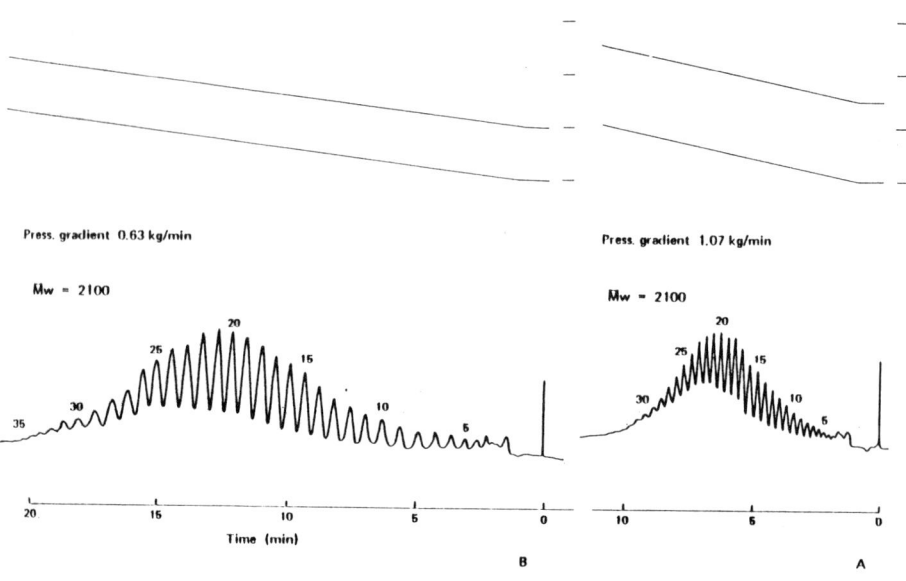

Figure 25. Analysis of polystyrene oligomer by SFC on pressure gradient operation. The same condition as in Fig. 23, except the pressure gradient is used; A: pressure gradient 1.07 kg/min: B: pressure gradient 0.63 kg/min.

of polyaromatic compounds to the supercritical CO_2 whose density increases with increased pressure and flow rate.

This can be confirmed by the chromatogram C which was obtained by an experiment with the same flow rate as C, but at a lower temperature, 36°C. Irrespective of the flow rate, the elution time in C is far shorter than that in B. The shorter elution time at lower column temperature is opposite to the usual result in HPLC. This fact can be explained by increased solubility with higher density at lower temperature. Because in cases C and B, the linear velocity of flow in the column at high temperature (40°C) in B may be faster

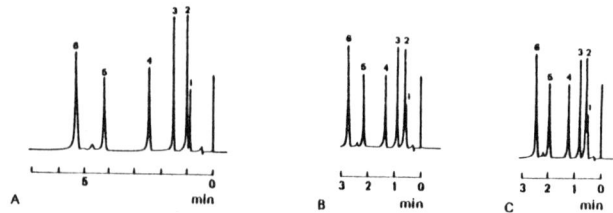

Figure 26. Analysis of polyaromatic hydrocarbons by SFC in isobaric condition of CO_2. Column: 1.7 mm i.d., 150 mm long, of stainless steel, packed with Deverosil ODS 5 µm. Column temperature is 40°C for A and B, and 36°C for C. Pressure at the top of column is 119 kg/cm² for A, 140 kg/cm² for B and 143 kg/cm² for C. Pressure at the end of column is 88 kg/cm² for A, 94 kg/cm² for B and 96 kg/cm² for C. Assignments of the peaks are 1: naphthalene, 2: biphenyl, 3: fluorene, 4: anthracene, 5: fluorantene and 6: pyrene.

than that at low temperature (36°C), considering of their density. Faster linear velocity promises faster elution in the usual sense, but here the effect of solubility increase with the increase of density dominates.

Figure 27 shows the retention profile of each component of polyaromatic hydrocarbons against the density plotted from the data of A, B and C in Fig. 26. The dotted line shows the retention time corrected to that of B and C, assuming that the linear velocity of A is kept constant.

According to the increase in density of CO_2, the retention time of polyaromatics decrease remarkably. This kind of relationship is similar to the profile against solvent strength in HPLC.

Reproducibility measurements of retention time for polyaromatic hydrocarbons have been done, on the isobaric condition of average pressure, 115 kg/cm^2. The results obtained by ten consecutive runs are listed in Table 2, which shows fairly good reproducibility had been attained as the result of stable operation of the intelligent cascade pump.

The density gradient programming is quite useful for obtaining a chromatogram keeping good resolution over the whole range in a short time.

Figure 28 shows a comparison between the chromatogram obtained on the isobaric condition and that on the pressure gradient condition for analysis of polyaromatic hydrocarbon compounds. The linear pressure gradient programming for density changing is adopted in the CO_2 delivery pump. The second pump delivers the liquid (here methanol) for the second column as shown in Fig. 10. As Fig. 28 shows, the resolutions of weak retained components such as naphthalene and biphenyl, become poor when elution

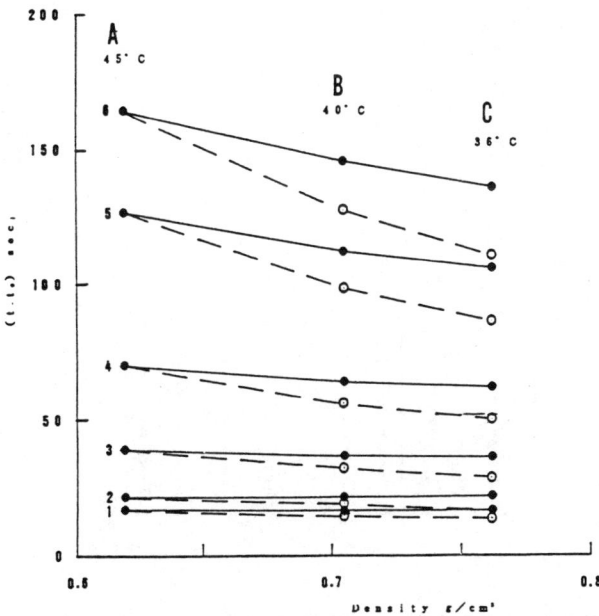

Figure 27. Retention time of each component versus the density of CO_2 predicted from Fig. 26. The dotted lines are the corrected retention times of B and C, assuming the same linear velocity as that of A.

time is short; however, in the pressure gradient condition this difficulty disappears.

In general pressure gradient elution in SFC is comparable with to solvent gradient elution in HPLC.

Table 2.
Reproducibility of retention time (min) of polyaromatic hydrocarbon analysis by SFC in isobaric condition of liquefied CO_2. Column: 1.7 mm i.d., 150 mm long, of stainless steel, packed with Deverosil ODS 3 μm. Pressure: 150 kg/cm^2 (top of column), 80 kg/cm^2 (end of column). Column temperature: 40°C

No.	Naphthalene	Biphenyl	Fluorene	Anthracene	Fluoranthene	Pyrene
1	0.970	1.093	1.613	2.497	4.120	5.200
2	0.970	1.093	1.613	2.498	4.113	5.188
3	0.970	1.098	1.600	2.468	4.098	5.167
4	0.970	1.088	1.600	2.472	4.078	5.158
5	0.981	1.103	1.620	2.505	4.103	5.192
6	0.973	1.097	1.620	2.497	4.102	5.167
7	0.987	1.110	1.620	2.483	4.082	5.150
8	0.958	1.080	1.598	2.463	4.080	5.142
9	0.967	1.087	1.603	2.477	4.080	5.147
10	0.963	1.085	1.603	2.465	4.070	5.138
$\bar{X}$	0.971	1.094	1.609	2.483	4.094	5.165
SD	0.008	0.009	0.009	0.015	0.015	0.021
CV%	0.81	0.80	0.55	0.60	0.38	0.64

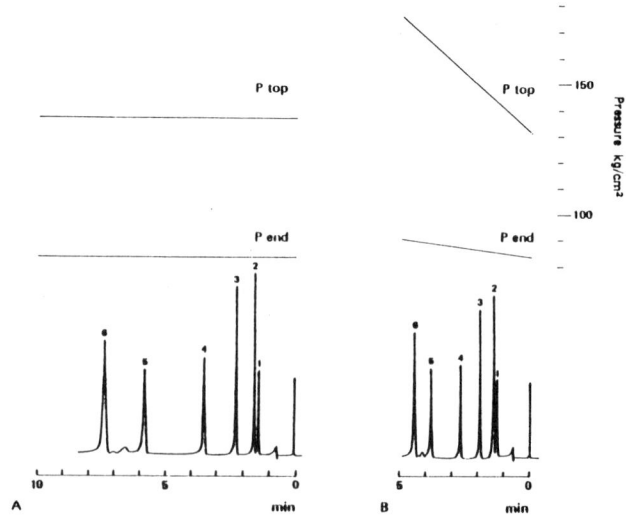

Figure 28. Analysis of polyaromatic hydrocarbons by SFC in pressure gradient condition. The same conditions as Fig. 26, except pressure gradient are used; A: pressure gradient 0.65 kg/cm^2, B: pressure gradient 1.05 kg/min.

DISCUSSION

Effect of thermal expansion of liquid

We have not so far discussed the effect of the thermal expansion coefficient of the liquid or the construction materials of the pump. It must be considered in actual experiment that the effect of thermal expansion of the liquid affects the precision of flow rate when the pump is operated at ambient temperature.

Recently, a displacement-type pump system has been used frequently for microcolumn experiments including SFC (Fjeldsted *et al.*, 1983). This type of pump basically has a large plunger volume (e.g., 5−10 ml), which inevitably makes the dead volume large. The large plunger volume and large dead volume cause a large volume change by the change of temperature.

As Fig. 9 shows, the thermal expansion coefficient of liquid, β is of the order 10^{-3}, and the change from one liquid to the other is of the same order. This means that if the total volume inside the pump is 10 000 µl, and the average thermal expansion coefficient, β, is 10^{-3}, 10 µl of the lost volume appears for a temperature change of 1°C, and the same order of change appears with change of eluent. This may be the effect of the change of the order of the flow rate used in the microcolumn experiment.

If the intelligent cascade pump uses a 3 µl plunger volume for operating with a flow rate of 10 µl/min; the lost volume due to the temperature change will be 3×10^{-3} µl per 1°C. Therefore the effect can be neglected.

Selection of 'auto mode' or 'lock mode'.

As described before (describing the proper operation of a cascade pump, in Experimental results), the signal P (pressure gradient during the period of the metering pump's discharging process) is fed for attaining complete compensation of the lost volume due to compression. This function is always 'on' in 'auto mode' operation of the intelligent cascade pump. However, sometimes this function causes fluctuation of the flow rate or of pressure enhanced, when the load of the pump is switched one to the other as in the case of sample injection. In such a case 'lock mode' operation is recommended. 'Lock mode' means fixing the parameters of the pump running just before switching on 'lock'.

As mentioned before, describing the experiment of SFC, we used two intelligent cascade pumps, one for delivery of eluent and the other for back pressure control. We have found irregular operation occurred when the two pumps were set to 'auto mode' at the same time, especially when both pumps were operating with pressure programming. Interference between the two pumps is cut by adopting 'lock mode' in one of them.

The experiment of SFC with *n*-hexane worked well either by adopting 'lock mode' in the eluent delivery pump, and 'auto mode' with pressure programming in the back pressure control pump, or adopting 'lock mode' in the back pressure control, and 'auto mode' 'in the delivery with pressure programming. With CO_2, the case of adopting 'auto mode' with pressure programming in the eluent delivery pump, and 'lock' on back pressure pump, was the only system which worked well.

Usefulness of real-time display of operation parameters

Usual requirements of real-time monitoring of an ordinary pump operation may be the display of the selected flow rate, attained pressure and pressure limit. In addition to the above, information concerning the temperature of the head, the status of operation, such as 'auto' or 'lock' and 'manual' or 'program', and repetition times of programming operation, is also useful.

Moreover, we found it useful to display the parameters in real time, together with the movement of the plunger of the intelligent cascade pump. With every movement of the plungers, the plunger volume of the metering pump, the plunger volume of the compression pump used for delivery of liquid — and for compression of liquid separately, as well as the pressure gradient with the selected flow rate, are displayed. By consulting these displayed data we can confirm that the pump is operating properly and know what order of compressibility of eluent is being used. This information may be used when the operation mode is switched from 'auto' to 'lock'. On the first trial operation, with the first load, or first liquid with a given flow rate, the attainable pressure is not known at the start of the operation.

The intelligent cascade pump has a built-in self-learning function for achieving a parameter as the plunger volume of the compression pump used for compressing liquid. This function is useful for flow-rate programming and pressure programming.

CONCLUSION

A new pumping system, which can produce non-pulsating flow independent of column pressure and compressibility of fluid, has been developed. The system consists of two reciprocating plunger pumps, serially combined in cascade mode. The first, the 'compression pump', completely compensates for volume of fluid lost due to compressibility; the second, the 'metering pump', delivers the compressed fluid accurately. Readjusting parameters so as to make the gradient of outlet pressure in a stroke zero, an accurate flow rate ranging from 1 to 2000 μl/min can be obtained, and kept at any pressure below 500 kg/cm^2 and for various liquids including liquefied CO_2.

ACKNOWLEDGEMENTS

The authors wish to acknowledge the help of Mr T. Fujisaki and his colleagues at Sigma Electronics Co. Ltd, with the intelligent control system of the pump. We appreciate the helpful advice and encouragement of Dr K. Jinno of Toyohashi University of Technology. Continuous help and support from many employees of Jeol Ltd, and many others related to this project, is greatly appreciated.

REFERENCES

Fjeldsted, J. C., Jackson, W. P., Peaden, P. A., and Lee, M. L. (1983). *J. Chromatogr. Sci. 21*, 222.
Hirata, Y. and Nakata, F. (1984). *J. Chromatogr. 295*, 315–322.
Klesper, E. (1978). *Angew. Chem. Int. Ed. Engl. 17*, 738–746.
Patternson, W. G. (1982). United States Patent, 4,352,639, October.
Saito, T., and Takeuchi, M. (1986). *Seventh International Symposium on Capillary Column Chromatography*, Nagoya, Gifu, Japan, 11–14 May.

Progress in HPLC, Vol. 4, pp. 53—71.
Yoshioka *et al.* (Eds)
© 1989 VSP.

Current use of packed columns in SFC

TYGE GREIBROKK, JO DOEHL and ELSA LUNDANES

Department of Chemistry, University of Oslo,
PO Box 1033, Blindern, 0315 Oslo 3, Norway

INTRODUCTION

After almost two decades of slow progress in supercritical fluid chromatography (SFC) the introduction of open tubular columns by Novotny *et al.*
(1981) suddenly induced new interest in the technique. Today most new
applications are performed with open tubular columns. The reasons for this
are many, but a major point, emphasized by Novotny *et al.* (1981), is the
pressure drop over the column. Even if the pressure drop in supercritical
fluids is much lower than in liquids, due to the low viscosities, the pressure
drop over a packed column is undeniably higher than over an open tube.

The pressure drop, which increases linearly with column length, will cause
an increase of the capacity factor along the column. With long packed
columns, k' values outside the optimum range may occur with accompany-
ing performance loss with respect to resolution and/or analysis time. This
can be only partially compensated by pressure programs. With short packed
columns, however, the pressure drop is not of much practical interest. Open
tubular columns also reduce the adsorption problems of packed columns,
and the low flow rates of capillary columns make them ideal for connection
to detectors with limited mobile phase acceptance.

However, the advantages of packed columns should also be emphasized:
higher loadability, wider commercial choice of column selectivities and higher
efficiency per unit time (of column dimensions available with current technology). The loading capacity of a 50 μm inner diameter thin film open
capillary column is as low as 10 ng per component. Split-less injection
techniques in SFC have been developed, but most commonly concentrated
sample solutions are required with open tubular columns. Since the injectors
currently in use need $1-10$ μl of sample solution to fill and flush the interior
dead volumes, and since the columns can often accept only a few nanoliters,
only a small part of the sample is ordinarily utilized. This illustrates the need
for better injection techniques with open tubular columns in SFC. With
packed columns current micro-HPLC injectors can be used without splitting.

The number of stationary phases commercially available for open tubular SFC have been limited, but are now increasing. Still a wider selectivity range is obtained on packed columns.

The efficiency of open tubular SFC per unit time was shown by Peaden et al. (1982) to be comparable with packed column HPLC (~40 plates/s). Since a major argument for the use of supercritical fluids in chromatography has been the high diffusion coefficients, the above fact should be a reminder that really high efficiencies per unit time in SFC can only be obtained with short columns packed with small particles, or alternatively with open tube diameters much smaller than the current ones.

MATERIALS AND METHODS

Fluids

Pressurized CO_2 (standard grade) and N_2O (medical grade) were obtained from AGA Norgas, Oslo, Norway. Liquid CO_2 was transferred from the tank to the pump by an eductor tube and a stainless-steel transfer line. The N_2O tank, which could not be obtained equipped with dip tube (for medical safety reasons), was mounted upside-down. A 2 μm filter was inserted in the transfer lines. The upside-down mounting of the N_2O tank, which is needed to transfer the liquid phase, creates occasional problems with the outlet valve becoming blocked with corrosion particles from the tank. Thus, in countries where this is allowed, the tanks should preferably be equipped with eductor tubes. The fluids were purified on a purifier column (500 × 7 mm), filled with active carbon (Alfa Products, Danvers, MA, USA). The purifier was replaced after purifying 10−15 kg of fluid, or whenever another fluid was installed.

The modifiers were of HPLC Grade S quality from Rathburn (Walkerburn, UK).

Pumps

Syringe pump. The syringe pump, Isco μ-LC 500 from ISCO (Lincoln, NB, USA) had a flow rate of 0.02−600 μl/min and a cylinder volume of 50 ml, which made it suitable for narrow bore columns. In order to fill the cylinder completely with liquid CO_2 at each refill operation, a cooling coil was mounted around the pump cylinder. Cold methanol at −15°C was circulated through the coil with a Julabo-F 20-VL circulation bath (Julabo Labortechnik GMBH, Sulbach, FRG) during filling. With the original pump seals the cooling had to be shut off after filling, due to substantial leaks of CO_2 at the cold seals. With more recently obtained seals no leaks have been observed, and cooling is maintained at all times in order to keep constant temperature.

The pump was operated at constant pressure with good stability. Pressure gradients were constructed manually or with a home-made programming device. The filling procedure is rather slow (45 min). Shut-off valves (SSI 02-0120, Scientific Systems Inc., State College, PA, USA), mounted at the inlet and outlet of the pump, allowed the rest of the system to be held at equilibrium pressure during pump filling, column replacement, purifier replacement, tank replacement etc.

Reciprocating piston pumps. The modified Waters 6000 A and Waters 590 pumps were utilized mainly with 1 and 2 mm (i.d.) packed HPLC columns. The Waters 6000 A pump was modified in the following way:

(1) Additional check valves (SSI-02-0129 from Scientific Systems, Inc.) were mounted at the pump head inlets and outlets. The four valves were surrounded by an aluminum block containing cooling channels.
(2) Cooling channels were drilled in the solid part of both pump heads.
(3) The cooling channels in the pump heads and the check valves were connected with PVC tubing, and cold methanol at −5°C was circulated through the system.

The electronics module was removed from the pump housing in order to avoid condensation problems.

The Waters 590 pump was modified in the following way. Since the cylindrical pump heads did not contain room for drilling cooling channels, the cooling was obtained by a clamp-on unit, made from aluminum; for details of construction see Greibrokk *et al.* (1986) and (1984). The check valves, the connections and the circulation were identical with the modified 6000 A pump described above. In contrast to the 6000 A pump the 590 pump could be operated in constant pressure mode, but the baseline noise was significantly higher at constant pressure than at constant flow. Thus the pump was used either at constant flow or with flow gradients using the flow programmer contained in the pump control.

The soft-seat check valves of the modified pumps malfunctioned quite often, and the O-rings then had to be replaced. Since some O-rings lasted for months and some only for a few days, it would appear that the reproducibility of the O-ring production is not satisfactory for this purpose.

With an experienced operator, and careful treatment, the modified reciprocating piston pumps can be used at controlled flow for long periods without problems. It is evident, however, that better check valves (or active valves) are needed in order to be able to compete seriously with the syringe pumps.

Injectors

Two different injectors were used. One was a Valco CI 4 W from VICI AG, Sehenkon, Switzerland, equipped with a 60 nl rotor. The other was a Rheodyne 7520 from Rheodyne Inc., Cotati, CA, USA, equipped with a 200 nl rotor. Both were mounted on the top of the column oven. The injector should *not*, contrary to many recommendations, be mounted inside the column oven. With the injector inside the oven the increased column temperatures which are sometimes needed to optimize a separation do not allow reproducible injection of samples in volatile solvents. If the injector needs heating a separate heating device should be used.

Columns

25 cm × 1.3 mm i.d., packed with 8 μm CP-Spher C_{18}, were obtained from Chrompack (Middelburg, The Netherlands); 3 cm × 2.1 mm i.d. Spheri-5 cyano (5 μm) MPLC cartridge columns were obtained from Brownlee Laboratories (Santa Clara, CA, USA) and a 15 cm × 1 mm i.d. Novapak C_{18} column was received as a gift from Millipore/Waters (Milford, MA, USA).

The fused silica columns were slurry packed with 4 μm Novapak C_{18} and with 3 μm Spherisorb ODS using a slurry vessel with a magnetic stirring bar, as described by Folestad (1985). The empty 0.2 mm and 0.3 mm columns were obtained from Chrompack.

Restrictors

With the flame ionization detector tapered fused silica restrictors were obtained by drawing a 10 μm or a 25 μm capillary in a flame. A batch of several restrictors were produced, the flow at 200 bar was recorded and the suitable restrictors were selected.

In the thermionic detector, platinum restrictors were made by crimping the end of 100 μm i.d. tubes, from Goodfellow Metals (Cambridge, UK).

Column oven and detectors

Hewlett-Packard 5700 gas chromatographs were used without modification as column ovens and with flame ionization detector (FID) and thermionic detector (NPD). The FID was kept at 325°C and the NPD at 350°C.

UV detection was obtained with a Perkin-Elmer LC-55 detector where all built-in tubing at the flow-cell inlet was removed. The quartz windows were tightened with an extra set of springs, and withstood peak pressures above 300 bar.

The packing procedures of the fused silica columns were controlled by on-column fluorescence detection with a Kontron LC-25 fluorescence detector.

COLUMN PACKING MATERIALS

In the first years of SFC, long columns packed with large particles were utilized. Even at high flow rates the analysis times were long. The combination of long analysis times and complicated instrumentation is probably one of the reasons for the slow development of SFC.

When conventional HPLC columns, packed with small particles, were introduced in SFC, by Gere et al. (1982) the actual, and not only the theoretical, speed of analysis was strongly emphasized by the authors. Later, Hirata and Nakata (1984) showed that packed fused silica capillaries could be used to separate polystyrenes with molecular weights up to approximately 8000 on a column with only 8000 theoretical plates.

Conventional HPLC columns can be utilized with detectors which are compatible with the high flow rates of these columns. Most gas-phase detectors (GC-type detectors), however, require the moderate to low flow rates delivered by narrow-bore columns. A compromise is to use a conventional column and a splitter after the column. In general the split systems cannot be recommended as permanent solutions. The reasons are that sample is wasted, fluid is wasted, the peak dilution is higher on a wide-bore column and the split-ratios are often difficult to reproduce, particularly small split-ratios. At low flow rates through the waste capillary, sample components may precipitate in the decompression process and slowly close the hole. With high split-ratios the sample components are more efficiently flushed out through the capillary. As a consequence the choice is then between narrow-bore steel columns, which are commercially available with internal diameters of approx. 1 mm, and fused silica columns of 0.2−0.5 mm diameter. Packed fused silica columns have recently become available commercially, but reasonably good columns can be obtained by relatively simple packing procedures.

In order to obtain short analysis times the particle diameters should be small (Fig. 1). With 2 μm particles Yang (1986) packed fused silica columns that produced 268 000 plates/m ($h = 2$). For applications which require selectivity more than efficiency, and for applications which require higher pore volumes, larger particles may be advantageous. Efficient short columns packed with small particles are also more vulnerable as regards pressure pulses. In HPLC the so-called 3 × 3 columns are recommended only in

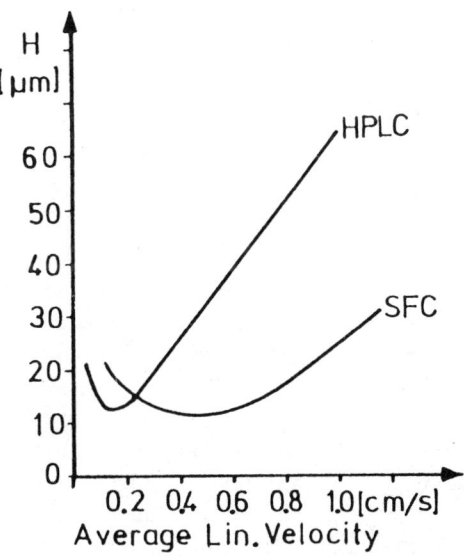

Figure 1. Plate height as a function of mobile phase velocity. Based on data from Gere *et al.* (1982).

combination with injectors equipped with a bypass loop, in order to avoid pressure pulses. Due to the higher compressibility of the mobile phase there are reasons to believe that the problem is smaller in SFC. But for similar reasons, open tubular columns may need protection against sudden pressure variations, as recommended by Lee and Markides (1986).

With a system based on pressure programs, the need for rigid packing materials is obvious. Accordingly, silica and silica-based packing materials with bonded stationary phases have dominated in SFC.

Silica

As in HPLC, silica is not compatible with solutes with polar groups unless polar modifiers are added to fluids such as carbon dioxide or hexane. Silica has, however, been extremely useful for separation of polystyrenes, with 5% ethanol in hexane, as shown by Hirata and Nakata (1984), or with mixtures of alkanes and dioxane, as shown by Schmitz et al. (1983). The latter authors found that the addition of methanol, which was needed to separate oligo (2-vinylpyridines), could be detrimental to the columns by partial dissolution of the silica matrix. The effect was less with ethanol than with methanol, and could apparently be disregarded at lower alcohol concentrations (5–10%), according to the authors. Anyway, this demonstrates that hydrophilic solvents should be used with care as additives to fluids with high critical temperatures, on silica columns. Without deactivation agents unmodified silica is of little value as a column material for solutes with polar functional groups, due to strong hydrogen bonding, ionic interactions and dipole interactions. The polar alternatives to silica are the chemical-bonded silica-based packing materials containing diol groups, amino groups or cyano groups.

Polar-bonded packing materials

Bonded phases with polar functional groups have so far not found extensive use in SFC, but there is no reason to believe that this will not change. Amino columns, which separate polycyclic aromatic hydrocarbons by ring size in HPLC, were shown by Takeuchi et al. (1984) to be useful in SFC also. The retention of condensed aromatics was reduced compared to the retention on ODS materials. Berry et al. (1986) used an amino column to separate ergot alkaloids with carbon dioxide modified with methanol or methoxyethanol.

A general problem with many silica-based packing materials is the incomplete coverage of the silica matrix, leaving residual silanol groups open to strong interactions with basic groups and electron donors such as large condensed aromatics. So far the only surface function which appears to shield the residual silanol groups from such interactions is the amino group. The disadvantage of the amino columns is the strong interactions with acidic components and the danger of forming imines with aldehydes/ketones. Thus, amino columns should only be used with due respect to the reactivity of the functional groups of the solutes.

Cyano columns have been utilized to trap the resins fraction of crude oils (Fig. 2). After back-flushing with CO_2, without modifiers, a minor part of the resins remained adsorbed to the column. Levy and Ritchey (1986) used a cyano column for the separation of aromatic amines, and a diol column for the separation of aminoazo compounds. Addition of modifiers improved peak shapes on both columns.

Non-polar-bonded phases

As in HPLC, the most widely used packed columns in SFC are the reversed-phase materials, particularly the C_{18} materials. Due to their excellent properties in HPLC the reversed-phase columns are commercially available in many sizes down to 1 mm i.d. Due to the low polarity of most SFC mobile phases, the properties of the C_{18} columns will be more affected by the manufacturing process than is normally the case in reversed-phase HPLC. Figure 3 demonstrates the effects on retention and peak shape which can be obtained on columns with different amounts of residual silanol groups.

Compounds with hydrogen bonding substituents, such as amines, amides, acids, alcohols, cyano compounds, nitro compounds, aldehydes and ketones should be expected to have a tendency to elute as tailing peaks with strong retention on packings with residual silanol groups. Even polycyclic aromatic hydrocarbons, with their high π-electron density, interact with the remaining silanol groups (Fig. 4). As demonstrated in Figs 3 and 4, such interactions can be reduced considerably on packing materials which have been manufactured for separation of amines and similar compounds. Still, it is not realistic to expect that silica-based materials can be produced completely without residual silanol groups. The options to overcome such interactions include the addition of modifiers or changing to resin-based packing materials, porous carbon materials or open tubular columns.

Non-silica hydrophobic materials

Gere (1983) showed that octylphenoxy polyethylene glycol oligomers from the surfactant Triton X-100 separated much better on a PRP (polymer reversed-phase) column than on a silica-based C_{18} column. Fourteen oligomers were resolved in 8 min. With a modifier gradient the resolution was improved even further (Fig. 5). The column was utilized at pressures exceeding 300 bar, which is high for a resin-based column. While the resin-based materials are interesting candidates as hydrophobic materials in SFC, too few experimental data are available to evaluate their usefulness. Their ultimate use in SFC will to a large extent depend on resistance to high pressures.

Porous graphite carbon, a material developed by Knox *et al.* (1986) for HPLC applications is another packing which also has interesting properties for SFC. Porous graphitic carbon is the only truly hydrophobic adsorbent; it

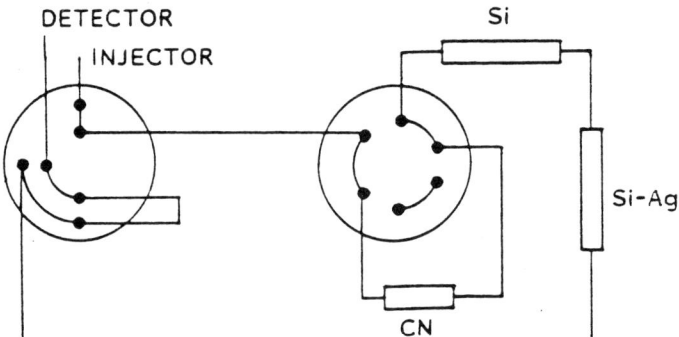

Figure 2. Column switching system for group characterization of crude oil. The cyano (CN) column is used for trapping the resins fraction and is back-flushed in the final step of the procedure. The silica (Si) and the silver nitrate on silica (Si−Ag) columns are used for separating aliphatics and aromatics.

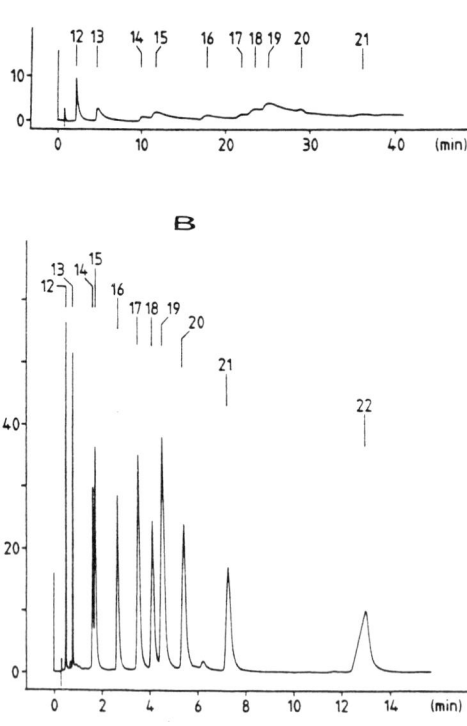

Figure 3. Nitro-PAH eluted with carbon dioxide at 40°C and 180 bar on CP Spher C_{18} (A) and on Nova-Pak C_{18} (B): 1-nitronaphthalene (12), 2-nitrofluorene (13), 3-nitrofluoranthene (14), 1-nitrotriphenylene (15), 2-nitrotriphenylene (16), 1-nitrobenzo[e]pyrene (17), 6-nitrobenzo[a]-pyrene (18), 9-nitrodibenz[a,c]anthracene (19), 3-nitrobenzo[e]pyrene (20), 7-nitrobenzo[ghi]-perylene (21), and 5-nitrobenzo[ghi]perylene (22).

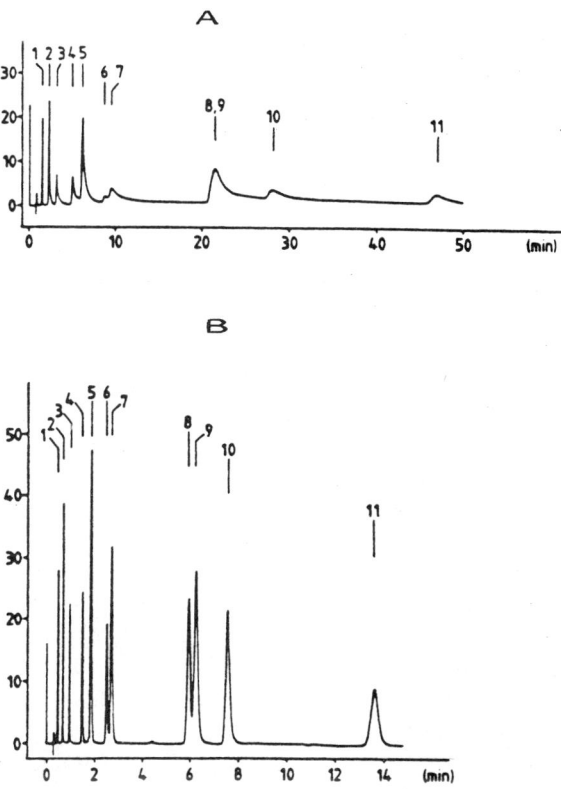

Figure 4. Polycyclic aromatic hydrocarbons with CO_2 at 175 bar and 40°C on CP Spher C_{18} (A) and on Nova-Pak C_{18} (B): naphthalene (1), fluorene (2), anthracene (3), fluoranthene (4), pyrene (5), benz[a]anthracene (6), chrysene (7), benzo[e]pyrene (8), benzo[a]pyrene (9), dibenz[a,c]anthracene (10) and benzo[ghi]perylene (11).

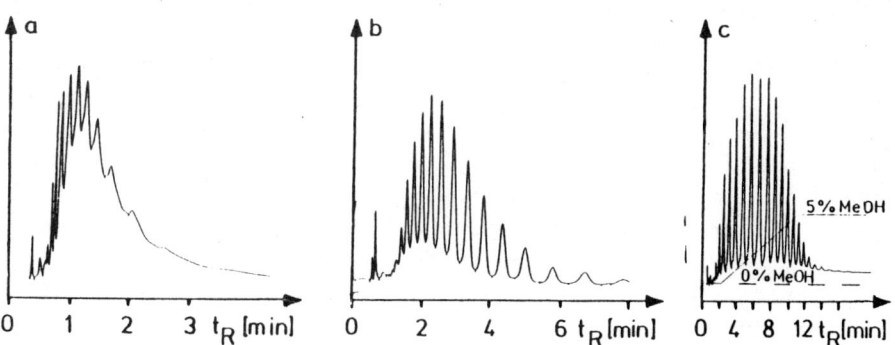

Figure 5. Separation of oligomers from the surfactant Triton X-100 on a reversed-phase column (a), on a PRP column (b), and on a PRP column with a modifier gradient (c). From Gere (1983), with permission.

is stronger than other carbon packings, but is still not as rigid as silica. A field of applications which look promising for porous graphitic carbon is the chemistry of petroleum/crude oils, providing the packings are rigid enough to withstand the pressures without being deformed.

MODIFIERS

The addition of solvents to the mobile phase can affect the properties of the mobile phase and the stationary phase. By adding limited amounts, around 1%, of polar solvents, the stationary phase can be deactivated while maintaining the desired critical conditions close to those of the pure fluid. In the simplest binary fluid systems the critical temperatures and pressures are approximately proportional to the mole fraction of polar modifiers, at limited levels of the modifier. Methods to calculate the critical loci of such simple systems were discussed briefly by Yonker and Smith (1986). With 2-propanol in carbon dioxide as an example, at a mole fraction of 0.2, the critical temperature of the mixture increased to above 100°C and the critical pressure increased to above 120 bar. With open tubular columns and modifier concentrations of 2% (w/w), the authors reported that the modifiers reduced peak tailing, and postulated a connection with active sites in the silicone stationary phase. Another property of the silicones, which was discovered recently, is the pronounced swelling, particularly of thick films — as shown by Novotny and David (1986).

Yonker and Smith (1986) also reported that 2-propanol in carbon dioxide (at a mole fraction of 0.2) reduced the retention of various aromatic compounds more than methanol in carbon dioxide. On a packed C_{18} column Blilie and Greibrokk (1985a) found slightly different results with small amounts (2% v/v) of modifiers. Since the solubilities of many of the aromatic compounds studied are higher in higher alcohols than in methanol, this shows that a differentiation between adsorption and solubility effects may be difficult to obtain, but that the major effect on silica-based packed columns is related to adsorption.

The other effect of the modifiers is to alter (usually increase) the solvating power of the mobile phase when adding substantial amounts of modifiers to the fluid. Berry et al. (1986) used up to 20% of alcohols in CO_2 in order to separate xanthines and sulfonamides. The modifier can be added at constant rate or at an increasing rate, as shown by Schmitz et al. (1983) and by Blilie and Greibrokk (1985b). Modifier gradients can be compared with density gradients, such as in Fig. 6 where a combined modifier/density gradient was determined to be the best choice for separation of nitropolycyclic aromatic hydrocarbons.

The most widely used modifier is methanol. The reason for this is probably the well-known efficiency of methanol as a deactivating agent in adsorption chromatography. The retention of nitropolycyclic aromatic hydrocarbons as a function of methanol content in the mobile phase is shown in Fig. 7. Not only the retention is affected, but the peak shape and thereby the detect-

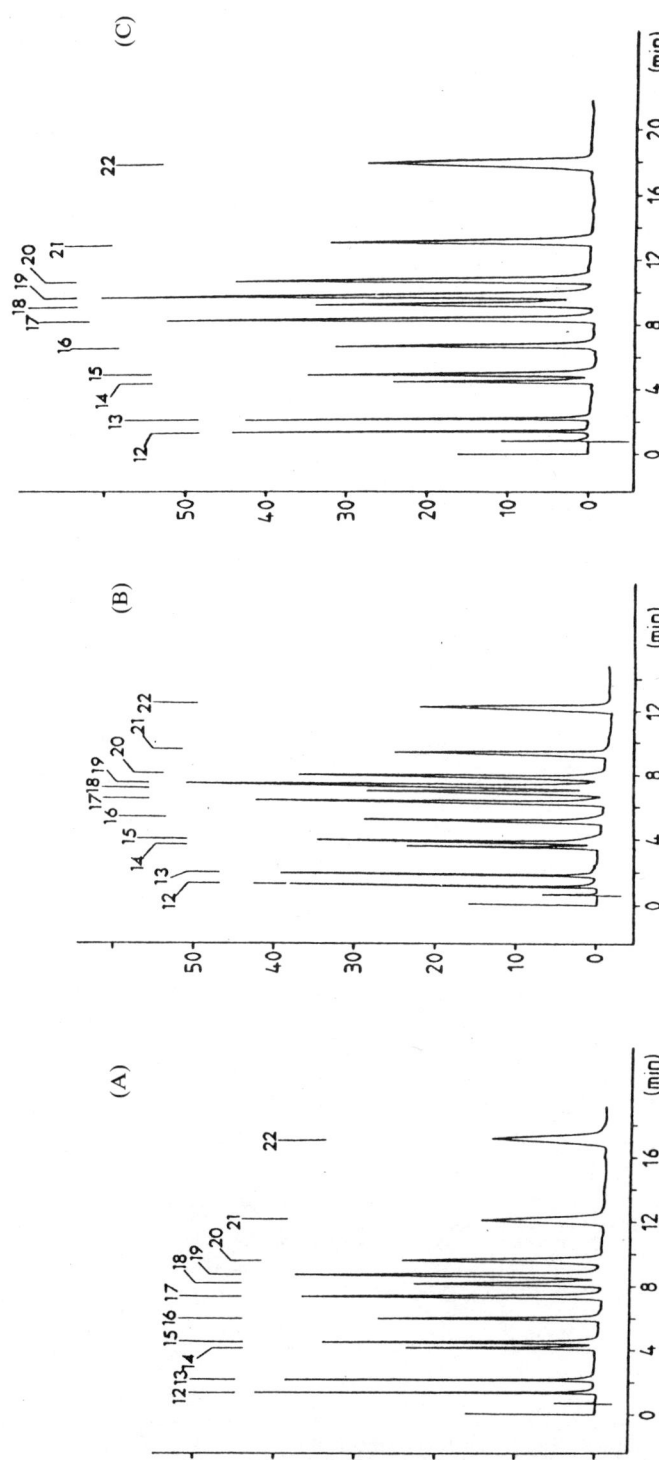

Figure 6. Improved separation of nitro-PAH with pressure gradient from 150 to 270 bar (A), with modifier gradient from 4% to 10% methanol in CO_2 (B) and with a combined pressure/modifier gradient from 160 to 270 bar and from 3% to 10% methanol in CO_2 (C). For identification of the individual compounds see Fig. 3.

ability can be significantly improved, as shown in Figs 8 and 9. Among the alcohols, however, methanol is not necessarily the most efficient modifier, as shown by Blilie and Greibrokk (1985a) and by Levy and Ritchey (1986). The higher alcohols had the strongest effect on polycyclic aromatic hydrocarbons (Fig. 10), nitropolycyclic aromatic hydrocarbons and on polystyrenes. Hexanol reduced the retention of octylbenzene, hexachlorobenzene, didodecyl phthalate and 4-nitroaniline more than methanol, while methanol had a stronger effect on aminoazo compounds. The most efficient modifier of all, however, was shown to be propylene carbonate (Levy and Ritchey, 1986).

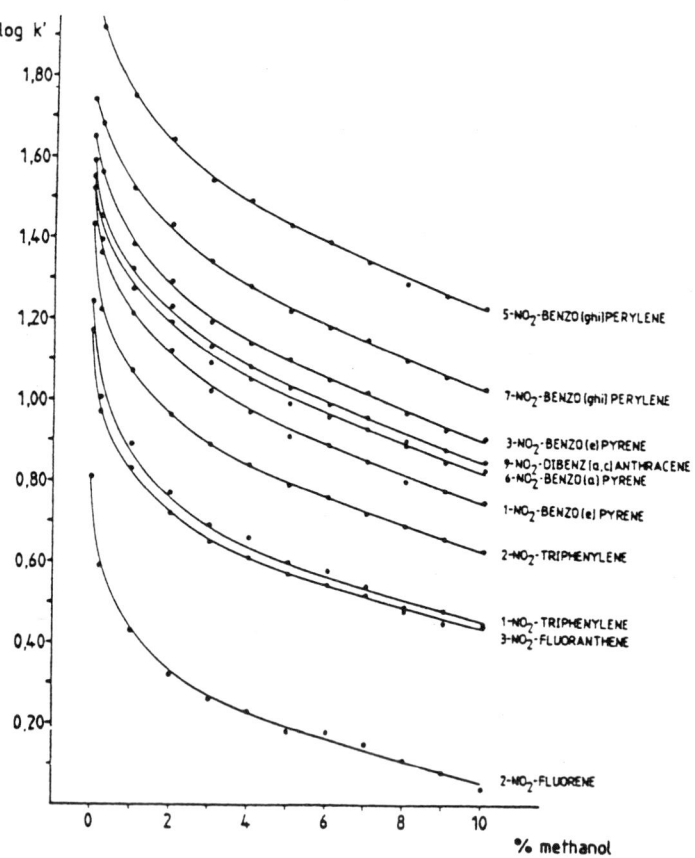

Figure 7. The retention of nitro-PAH as a function of the amount of methanol in CO_2, at 170 bar and 40°C on CP Spher C_{18}.

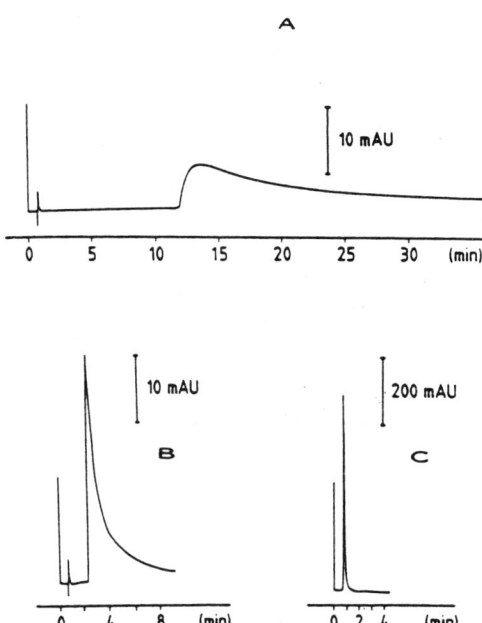

Figure 8. Benzoic acid injected on CP Spher C_{18} at 160 bar and 40°C with 0% (A), 1% (B) and 10% (C) methanol in CO_2.

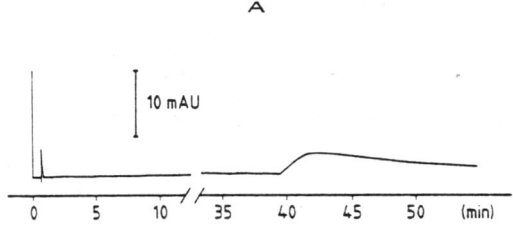

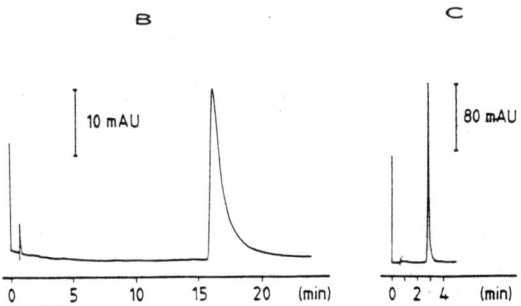

Figure 9. 13 H-dibenzo[a,i]carbazol injected on CP Spher C_{18} at 160 bar and 40°C with 0% (A), 1% (B) and 10% (C) methanol in CO_2.

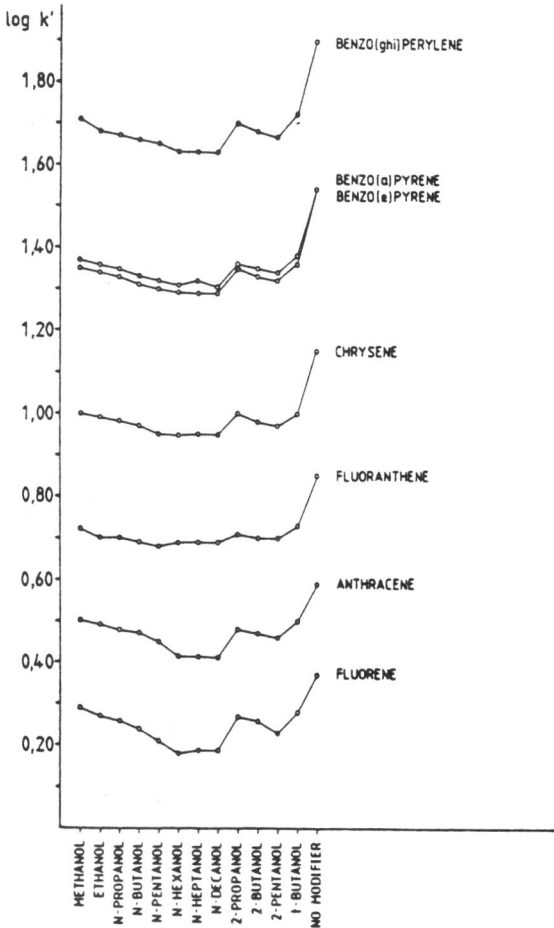

Figure 10. Retention of polycyclic aromatic hydrocarbons as a function of alcohols (2%) in CO_2 at 160 bar and 40°C on CP Spher C_{18}.

Modifier mixing

The simplest mixing process is to purchase pre-mixed fluids, but the selection is limited. Two pre-mixed fluids and two SFC pumps are needed to obtain gradients. The fluid may also be mixed in a separate tank or in the pump cylinder of syringe pumps with relatively large cylinder volumes. A known volume of modifier is first transferred to the cylinder, followed by the fluid in the liquid state. Batch-to-batch reproducibility is directly connected to the extent to which the cylinder is filled with fluid in the liquid state. This system also needs two SFC pumps for gradient formation.

After a syringe pump has been used with a mixed fluid the system cannot be changed without thorough cleaning. In our experience removal of the last traces of solvents needs two to four refilling and flushing cycles. With certain pumps this is a highly time-consuming task, and reminds one of the early days of HPLC when syringe pumps were common for solvent delivery.

The most flexible way of mixing the modifier with the fluid is to pump the fluid and the modifier into a mixer with two different pumps. In order to be able to control the mixing rate the fluid pump should be on flow control, not on pressure control. If very low flow rates require a syringe pump to supply the modifier, this pump should be connected to the mixer with a piece of narrow tubing in order to avoid back-diffusion of fluid from the SFC pump into the cylinder of the modifier pump. By using reciprocating piston pumps for modifier delivery, no back-diffusion into the solvent reservoir will occur. With capillary columns it is difficult, however, to find reciprocating pumps with flow rates and flow-rate reproducibilities useful for modifier supply (at low percentages of modifier). Thus the eventual design of the modifier delivery unit and the mixing system is column-dependent. Even if simple T-tubes have been used with good results, more efficient mixers, with minimum dead volumes, are recommended. A simple mixer can be obtained by filling a small tube with glass beads. Post-column HPLC mixers should be used with care, since they do not necessarily withstand the pressures of SFC.

PACKED FUSED SILICA COLUMNS

Column dimensions

Considering the wall thickness, the strength of the polyimide fused silica tubes is quite impressive. Few data are published, however, on maximum allowable pressures. In our experiments with commercially available standard sizes of tubing we found that tubes with 0.4 mm outer diameter and 0.3 mm inner diameter ruptured at pressures of approximately 400 bar. With an inner diameter of 0.2 mm and outer diameter of 0.4 mm all the columns withstood pressures up to 600 bar. The rupturing usually started at the column inlet, where the highest pressure is found, and where the connection puts an extra strain on the wall. The rupturing process varied from a clean break to a long smoldering process. Since pressures of 400–600 bar may be required to pack fused silica columns efficiently, 0.2 mm columns may seem the natural choice. Unfortunately the 0.2 mm columns are not practical for connections with other fused silica tubes, since most connecting lines have outer diameters larger than 0.2 mm, not allowing efficient internal connections.

One alternative is to reinforce 0.3 mm columns. Takeuchi *et al.* (1984) cemented the fused silica tube into a stainless-steel tube with an epoxy glue. We have used a similar reinforcement, but with an outer fused silica tube (i.d. 0.53 mm, o.d. 0.8 mm). Since the heat conductivity is reduced by an extra wall of fused silica, aluminum-clad columns may possibly have certain advantages. Depending on the choice of reinforcement, the column outer diameter may end up as 1.6 mm (1/16 in.), 0.8 mm (1/32 in.) or 0.4–0.5 mm, respectively.

With fused silica columns the high-volume pumps used to pack standard HPLC columns are not needed. Due to the small volumes almost any pump can be utilized, provided sufficiently high pressures can be delivered.

Column connections

All packed columns need to be terminated by a particle filter. It is, however, equally important to protect the inlet of the SFC column with a filter. Without inlet filter the momentary pressure reduction during injection may blow packing material back into the injector.

With the 1/16 in. stainless-steel tube reinforcement design of Takeuchi *et al.* (1984), the double tube ended up against a stainless-steel frit in a union in both ends (Fig. 11a). Whether this system produce extra band broadening depends on the finish of the column ends and the size of the metal frit.

In other designs the filter is placed inside the column in order to reduce band broadening. For HPLC applications Gluckman *et al.* (1983) used a thin porous teflon disc. Einarsson *et al.* (1986) used a glass filter plug with 2 µm porosity (Fig. 11b). Alborn and Stenhagen (1985) used a coarse packing placed at the end of the column which was drawn in the flame to a fine tip. The latter design was constructed for LC−MS, but cannot be transferred to SFC and SFC−MS due to the lack of restrictor, unless the tip is drawn as restrictor. With this design the column is destroyed at the same time as the tip is plugged or damaged. We have used a modification of the method of Gluckman *et al.* (1983), but with a 0.8 µm porosity glass-fiber filter for 3 and 4 µm packing materials.

At the outlet of the column the most efficient way of connecting two tubes is to use connector tubing (or restrictor tubing) with an outer diameter small

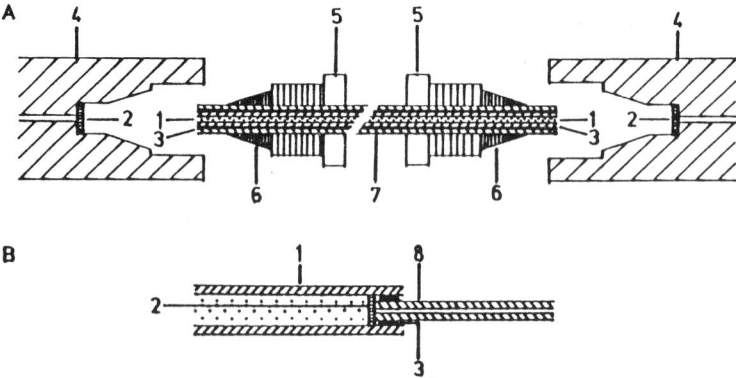

Figure 11. Fused silica column in stainless-steel tubing, terminated with metal frits in union connectors (A). Fused silica column terminated with glass-fiber plug inside the column and fused silica connector (B). 1 = Column, 2 = filter, 3 = glue, 4 = union connectors, 5 = compression screw, 6 = ferrule, 7 = stainless-steel tubing and 8 = fused silica connector.

enough to be inserted into the end of the column (Fig. 11b). The two tubes are fixed with a standard epoxy glue which is applied on the outside of the internal tube, almost, but not completely, to the end. If the restrictor tubing is too wide to be inserted into the column there are two alternatives. One is to use a short connector tubing which can be inserted into and glued to both the column and the restrictor tubing (Fig. 12a). The other alternative is to use a butt-to-butt connection between the column and the restrictor. Since the standard butt-to-butt connectors are too large to pass through the narrow openings in many detectors, a much slimmer design is needed, whereby the two tubes and a glass filter plug are cemented butt-to-butt into an outer fused silica tube, fitting like a glove on the two capillaries (Fig. 12b). Thus the choice of restrictor/detector connection is totally dependent of the kind of restrictor used.

At the column inlet side the alternatives are fewer: direct connection of the column to the injector or indirectly via a connecting tube. The latter is inserted with a glass filter plug into the column and fixed by glue. The design of the direct connection to the injector depends on the internal bore of the injector and on the size of the connecting capillary. With a 1/16 in. bore injector the column can be connected directly with a metal frit (Fig. 13a) or via a connector (Fig. 13b). In order to obtain a tight fitting of the tube and avoid dead volumes, Springston and Novotny (1984) used a Teflon insert and a Vespel ferrule drilled to fit over the insert, which was suitable up to at least 240 bar. We have preferred to use a Teflon filler which is not inserted through the ferrule (Fig. 13b), since the friction between the capillary and Teflon is less than between the capillary and Vespel, reducing the risk of the capillary sliding out at high pressures.

The metal frit connection, shown in Fig. 13a, should be avoided with highly efficient columns in order to reduce the extra-column band broadening.

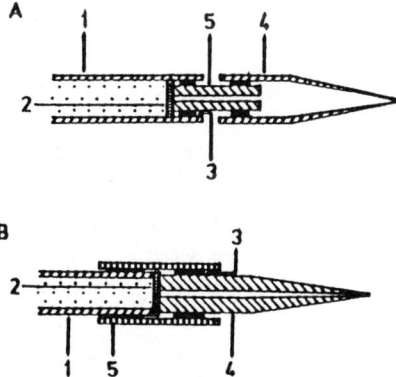

Figure 12. Internal connection of tapered restrictor to fused silica column (A). Butt-to-butt connection of tapered restrictor and fused silica column (B). 1 = Column, 2 = filter, 3 = glue, 4 = restrictor and 5 = connecting tube.

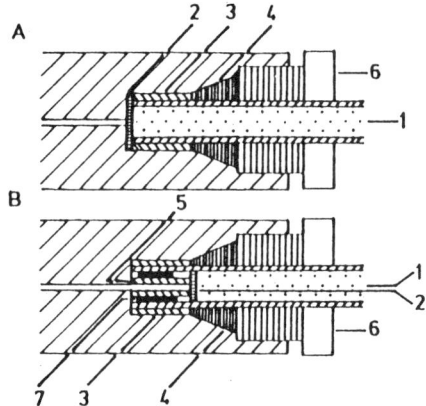

Figure 13. Direct connection of fused silica column to 1/16 in. bore injectors and detectors with reducing Vespel ferrule and Teflon sleeve, terminated with metal frit (A) or with glass-fiber plug and internal capillary (B). 1 = Column, 2 = filter, 3 = Teflon tube, 4 = Vespel ferrule, 5 = internal capillary, 6 = compression screw and 7 = glue.

ACKNOWLEDGEMENTS

Financial support was received from Statoil, Norway, through the Norwegian Academy of Sciences, and from the Norwegian Council for Scientific and Industrial Research (NTNF).

REFERENCES

Alborn, H. and Stenhagen, G. (1985). Direct coupling of packed fused-silica liquid chromatographic columns to a magnetic sector mass spectrometer and application to polar thermolabile compounds. *J. Chromatogr. 323*, 47–66.

Berry, A. J., Games, D. E., and Perkins, J. R. (1986). Supercritical fluid chromatographic–mass spectrometric studies of some polar compounds. *J. Chromatogr. 363*, 147–158.

Blilie, A. L. and Greibrokk, T. (1985a). Modifier effects on retention and peak shape in SFC. *Anal. Chem. 57*, 2239–2242.

Blilie, A. L. and Greibrokk, T. (1985b). Gradient programming and combined gradient–pressure programming in supercritical fluid chromatography. *J. Chromatogr. 349*, 317–322.

Einarsson, S., Folestad, S., Josefsson, B., and Lagerkvist, S. (1986). High-resolution reversed-phase liquid chromatography system for the analysis of complex solutions of primary and secondary amino acids. *Anal. Chem. 58*, 1638–1643.

Folestad, S. (1985). Detection and column technology in miniaturized liquid chromatography. Doctoral thesis, Department of Analytical and Marine Chemistry, University of Gothenburg, Sweden, p. 31.

Gere, D. R., Board, R., and McManigill, D. (1982). Supercritical fluid chromatography with small particle diameter packed columns. *Anal. Chem. 54*, 736–740.

Gere, D. R. (1983). Oligomer separations by supercritical fluid chromatography. Application Note AN 800-3. Hewlett-Packard Co., Avondale, PA.

Gluckman, J. C., Hirose, A., McGuffin, V. L., and Novotny, M. (1983). Performance evaluation of slurry-packed capillary columns for liquid chromatography. *Chromatographia 17*, 303–309.

Greibrokk, T., Blilie, A. L., Johansen, E. J., and Lundanes, E. (1984). New system for delivery of the mobile phase in supercritical fluid chromatography, *Anal. Chem. 56*, 2681–2684.

Greibrokk, T., Doehl, J., Farbrot, A., and Iversen, B. (1986). Mobile-phase delivery in supercritical fluid chromatography. *J. Chromatogr. 371*, 145–152.

Hirata, Y. and Nakata, F. (1984). Supercritical fluid chromatography with fused-silica packed columns. *J. Chromatogr. 295*, 315–322.

Klesper, E., Corwin, A. H., and Turner, D. (1962). High pressure gas chromatography above critical temperatures. *J. Org. Chem. 27*, 700–701.

Knox, J. H., Kaur, B., and Millward, G. R. (1986). Structure and performance of porous graphitic carbon in liquid chromatography. *J. Chromatogr. 352*, 3–25.

Lee, M. L. and Markides, K. E. (1986). Capillary supercritical fluid chromatography: sample introduction, flow restriction, column stability and migration of polar solutes. *J. Chromatogr.*, in press.

Levy, J. M. and Ritchey, W. M. (1986). Investigation of the uses of modifiers in supercritical fluid chromatography. *J. Chromatogr. Sci. 24*, 242–248.

Novotny, M., Bertsch, W., and Zlatkis, A. (1971). Temperature and pressure effects in supercritical fluid chromatography. *J. Chromatogr. 61*, 17–28.

Novotny, M., Springston, S. R., Peaden, P. A., Fjeldsted, J. C., and Lee, M. L. (1981). Capillary supercritical fluid chromatography. *Anal. Chem. 53*, 407A–414A.

Novotny, M. V. and David, P. A. (1986). The mobile-phase effects on stationary-phase properties in capillary supercritical fluid chromatography. *J. Chromatogr.*, in press.

Peaden, P. A., Fjeldsted, J. C., Lee, M. L., Springston, S. R., and Novotny, M. (1982). Instrumental aspects of capillary supercritical fluid chromatography. *Anal. Chem. 54*, 1090–1093.

Schmitz, F. P., Hilgers, H., and Klesper, E. (1983). Gradient elution in supercritical fluid chromatography. *J. Chromatogr. 267*, 267–275.

Springston, S. R. and Novotny, M. (1984). Mobile-phase solute mass transfer in supercritical fluid chromatography. *Anal. Chem. 56*, 1762–1766.

Takeuchi, T., Ishii, D., Saito, M., and Hibi, K. (1984). Supercritical fluid chromatography with micro packed columns and carbon dioxide as a mobile phase. *J. Chromatogr. 295*, 323–331.

Yang, F. J. (1986). Microbore column chromatography: a review of techniques and applications of CGC, capillary–SFC and capillary–HPLC. *J. Chromatogr.*, in press.

Yonker, C. R. and Smith, R. D. (1986). Study of retention processes in capillary supercritical Luid chromatography with binary fluid mobile phases. *J. Chromatogr. 361*, 25–32.

Progress in HPLC, Vol. 4, pp. 73—85.
Yoshioka *et al.* (Eds)
© 1989 VSP.

Oligomer separation by supercritical fluid chromatography using gradient elution

FRANZ P. SCHMITZ and BERND GEMMEL

Lehrstuhl für Makromolekulare Chemie, RWTH Aachen,
Worringerweg, D-5100 Aachen, Federal Republic of Germany

INTRODUCTION

Separation of polymers is most frequently done by means of size exclusion chromatography (SEC). This applies also to low molecular weight polymers, where it is possible to separate adjacent oligomers, i.e. to separate members of the homologous series which differ in length only by one repetitive unit. Although method development has considerably improved the separation power of SEC, oligomer separations by SEC are rather time-consuming. Furthermore, SEC techniques are restricted to separation based on molecular size, and are unable to differentiate according to chemical factors. Hence, parallel homologous series which may be present in polymer samples can be identified by SEC only when the size of the molecules in solution differs.

Separations of higher efficiency than by SEC can be achieved by retention chromatography techniques, where the separation mechanism is based on the interaction of dissolved molecules with the stationary phase. This was first applied by Jentoft and Gouw (1969), separating styrene oligomers via pressure-programmed supercritical fluid chromatography (SFC). Some years later, van der Maeden *et al.* (1978) used gradient elution high-performance liquid chromatography (HPLC) for the separation of oligomers. For eluting a homologous series by retention chromatography, the application of a gradient technique is essential since, as has frequently been observed, a linear correlation exists between the logarithm of the capacity ratio and the degree of polymerization, i.e. retention time increases exponentially with increasing molecular mass. Therefore, the solvent strength of the mobile phase has to be increased steadily during a chromatographic run in order to obtain equally spaced peaks.

While in liquid chromatography (LC) the enhancement of the solvent strength is increased by changing the eluent composition, reduction of reten-

tion times is achieved in gas chromatography (GC) by increasing the column
temperature, thereby increasing the solutes' vapor pressures. In SFC a third
gradient method is applied, making use of the increase of solvent strength
with increasing mobile phase density, i.e. by increasing the column pressure.
All three gradient methods have been shown to be applicable to the SFC
separation of oligomers (Jentoft and Gouw, 1967; Schmitz and Klesper,
1981a; Schmitz et al., 1987b), with pressure/density gradients being most
popular. In this communication we shall focus on gradients of eluent composition, the importance of which has recently been seen to be increasing.

MATERIALS AND METHODS

Apparatus

Separations were performed using a modified HPLC instrument which is
schematically shown in Fig. 1. It is based on a model 1084 B apparatus from
Hewlett-Packard, supplemented by an additional oven capable of attaining
temperatures up to 300°C, as well as by pressure control devices.

Mobile phases with boiling temperatures below 50°C are pre-pressurized
in steel storage tanks (1) by application of helium pressure. Alternatively,
mobile phases with sufficiently high boiling points can be fed without pre-
pressurization from the glass bottles (2) provided with the 1084 B. The
components are metered separately in the liquid state by the membrane-
type pumps (3) and consecutively pass the pulsation damping system
combined with a device for measuring the feed rate (4), and the mixing
chamber (5). For pumping low-boiling eluents (such as carbon dioxide) at
elevated ambient temperatures it may become necessary to cool the pump
by means of a cooling device attached to the pump head. After having
passed a filter (6) the mobile phase reaches the variable volume injection
system (7) or, if low-boiling eluents are used, a loop-type injector (8).

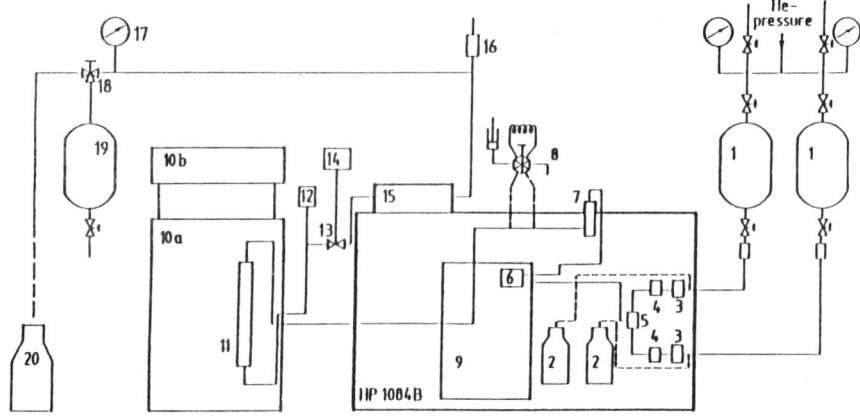

Figure 1. SFC instrument (schematic). For explanation see text.

Since the oven of the 1084 B (9) is usable only up to 100°C, the mobile phase is led directly out of the unheated oven to an external oven (10a). This external oven was either a laboratory oven with forced-air circulation (UT 5042 EK, W. C. Heraeus, Hanau, FRG) equipped with a temperature programmer (10b; Kelvitron TPG2, Heraeus), or a GC oven (Carlo Erba Fractovap 4200, Carlo Erba Instruments, Hofheim/Ts., FRG). The mobile phase is transferred to the supercritical state within the capillary inside the oven. The supercritical mobile phase then enters the separation column (11) and, after having left the column and the oven, is allowed to liquefy. Transferring the mobile phase back to the liquid state prior to detection has no significant detrimental influence on the quality of the chromatogram, as has also been observed by Johnson *et al.* (1985). The pressure at the column outlet is monitored by an electric pressure transducer (12; Meßumformer Teleperm D, M 56441, Siemens, Karlsruhe, FRG), while the pressure upstream of the column is measured by the pressure transducer of the 1084 B. The pressure is set by means of a regulating valve (13); and by combining pressure transducer (12) and valve (13) with a programmer and a motor (14), pressure programming facilities are obtained. Alternatively, for keeping the pressure constant, a back-pressure regulating valve (Tescom 26-3220-24-004, WSM, Eching, FRG) was used.

At a lowered pressure level the stream of the mobile phase is then led to the UV detector (15) of the 1084 B, which was modified to withstand higher pressures (cf. McManigill *et al.*, 1982) and is capable of working in a wavelength range of 190−600 nm. Alternatively, another UV detector (LC 75, Bodenseewerk Perkin-Elmer, Überlingen, FRG) with variable wavelength detection (190−600 nm) was used. With both detectors, chromatograms were recorded using the plotter of the 1084 B. An external metering valve (18) and a manometer (17) are used to raise the pressure in the detector cell to a level that prevents formation of gas in the mobile phase. To protect the detector an adjustable spring-actuated safety valve (16) can be provided. Frequently the column outlet pressure was regulated downstream of the detector; in these cases one valve and one manometer could be omitted. The effluent is finally collected in a pressure-resistant metal container (19) as a liquid under its own vapor pressure or, in the case of high-boiling eluents, in a glass bottle (20).

Mobile and stationary phases

Eluents which are liquids at ambient conditions (pentane, dioxane, acetonitrile) were dried over sodium, distilled and degassed. Carbon dioxide was purchased with sufficient purity of 99.995% (Messer-Griesheim, Hürth, FRG). The carbon dioxide containers were inverted for supplying liquid CO_2 to the pumps. Pentane and CO_2 were fed to the pumps by applying helium pressure to the containers; a filter cascade of four sintered metal frits having pore diameters of 20 to 5 μm were provided between the eluent containers and the pumps. Critical data for the pure eluent components are shown in Table 1.

Table 1.
Critical data for eluents (according to Reid *et al.*, 1977)

Compound	$T_c/°C$	P_c/bar
Carbon dioxide	31.1	73.8
Pentane	196.5	33.7
1,4-Dioxane	313.9	52.1
Acetonitrile	275	48.3

As the stationary phase, bare silica (LiChrosorb Si 60 or LiChrosorb Si 100) with an average particle diameter of 10 μm was used. This silica material is modified 'on-column' by dioxane, either during column conditioning with dioxane-containing mobile phases or by separate treatment with dioxane at elevated temperatures (Schmitz *et al.*, 1987a). The stationary phase was packed into the stainless steel columns (length 25 cm, internal diameter 4.6 mm) using a slurry technique. The slurry, consisting of a suspension of the stationary phase (2.5 g) in 20 ml of toluene and 30 ml of cyclohexanol, is filled into a slurry reservoir connected to the column. The remaining volume of the reservoir is then filled with the packing medium (heptane, pentane or methanol) and connected to a membrane-type reciprocating pump (AE 10-4.4, Orlita, Gießen, FRG). Then the packing medium is pumped onto the slurry, and the compression of the column packing is carried out at a pressure of 300–400 bar. Finally, the columns are conditioned with the desired mobile phase at supercritical conditions.

Samples

The oligostyrene sample (PS 800) was purchased from Pressure Chemical Co. (Pittsburgh, PA) and had a $\bar{M}_w/\bar{M}_n$ value of ≤1.30 (according to the manufacturer's data). The polycaprolactonediol sample was purchased from Aldrich (Steinheim, FRG) and should have an average molecular weight of 530 g/mol (according to the manufacturer). The preparation of the vinylbiphenyl oligomers has been described elsewhere (Schmitz *et al.*, 1985). Sample solutions were prepared using the secondary component of the eluent system as the solvent.

RESULTS AND DISCUSSION

The use of gradients of eluent composition in supercritical fluid chromatograpy was first proposed by Klesper (1978). This gradient technique, being of eminent importance in HPLC, was first applied some years later (Schmitz and Klesper, 1981a, b) with the separation of styrene oligomers. In these studies, adjacent styrene oligomers could be separated up to degrees of polymerization, *n*, of *n* = 54.

As in liquid chromatographic separations, increasing the amount of a component having a higher solvent strength for the sample under investigation in the mobile phase (eluent B) leads to reduced retention. Hence, higher molecular weight sample components, which do not elute at the initial composition of the mobile phase, will leave the separation column when the solvent strength has been sufficiently ameliorated by the addition of eluent B. According to the decreasing solubility difference between adjacent homologues with increasing molecular weight, a linear composition gradient is frequently not appropriate. As a simple nonlinear gradient it has been found suitable to increase the ratio Q_B linearly.

$$Q_B = \frac{p_B}{1 - p_B} \tag{1}$$

Here, p_B is the volume fraction of eluent B in the eluent mixture.

As an example of an almost linear gradient, Fig. 2 shows the separation of styrene oligomers with a pentane–dioxane mobile phase. The amount of dioxane in the mobile phase was increased according to the gradient shown in the figure. Similar results were obtained replacing pentane as the primary component (eluent A) of the mobile phase by hexane, butane and propane (Schmitz *et al.*, 1983). Using a pentane–dioxane mobile phase, and applying higher pressures, separations could also be achieved with polystyrenes of narrow molecular weight distribution (Schmitz and Klesper, 1983). Furthermore, additional vinyl arene oligomers were separated using the pentane–dioxane mobile phase (Schmitz *et al.* 1985; Schmitz and Hilgers, 1986; Schmitz *et al.*, 1986).

A variety of eluent pairs has been tested with respect to their applicability for the separation of oligomers (Table 2). In most cases the oligomers were of the vinylarene type, because such oligomers can easily be detected by means of a UV detector. For these oligomers, dioxane was suitable as the B component of the eluent system, the A component being an alkane, diethyl ether, trifluoromethane, or carbon dioxide. Dioxane could be replaced by cyclohexane or by methylene chloride, which are also better solvents for vinyl arene oligomers than are any of the above A components. Methylene chloride, being thermally less stable and reported to decompose when used as a single supercritical fluid (Asche, 1978), can be applied in binary mixtures when the A component has a low critical temperature and at fairly low

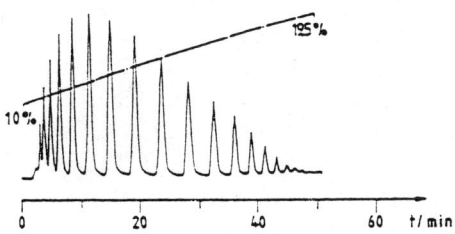

Figure 2. SFC-separation of styrene oligomers (PS 800). Mobile phase: pentane–dioxane. Pressure at column outlet (p_e) at the start of the chromatogram: 50 bar; temperature: 240°C; flow rate: 1 ml/min (measured at the pumps in the liquid state); stationary phase: LiChrosorb Si 100, 10 μm; detection at 254 nm. The gradient is as shown, percentage given as volume ratio.

Table 2.
Eluent pairs applied for oligomer separation

Eluent B	Eluent A	Eluent B
1,4-Dioxane	Hexane — Pentane — Butane — Propane — Ethane — Diethylether — CHF₃ — CO₂	Cyclohexane — Methanol — Dichloromethane — Ethanol — Acetonitrile

percentages of methylene chloride. The critical temperature of the mixture is then relatively low. The same observations also apply to acetonitrile. Lower alcohols such as methanol and ethanol are not useful as a component of the eluent system when a silica stationary phase is used, since at supercritical conditions and alcohol concentrations exceeding 10% the silica stationary phase dissolves in the mobile phase (Schmitz *et al.*, 1983). However, a polymer stationary phase should provide sufficient stability towards alcohols which is indicated by the successful gradient separation of oxyethylene oligomers with CO_2−methanol on a polystyrene stationary phase reported by Gere (1983).

Comparisons which were performed with separations of vinylarene oligomers, using pentane−dioxane eluents at liquid and at supercritical conditions, demonstrate that better separations can be obtained when the mobile phase is in the supercritical state (Schmitz *et al.*, 1985, 1986). This can be ascribed to the diffusion coefficients being higher, and the viscosity lower, at supercritical compared to liquid conditions. It was furthermore demonstrated that by changing pressure and temperature with supercritical binary fluids, the separation can be tailored to specific requirements (Schmitz *et al.*, 1985). A study on the influence of pressure on the oligostyrene separation with pentane−dioxane as shown in Fig. 2 revealed that the overall resolution reached a maximum value when the column outlet pressure was set at about 42 bar at the start of the chromatogram (Schmitz, 1986). Hence, SFC with gradient elution possesses great variability concerning method optimization.

In most cases, and this also applies to the separation shown in Fig. 2, the column outlet pressure was not maintained constant during a run, but instead was allowed to increase according to the higher viscosity of the eluent component B. Such an increase of pressure was shown to support the composition gradient (Schmitz *et al.*, 1983). With such an increase of pressure the gradient is actually of a binary type, being composed of a composition gradient and an 'inherent' pressure gradient. When the pressure gradient is increased still further, e.g. by increasing the overall flow rate during a run, it becomes possible to elute compounds which can be eluted neither by composition programming nor by pressure programming; this was demonstrated with separations of *N*-vinyl carbazole oligomers (Schmitz *et al.*, 1985, 1986).

While the combination of composition and pressure program can be used for accelerating a separation, combination of either one of these techniques with temperature programming can be used for improving the resolution of a separation. As is shown in the Schmitz *et al.* chapter in this book, maxima of resolution are observed in dependence of pressure and/or eluent composition. The temperature T_R^m where these maxima occur shifts to higher values with increasing pressure and increasing amount of component B in the eluent mixture, which is shown in Fig. 3 for the CO_2−dioxane eluent pair. This indicates that, for maintaining a maximum resolution throughout a separation, the temperature should be increased during the separation according to the shift of the temperature of the resolution maximum with eluent composition and/or pressure. In fact, some preliminary studies demonstrated the improvement of resolution upon superimposing a temperature gradient on a composition gradient (Leyendecker *et al.*, 1986; Schmitz *et al.*, 1987b) or on a pressure gradient (Schmitz *et al.*, 1987b). Some examples of this multiple gradient technique are given below.

Figure 4 shows four separations of styrene oligomers. Chromatograms *a* and *c* were obtained isothermally at 145°C, the difference being a more steeply sloped nonlinear gradient for chromatogram *c*. Employing the same composition gradient as for *a*, but superimposing the corresponding temperature program, as determined for the increase of both dioxane content and pressure from Fig. 3, chromatogram *b* was obtained. The same relation applies to chromatograms *c* and *d*. In both cases the resolution is improved. This can also be seen from Table 3, giving the resolution values for adjacent peak pairs as well as the average resolution, R_m. The improvement for R_m is found to be almost identical in both cases; however, a closer inspection reveals differences for the resolution of individual peak pairs. On one hand, high resolution values are reduced ($R_{2,3}$, $R_{3,4}$), or remain constant ($R_{4,5}$, $R_{5,6}$) when chromatogram *a* is compared to chromatogram *b*. On the other hand, the lower resolution values found for the higher homologues are generally improved. Comparing chromatograms *c* and *d*, both of which were obtained at higher eluent gradient ramps, the resolution is more uniformly improved throughout the chromatograms than with the slower eluent composition gradient (chromatograms *a* and *b*).

Figure 5a shows a separation of a 4-vinyl biphenyl oligomer sample, obtained with the same eluent pair CO_2−dioxane, applying a dioxane gradient isothermally at 135°C with an inherent pressure gradient. Using similar conditions with an additional temperature program leads to the chromatogram of Fig. 5b. With the latter chromatogram, where the pressure gradient was more steeply ramped, more compounds are eluted at an even smaller analysis time compared to the isothermal chromatogram, but due to the temperature gradient most of the resolution is retained.

As already mentioned, the use of a low-boiling compound as the component A of an eluent system allows one to choose as the component B a compound of relatively low thermal stability. This has been demonstrated for gradient elution with the pair CO_2−CH_2Cl_2 (Schmitz *et al.*, 1987b). An

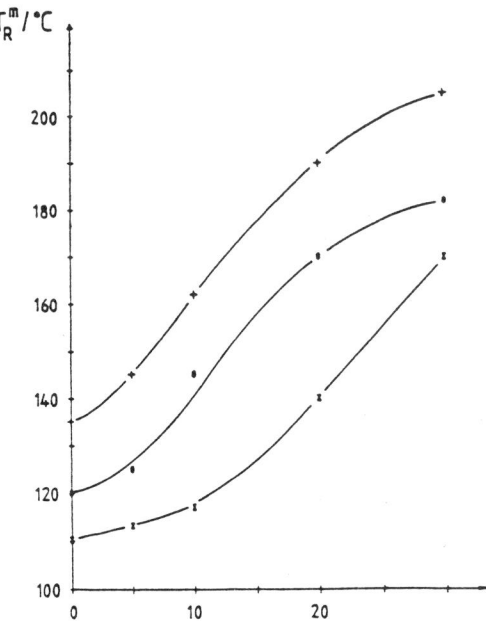

Figure 3. Variation of the temperature of the resolution maximum, T_R^m, on the amount (w/w) of dioxane in CO_2, p_e = 150 bar (×), 200 bar (○), 250 bar (+). Stationary phase and flow rate as for Fig. 2.

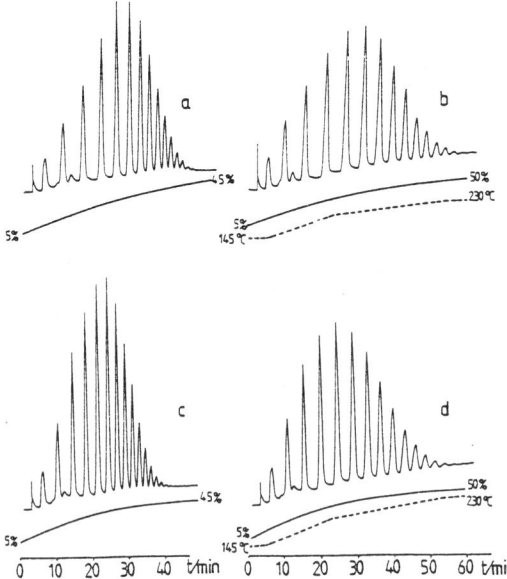

Figure 4. SFC separations of styrene oligomers (PS 800) with gradient elution isothermally at 145°C (*a*, *c*) and with superimposed temperature-programming (*b*, *d*), p_e increasing from 250 to 280 bar during each chromatogram. Mobile phase: CO_2–dioxane. Gradients as shown (——, composition gradient, – – –, temperature gradient). Other conditions as for Fig. 2.

Table 3.
Resolution data for the oligostyrene separations of Fig. 4

Resolution[a] $R_{i,j}$	Chromatogram			
	a	b	c	d
$R_{2,3}$	2.74	2.45	2.30	2.60
$R_{3,4}$	3.25	3.09	2.77	3.17
$R_{4,5}$	3.30	3.31	2.97	3.29
$R_{5,6}$	3.16	3.16	2.64	2.96
$R_{6,7}$	2.61	2.80	2.31	2.61
$R_{7,8}$	2.05	2.25	2.04	2.23
$R_{8,9}$	1.74	2.08	1.92	1.88
$R_{9,10}$	1.54	1.77	1.68	1.70
$R_{10,11}$	1.34	1.64	1.57	1.51
$R_{11,12}$	1.06	1.40	1.25	1.32
$R_{12,13}$	1.09	1.39	1.20	1.24
$R_{13,14}$	0.88	1.24	1.05	1.19
$R_{14,15}$	0.91	1.11	1.05	1.14
$R_{15,16}$	0.78	0.89	0.97	1.09
R_m	1.89	2.04	1.84	2.00

[a] $R_{i,j}$ is the resolution between neighbouring peaks i and j.

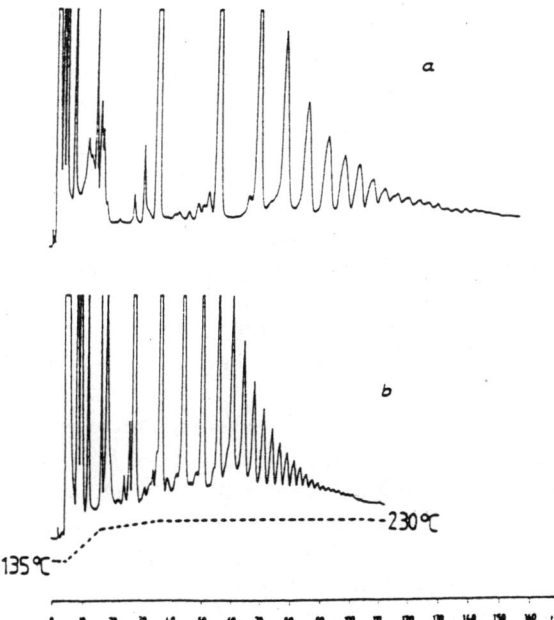

Figure 5. Gradient elution SFC separation of 4-vinylbiphenyl oligomers isothermally at 250°C
(a) and with superimposed temperature program (b). Mobile phase: CO_2–dioxane; composition gradient: 5 to 60% dioxane (v/v) within 160 min, according to eqn (1); inherent pressure gradients: p_e increasing from 247 to 310 bar (a) and increasing from 245 to 395 bar (b). Other conditions as for Fig. 2.

additional compound which should be used as a cofluid only with low-boiling primary components due to its low thermal stability (Asche, 1978) is acetonitrile. This compound is also frequently used as a second component in HPLC. The influence of the addition of acetonitrile to a CO_2 primary component on the retention of aromatic analytes in SFC has been studied isocratically with packed columns (Levy and Ritchey, 1985, 1986; Blilie and Greibrokk, 1985) as well as with capillary columns (Yonker and Smith, 1986). For the separation of aromatics (Mourier et al., 1985a, b) and of K-vitamins (Hondo et al., 1986), acetonitrile concentrations of 4% and 10%, respectively, have been used. Acetonitrile is also of interest because of its high spectral transparency: mixtures with CO_2 are transparent even at wavelengths below 200 nm (Hondo et al., 1986). Hence, acetonitrile is of high interest as a cofluid for gradient elution separations of oligomers which show absorption only at low wavelengths.

In Fig. 6a the gradient elution separation of caprolactonediol oligomers is shown using a CO_2−acetonitrile mobile phase at a detection wavelength of 220 nm. Although not being optimized, the chromatogram shows a fair separation of the homologous series as well as the occurrence of at least one parallel oligomer series. During the separation the mobile phase can be assumed to be in the supercritical state for the first half of the chromatogram, but in the subcritical (liquid) state for the second half. This can be estimated from the critical data of the pure components (cf. Table 1). The assumption could be made that a similar separation would be obtained when the mobile phase is in the liquid state during the whole separation. Accordingly, a chromatogram was recorded at ambient temperature (Fig. 6b). This chromatogram shows only a minor separation of a homologous series during the first 30 min. The remainder of the sample is then eluted as one large peak without any significant resolution. As the reason for the difference between the chromatograms in Fig. 6a and Fig. 6b, it can be assumed that the sample components had already been transported to a significant extent through the column in the supercritical state before being eluted from the column at subcritical conditions. The higher separation efficiency brought about by the supercritical eluent would then be responsible for the reasonable separation seen in Fig. 6a.

Additionally, using carbon dioxide−acetonitrile, gradient elution SFC separations of neopentyl carbonate oligomers have been performed. These oligomers, prepared via anionic initiation (Keul et al., 1986), show an absorption maximum at 230 nm, and this wavelength has been applied for detection. With the separation of these oligomers we have observed that, by changing the column outlet pressure, the selectivity for some compounds present in the oligomer sample was altered.

CONCLUSION

Gradients of eluent composition (gradient elution) in SFC are shown to be an interesting alternative to the frequently used pressure/density gradients

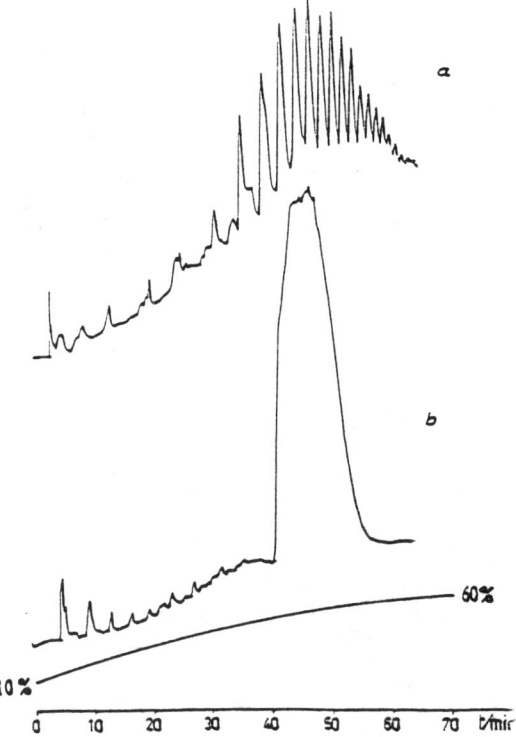

Figure 6. (a) Gradient elution SFC separation of caprolactonediol oligomers. Mobile phase: CO_2–acetonitrile. The gradient is shown. Temperature: 135°C; $p_e = 232$ bar at the start of the chromatogram; detection wavelength: 220 nm. Other conditions as for Fig. 2. (b) Gradient elution of caprolactonediol oligomers at LC conditions. Temperature: ambient. Other conditions as for (a).

which is demonstrated for the separation of different oligomers. Not only can this gradient technique be applied in a rather simple way, but retention and resolution depend sensitively on the gradient, offering variability for optimizing separations. The possibility of employing gradient elution as part of multiple programming techniques may become of especial importance, since this provides the chromatographer with even more versatile tools for method development.

SUMMARY

This chapter has demonstrated that gradient elution is a viable alternative to pressure/density programming in supercritical fluid chromatography. For the separation of vinyl arene oligomers, combinations of 1,4-dioxane with alkanes, ethers, carbon dioxide, or trifluoromethane, increasing the amount of dioxane in the eluent mixture during the separation, have been found most suitable. Furthermore, the eluent pair carbon dioxide–acetonitrile turned out to be of special interest for the separation of oligomers absorbing only at low wavelength, since CO_2—acetonitrile offers high transparency even below 200 nm.

Additionally, the usefulness of multiple-gradient separations was demonstrated. By superimposing a temperature gradient on a gradient of eluent composition the resolution can be improved. Further optimization was shown to be achieved by addition and variation of a pressure gradient. Multiple gradient separations, being a unique feature of SFC, improve the possibilities for optimization of separations considerably.

ACKNOWLEDGEMENT

Thanks are expressed to Prof. Dr E. Klesper for stimulating discussions. Financial support by the Deutsche Forschungsgemeinschaft (DFG) is gratefully acknowledged.

REFERENCES

Asche, W. (1978). Mobile phases for supercritical fluid chromatography. *Chromatographia 11*, 411–412.

Blilie, A. L. and Greibrokk, T. (1985). Modifier effects on retention and peak shape in supercritical fluid chromatography. *Anal. Chem. 57*, 2239–2242.

Gere, D. R. (1983). Oligomer separations by supercritical fluid chromatography. Appl. Note 800-3, Publ. No. 43-5953-1692, Hewlett Packard Co., Avondale, PA.

Hondo, T., Saito, M., and Senda, M. (1986). Analysis of vitamin K by directly coupled supercritical fluid extraction/supercritical fluid chromatography (in Japanese). *Bunseki Kagaku 35*, 316–319.

Jentoft, R. E. and Gouw, T. H. (1969). Supercritical fluid chromatography of a 'monodisperse' polystyrene. *J. Polym. Sci. B7*, 811–813.

Johnson, C. C., Jordan, J. W., Taylor, L. T., and Vidrine, D. W. (1985). On-line supercritical fluid chromatography with fourier transform infrared spectrometric detection employing packed columns and a high pressure lightpipe flow cell. *Chromatographia 20*, 717–723.

Keul, H., Bächer, R., and Höcker, H. (1986). Anionic ring-opening polymerization of 2,2-dimethyltrimethylene carbonate. *Makromol. Chem. 187*, 2579–2589.

Klesper, E. (1978). Chromatographie mit Überkritischen fluiden Phasen. *Angew. Chem. 90*, 785–793; *Angew. Chem., Int. Ed. Engl. 177*, 738–746.

Levy, J. M. and Ritchey, W. M. (1985). The effects of modifiers in supercritical-fluid chromatography. *J. High Res. Chromatogr. Chromatogr. Comm. 8*, 503–509.

Levy, J. M. and Ritchey, W. M. (1986). Investigations of the uses of modifiers in supercritical fluid chromatography. *J. Chromatogr. Sci. 24*, 242–248.

Leyendecker, D., Leyendecker, D., Schmitz, F. P., and Klesper, E. (1986). Multiple gradients in supercritical fluid chromatography (SFC). Ternary gradients by programming eluent composition and temperature at varying pressure. *J. High Res. Chromatogr. Chromatogr. Comm. 9*, 525–527.

McManigill, D. Board, R., and Gere, D. R. (1982). Hardware adaptations to HPLC apparatus to enable operation as a supercritical fluid chromatograph. Techn. Paper, Publ. No. 43-5953-1647, Hewlett-Packard Co., Avondale, PA, pp. 21–23.

Mourier, P., Caude, M., and Rosset, R. (1985a). Rétention, sélectivité et efficacité en chromatographie en phase supercritique. *Analusis 13*, 299–311.

Mourier, P., Eliot, E., Caude, M. H., Rossett, R. H., and Tambute, A. G. (1985b). Supercritical and subcritical fluid chromatography on a chiral stationary phase for the resolution of phosphine oxide enantiomers. *Anal. Chem. 57*, 2819–2823.

Reid, R. C., Prausnitz, J. M., and Sherwood, T. K. (1977). *The Properties of Gases and Liquids*. McGraw-Hill, New York, 3rd edn.

Schmitz, F. P. (1986). Chromatography with sub- and supercritical eluents. Influence of the separation conditions on selectivity, plate number and resolution. *J. Chromatogr. 356,* 261–269.

Schmitz, F. P. and Hilgers, H. (1986). Separation by means of supercritical fluid chromatography of 1-vinyl- and 2-vinylnaphthalene oligomers prepared through radical and anionic initiation. *Makromol. Chem., Rapid Commun. 7,* 59–63.

Schmitz, F. P. and Klesper, E. (1981a). Separation of styrene oligomers by supercritical fluid chromatography (SFC) using a modified HPLC-instrument. *Polym. Bull. 5,* 603–608.

Schmitz, F. P. and Klesper, E. (1981b). Supercritical fluid chromatography of oligostyrenes by means of eluent gradients. *Makromol. Chem., Rapid Commun. 2,* 735–739.

Schmitz, F. P. and Klesper, E. (1983). Polystyrene separation by supercritical fluid chromatography (SFC). *Polymer Commun. 24,* 142–144.

Schmitz, F. P., Hilgers, H., and Klesper, E. (1983). Gradient elution in supercritical fluid chromatography. *J. Chromatogr. 267,* 267–275.

Schmitz, F. P., Hilgers, H., Lorenschat, B., and Klesper, E. (1985). Separation of oligomers with UV-absorbing side groups by supercritical fluid chromatography using eluent gradients. *J. Chromatogr. 346,* 69–79.

Schmitz, F. P., Hilgers, H., and Gemmel, B. (1986). High-performance liquid and supercritical fluid chromatographic separation of vinyl oligomers by gradient elution. *J. Chromatogr. 371,* 135–144.

Schmitz, F. P., Leyendecker, D., and Leyendecker, D. (1987a). Permanent 'on column' modification of a silica stationary phase with 1,4-dioxane. *J. Chromatogr. 389,* 245–250.

Schmitz, F. P., Leyendecker, D., Leyendecker, D., and Gemmel, B. (1987b). Possibilities for optimization of oligomer separation with supercritical fluid chromatography. *J. Chromatogr. 395,* 111–123.

van der Maeden, F. P. B., Biemond, M. E. F., and Janssen, P. C. G. M. (1978). Oligomer separations by gradient elution high-performance liquid chromatography. *J. Chromatogr. 149,* 539–552.

Yonker, C. R. and Smith, R. D. (1986). Study of retention processes in capillary supercritical fluid chromatography with binary fluid mobile phases. *J. Chromatogr. 361,* 25–32.

Progress in HPLC, Vol. 4, pp. 87—110.
Yoshioka *et al.* (Eds)
© 1989 VSP.

Microscale supercritical fluid extraction and coupling of microscale supercritical fluid extraction with supercritical fluid chromatography

MUNEO SAITO, TOSHINOBU HONDO, MASAAKI SENDA

Jasco, Japan Spectroscopic Co., Ltd, No. 2967-5 Ishikawa-cho, Hachioji City, Tokyo 192 Japan

and KENKICHI SUGIYAMA

Research Institute, Morinaga & Co., Ltd, 2-1-1 Shimosueyoshi, Tsurumi-ku, Yokohama 230 Japan

INTRODUCTION

Although Hannay and Hogarth (1879) reported more than 100 years ago that supercritical fluids showed solvating power, supercritical fluid extraction (SFE) was introduced by Zosel *et al.* only in the 1960s. Since then the method seems to have been developing mainly as an industrial-scale extraction technique, as reported by Zosel (1980), Hubert and Vitzthum (1980), Peter and Brunner (1980), Eggers (1980), Eggers and Tschiersch (1980), Stahl *et al.* (1980), Williams (1981), Coenen and Rinza (1981), Brogle (1982), Filippi (1982), Bott (1982), Vollbrecht (1982), Calame and Steiner (1982), Gardner (1982), and Brunner and Peter (1982).

In food, pharmaceutical, perfume and cosmetics industries, such an extraction medium, being non-toxic, non-polluting, non-flammable, and inexpensive, has long been in demand. Considering the extraction efficiency, the extraction medium must have high solvating power and selectivity for the target components, and a large difference of boiling points for easy recovery of extracts from the medium. Carbon dioxide, which has a relatively low critical pressure, 73 bar, and a low critical temperature, 31°C, so that a supercritical phase is easily obtained, is expected to meet the above requirements. For better performance in SFE the extract must be analyzed and checked if the target component is to be efficiently extracted, in order to optimize extraction conditions. For such a purpose, chromatography is most essential and various types of chromatography, such as thin-layer chromatography (TLC), gas chromatography (GC) and high-performance liquid chromatography (HPLC) are employed. Among them, HPLC is now the most widely utilized due to its wide application range.

HPLC is a separation analysis method not only contemporary with SFE, but also having a similar history of development. Chromatography originated with Tswett in the 1900s as liquid chromatography; however, rapid developments also took place in the 1960s. These two techniques have many things in common from the instrumentation aspect. Both use high-pressure pumps, sample introduction devices, packed or hollow separation columns, etc. Since the late 1960s, numerous reports on HPLC and SFE have been published. However, the systems seem to have been developing independently, and have little to do with each other from the viewpoint of instrumentation.

Supercritical fluid chromatography (SFC), which uses supercritical fluid as mobile phase, was originated, also in the 1960s, from high-pressure gas chromatography by several research groups — Klesper et al. (1962), Myers and Giddings (1965, 1966), Jentoft and Gouw (1970, 1972), Novotny et al. (1971), Bartmann and Schneider (1973), and others. In the early 1980s, advances in micro-HPLC renewed the interest in SFC. Rapid mass transfer in supercritical mobile phase attracted researchers, as it offers high-speed separation with high resolution on an open tubular capillary column and also on a packed capillary column. The small consumption of a fluid encouraged chromatographers to use inflammable and even toxic fluids under high pressure and high·temperature. Thus, extensive research work has been performed by several groups — Novotny et al. (1981), Peaden et al. (1982), Peaden and Lee (1982, 1983), Takeuchi et al. (1984), Hirata and Nakata (1984) and others. In addition to the use of a simple UV detector with a high-pressure cell, combinations of SFC with new detectors have also been investigated: an SFC/photodiode array multiwavelength UV detector by Saito et al. (1985); SFC/MS by Smith et al. (1982, 1984), and Crowther and Henion (1985); SFC/FTIR by Shafer and Griffiths (1983), Olesik et al. (1984), Johnson et al. (1985); SFC/IR based on buffer memory technique by Fujimoto et al. (1985); SFC/FID by Rawdon (1984), Norris and Rawdon (1984), and Chester (1984). SFC systems based on HPLC instrumentation were also reported by Gere et al. (1982), and Greibrokk et al. (1984). A sophisticated system was reported by Saito et al. (1985) which consists of an SFE system directly combined with SFC, a photodiode-array multiwavelength detector and its data processor, including capability of spectral data acquisition and peak deconvolution. Sophisticated application using the peak deconvolution technique of the system was reported by Jinno et al. (1986a, b).

In this chapter we shall first discuss microscale SFE systems, then direct coupling of SFE with SFC.

MICROSCALE SUPERCRITICAL FLUID EXTRACTION

Introduction
Supercritical Fluid Extraction (SFE) seems to have been examined and utilized for only industrial scale extraction, though SFE can be performed

on a microscale with an extraction vessel of only a few tenths to tens of milliliters in volume as the authors reported before (Saito *et al.*, 1985; Sugiyama *et al.*, 1985). Such SFE, we shall call this micro-SFE, has many advantages over a pilot plant SFE having an extraction vessel of one tenth to a few liters in volume, from several viewpoints. The advantages of micro-SFE are: (1) easy to build; (2) easy to operate; (3) small sample quantity; (4) low running cost; (5) on-line and/or off-line monitoring of extract with anlytical instruments such as UV, chromatograph, IR, NMR, MS, etc.; and (6) potential sample pre-treatment method for chromatographic analysis.

In SFE, carbon dioxide is generally the preferred extraction medium and is widely used, and as mentioned before, we shall also deal with carbon dioxide as both the extraction medium and the chromatographic mobile phase throughout all the sections in this chapter.

Instrumentation
In principle, an SFE system consists of a high-pressure pump, an extraction vessel, a back-pressure regulator and a separation vessel. Figure 1 shows schematic diagrams of different types of SFE system.

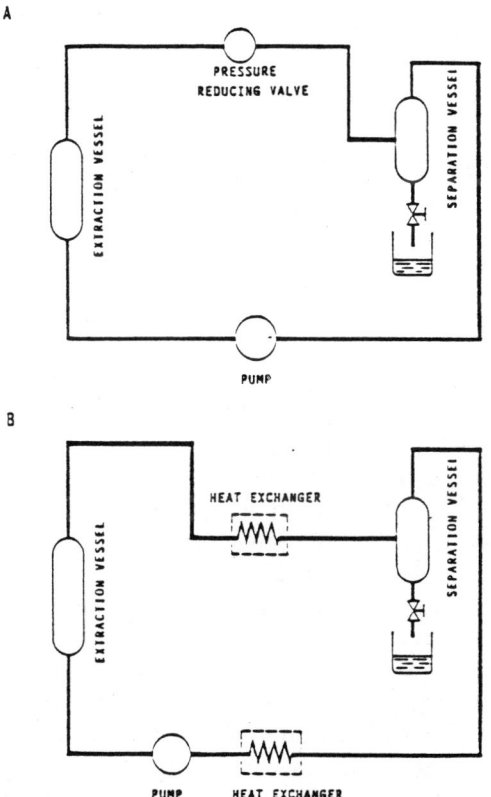

Figure 1. Schematic diagrams of different types of SFE systems. A type of SFE system having extract separation based on the pressure reduction (A). Another type based on the temperature change (B). After Hubert and Vitzthum (1980).

Recent advances in HPLC instrumentation technology permit us to build a micro-SFE system without any difficulty. As many researchers demonstrated (Gere et al., 1982; Greibrokk, 1984; Saito et al., 1985), an HPLC pump with a cooling jacket for the pump head can be used for delivery of liquefied carbon dioxide. A short empty column can be used as an extraction vessel, and a restrictor having a proper flow resistance can serve as a device which generates a back-pressure. A commercially available back-pressure regulator can conveniently be used for applying the suitable back-pressure without changing the mass flow rate of the fluid, though fractionation of the extract is difficult because the dead-volume of such a regulator is large — several tens of milliliters — in comparison with the volume of the extraction vessel — a few tenths to tens of milliliters — so that the extract is left in the regulator. However, a UV detector can be placed between the extraction vessel and the back-pressure regulator in order to perform on-line UV absorption monitering of the extract as a function of extraction time, or even UV spectra can be obtained as a series along the time axis, i.e. the three-dimensional UV spectrum, if a multiwavelength detector is employed. We shall call this type of data *extraction profile*. Such an arrangement is very useful to examine extraction conditions. In a pilot plant SFE system, examination of the conditions have generally been performed off-line by chromatography, spectrometry and other analytical methods after collecting extracts.

Application of micro-SFE to investigation of lemon peel oil extraction with supercritical carbon dioxide

Lemon peel oil is a fairly expensive material used in the perfume and flavor industries. Conventionally, lemon peel oil is isolated from the peel by using a sophisticated cold-press equipment. Since a delicate balance of the composition of the oil influences the quality of the product, extensive analytical works have been carried out. These works include: the investigation of analytical methods by Staroscik and Wilson (1982a), Analytical Method Committee (1984), Baaliouamer et al. (1985), Cotroneo et al. (1986a) and Takeoka et al. (1986); seasonal and regional variations in the composition of oils, by Staroscik and Wilson (1982b) and Cotroneo et al. (1986a, b); the variation of the composition by different extraction methods, by Cotroneo et al. (1986a) and Calame and Steiner (1982) examined extraction with supercritical carbon dioxide; and deterioration mechanism of flavor by Kimura et al. (1983), Shaw and Wilson (1982) and Klavons and Bennet (1985).

Calame and Steiner (1982) used a 4-l extraction pilot plant, and reported that the oil was obtained in 0.9% yield at 300 b and 40°C. The composition of the extracted oil they measured was a little different from that of a commercial cold-pressed oil, the major differences between the two oils were that the extracted oil contained less citral and more alcohol than the cold-pressed one.

Coppella and Barton (1987) examined the concentration of a cold-pressed lemon oil with carbon dioxide by measuring the lemon oil−carbon dioxide equilibrium. They proposed the conditions of 308–313°K (35–40°C) and 7.7–8.5 MPa (77–85 b) for considering the quality of the obtained oil.

In this paper, a simple extraction method with supercritical carbon dioxide and its application to the analysis of lemon peel oil will be presented.

Experimental. Materials and apparatus. A lemon was purchased from a grocery store, and its *flavedo* layer was carefully cut out from the fiber organism of the peel using a clean razor into 3 mm × 10 mm; each piece of the peel weighed about 100 mg. Carbon dioxide was purchased in a cylinder with a siphon tube from Toyoko Kagaku (Kawasaki, Japan).

We built two types of micro-SFE systems, schematic diagrams of which are shown in Figs 2 and 3. The JASCO Model BIP-1 (Tokyo, Japan) with a cooling jacket was used as the carbon dioxide delivery pump. A cartridge type extraction vessel was made of 4.6 mm ID × 1/4 in. OD × 35 mm long stainless steel tube with ordinary 1/4 in. HPLC column end fittings. A six-port valve for changing the flow line was the RHEODYNE Model 7000 (CA, USA). An HPLC column oven (Model TU-100, JASCO) was used for elevating the temperature of carbon dioxide and the extraction vessel above

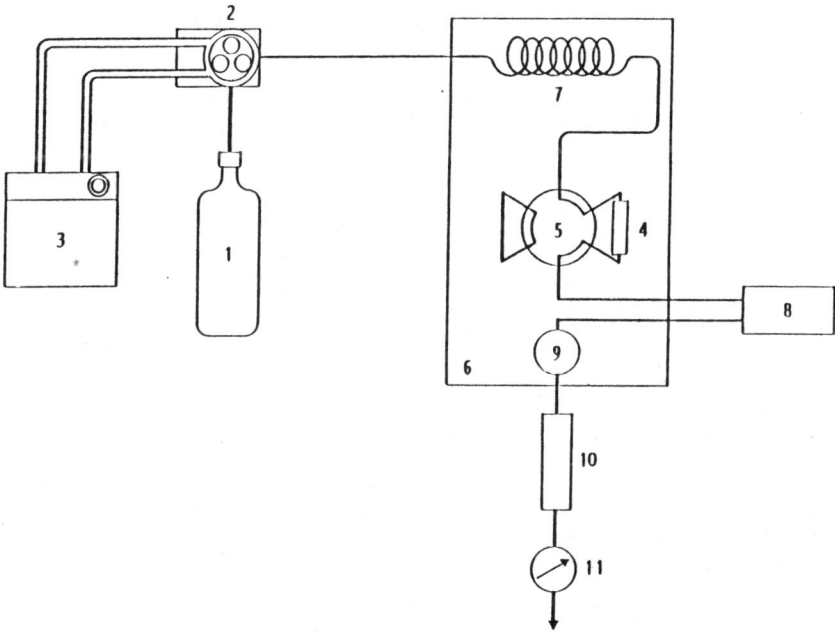

Figure 2. Micro-SFE system for extraction profile monitoring. Components: 1 = carbon dioxide cylinder, 2 = carbon dioxide delivery pump having head cooling jacket, 3 = coolant circulating bath, 4 = extraction cartridge of 4.6 mm ID × 1/4 in. OD × 35 mm length, 5 = six-port valve for changing the flow line, 6 = oven, 7 = 0.5 mm ID × 1/16 in. OD × 5 m long heat exchanger coil, 8 = multiwavelength UV detector with high pressure cell, 9 = back-pressure regulator, 10 = trap for mass flow meter, 11 = mass flow meter. Reproduced from Sugiyama and Saito (1988) with permission.

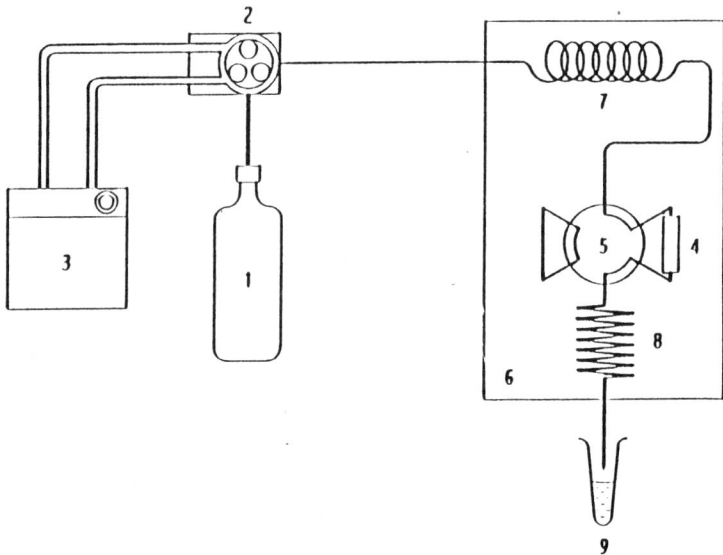

Figure 3. Micro-SFE system for collecting extract. Components: 1 through 7 are the same as in the caption for Fig. 2; 8 = flow restrictor, 9 = glass capillary tube. Reproduced from Sugiyama and Saito (1988) with permission.

its critical temperature. A 0.5 mm ID × 1/16 in. OD × 5 m long stainless steel tube was connected between the pump and the extraction vessel. The tube was coiled and kept in the oven serving as a heat-exchanger. The TESCOM Model 26-3200-24 (MN, USA) backpressure regulator was used to change the back-pressure while monitoring UV spectra of the extract with a multi-wavelength UV detector (JASCO MULTI-320) as shown in Fig. 2. The KOFLOC model 2500 CO_2 mass flow meter (Kojima Flow Instruments Corp., Kyoto, Japan) was used for monitoring the mass flow rate of carbon dioxide after the back-pressure regulator, reducing the pressure to atmospheric pressure. For collecting the extract, a 0.25 mm ID × 250 mm long stainless steel tube was used, instead of the back-pressure regulator, as a flow restrictor as shown in Fig. 3. The restrictor tube was pinched with a pair of pliers to increase the flow resistance.

For identification of components of the oil, a GC-MS system, consisting of the HP-5790 gas chromatograph (Hewlett-Packard, CA, USA), and the JMS-DX300 (JOEL, Japan) mass spectrometer was used.

Procedure. (1) Extraction profile monitering with multiwavelength UV detector.

The hydraulic system shown in Fig. 2 was used for extraction profile monitoring. The extraction vessel containing a piece of lemon peel was connected to the flow line. For equilibration of the system, carbon dioxide was first delivered from the pump into the detector, by-passing the vessel, then vented to atmospheric pressure, via the back-pressure regulator, which controls the extraction pressure, and the mass flow meter. The extraction temperature was controlled by the oven in which the vessel, back-pressure regulator, etc. were installed. Equilibration was checked with the detector which allowed real time monitoring of UV spectrum as a function of time. After equilibration, the 6-port valve was switched on to start extraction. The UV spectrum of the extract was displayed on CRT on real-time, and at the same time spectral data was stored on a floppy disc for later use.

(2) Collection of the extract.

The procedure for the collection of the extraction is very similar to that for the extraction profile monitoring except for the hydraulic system used which was shown in Fig. 3. The carbon dioxide pump was operated in the constant pressure mode at a preset extraction pressure. The pressure of the fluid containing the extract was released as it flowed along the restrictor and the lemon peel oil was collected in a glass capillary tube of 1.5 mm ID × 30 mm length. The extraction yield was calculated from the weight of the collected oil and the peel weight before extraction.

(3) Capillary GC–MS analysis.

Extracted oil at various conditions were chromatographed on a CBP-20 capillary column (Shimadzu, Kyoto, Japan) and monitored with a flame ionization detector (FID). Identification of each component was made by GC–MS analysis. The obtained results were compared with each other, and also with analytical results of a commercially cold-pressed oil by the same chromatographic method.

Results and discussion

Extraction profile. The extraction profile of the lemon peel is shown in Fig. 4. The extraction temperature was kept constant at 45°C, the pressure

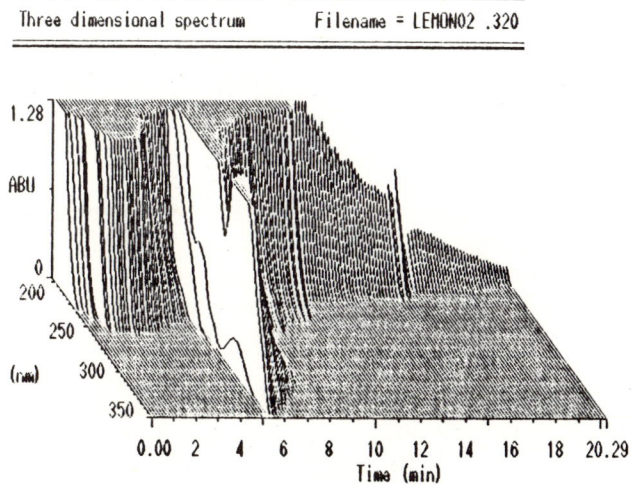

Figure 4. Extraction profile of lemon peel. SFE conditions: temperature kept constant at 45°C; pressure changed stepwise from 90, to 110, 140 and 170 kg/cm^2 at time intervals of 5.0 min. Reproduced from Sugiyama and Saito (1988) with permission.

was, however, changed stepwise from 90, to 110, 140 and 170 kg/cm^2 at time intervals of 5.0 min, while maintaining the mass flowrate of carbon dioxide constant. According to the profile, the extraction of the lemon peel oil started at a comparably low pressure of 90 kg/cm^2. Therefore, we decided to examine the extraction pressure in the range 100−250 kg/cm^2.

Percent yields of extracted oil under various conditions. A percent yield was obtained by a simple calculation; the mass of collected oil divided by the mass of sample piece before extraction. Figure 5 shows the relationship between the per cent yield and the extraction pressure using temperature as a parameter.

At 58°C, the oil was hardly extracted below 150 kg/cm^2 pressure, however, at 200 kg/cm^2 the extraction yield rapidly increased to 1.94%. Although the yield was the maximum among all the conditions we examined, the fragrance of the extract gave a slight off-flavor. Therefore, such a condition is not suitable from the viewpoint of the quality of the product.

The extraction at 45°C temperature showed satisfactory results from the viewpoints of both the quality of the flavor and the efficiency of extraction over the relatively wide range of pressure, 150−250 kg/cm^2.

It was found that the yield at 100 kg/cm^2 and 30°C, which is a slightly below the critical temperature of 31.3°C, is higher than that at 45°C and 100 kg/cm^2, and a little lower than that at 200 kg/cm^2. The quality of the oils extracted under the above conditions was comparable with the oils extracted at 45°C and 150 and 250 kg/cm^2. This suggests that the extraction

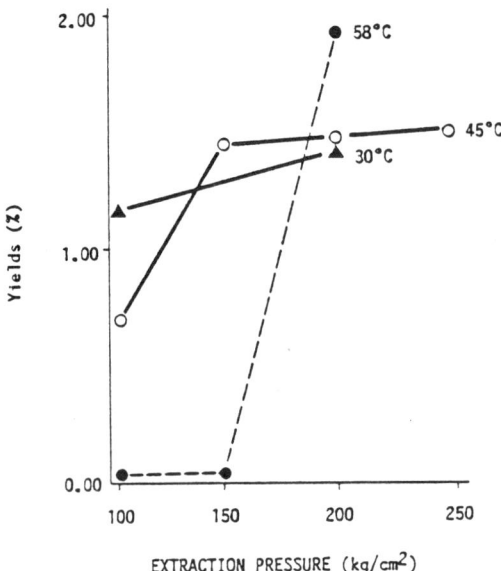

Figure 5. Extraction yields of lemon peel oil under various SFE conditions. Conditions: ▲ = 30°C, ○ = 45°C, ● = 58°C. Reproduced from Sugiyama and Saito (1988) with permission.

could be performed at room temperature with sub-critical or liquid carbon dioxide at comparatively low pressures of around 100 kg/cm^2. This condition is very similar to that proposed by Coppella and Barton (1987), that was investigated by precise lemon oil-carbon dioxide phase equilibrium experiment, though their material was not raw lemon peel but a cold-pressed oil.

Comparison of components of extracted oils and cold-pressed oil. Figure 6 shows a typical GC chromatogram of the oil extracted at 100 kg/cm^2 and

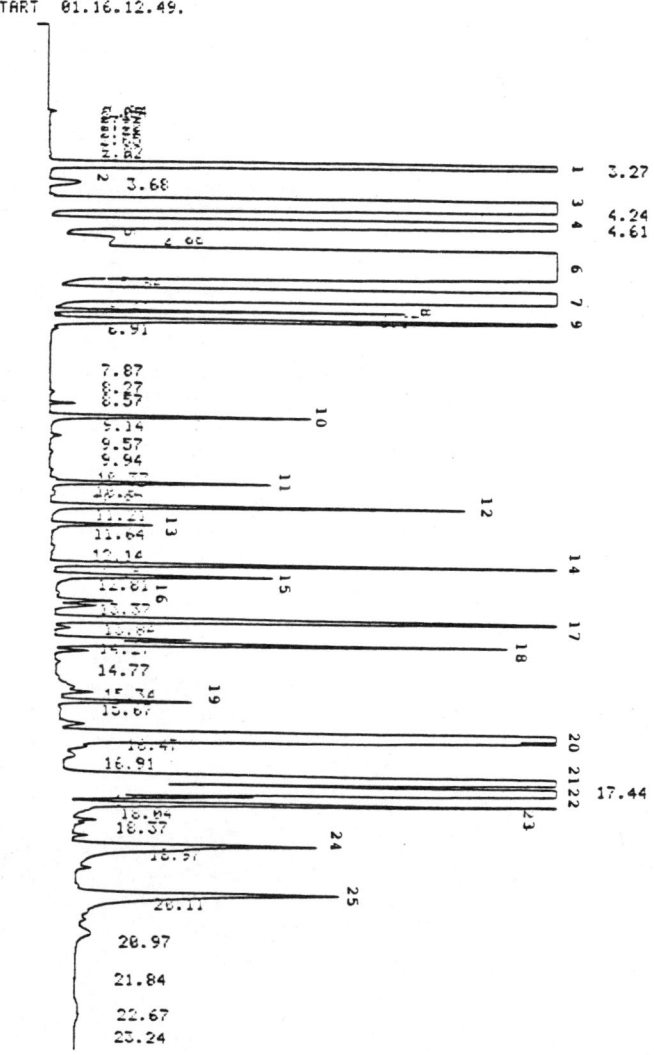

Figure 6. Typical capillary GC chromatogram of lemon peel oil. SFE conditions: temperature, 45°C; pressure, 100 kg/cm^2. GC conditions: column, 25 m long × 0.2 mm ID. CBP 20; carrier gas, He (1 ml/min); split ratio, 1/90; column temperature, 90°C for 2 min, 5°C/min to 175°C; detector, FID. Reproduced from Sugiyama and Saito (1988) with permission.

45°C. There are 25 major peaks in the chromatogram and each peak area and the total sum of the areas were taken into the calculation. Peak identification by GC−MS analysis was successful for 23 components, however, peak nos. 13 and 16 were not identified. Table 1 lists per cent contents of components of extracted oils under various conditions, those of a cold-pressed oil are also listed for comparison.

It is apparently seen that the limonene content in cold-pressed oil is higher than that in any of the CO_2 extracted oils. In order to facilitate the comparison, the component listed in Table 1 were grouped into several compound types, namely, hydrocarbons, aldehydes, alcohols, esters and oxygenated compounds. Total amount of these types of compounds and their major components are shown in Table 2, respectively.

Table 1.

Per cent amount[a] of component of extracted oil with carbon dioxide under various conditions.

	Pressure (kg/cm^2)/Temperature of CO_2						
	100	200	150	200	250	200	
Peak no.[b]/							Cold-
Component	30°C		45°C			58°C	pressed
1 α-Pinene	1.52	1.19	1.44	1.13	1.51	1.49	1.77
2 Camphene	0.04	0.03	0.03	0.03	0.04	0.04	0.03
3 β-Pinene	11.47	9.18	9.95	8.97	11.39	10.54	9.68
4 Myrcene	1.40	1.40	1.45	1.39	1.38	1.48	1.80
5 α-Terpinene	0.09	0.07	0.09	0.09	0.10	0.10	0.09
6 Limonene	64.28	68.38	68.23	68.63	64.15	68.59	71.80
7 γ-Terpinene	11.29	9.80	9.85	9.63	11.35	9.60	6.35
8 p-Cymene	0.12	0.07	0.08	0.07	0.11	0.07	0.29
9 Terpinolene	0.57	0.50	0.55	0.57	0.58	0.47	0.51
10 Nonanal	0.10	0.11	0.08	0.09	0.11	0.10	0.08
11 Limonene Oxide	0.11	0.10	0.09	0.09	0.11	0.09	0.02
12 trans-Sabinene	0.17	0.16	0.13	0.16	0.19	0.14	0.33
13 — hydrate	0.05	0.05	0.04	0.05	0.05	0.04	0.02
14 Citronellal	0.23	0.18	0.17	0.18	0.22	0.16	0.45
15 Octylacetate	0.15	0.11	0.12	0.12	0.14	0.10	0.08
16 —	0.07	0.06	0.04	0.06	0.07	0.05	—
17 Linalool	0.56	0.56	0.44	0.56	0.58	0.58	0.07
18 Linalylacetate	0.42	0.35	0.27	0.34	0.42	0.46	0.09
19 4-Terpineol	0.10	0.09	0.07	0.09	0.14	0.08	0.01
20 Neral	1.67	1.82	1.48	1.66	1.90	1.54	1.89
21 Citral	1.68	1.91	1.36	1.80	1.88	1.38	1.02
22 Nerylacetate	2.01	2.17	2.21	1.96	2.27	1.87	2.34
23 Geranylacetate[c]	0.39	0.44	0.37	0.50	0.36	0.35	0.51
24 Nerol	0.57	0.42	0.41	0.56	0.34	0.37	0.01
25 Geraniol	0.65	0.56	0.46	0.81	0.44	0.52	0.01

[a] Percent amount of each component was calculated as; (Peak area) × 100/(total sum of assigned 25 peak area).

[b] Peak numbers in this table correspond to those in Fig. 5.

[c] Peak 23 may have included citronellol.

Table 2.
Composition of compound type and major component

Compound type/ Major component	Pressure (kg/cm²)/Temperature of CO₂						
	100	200	150	200	250	200	Cold-pressed
	30°C		45°C			58°C	
Total hydrocarbons	91.00	90.83	91.84	90.72	90.85	92.56	92.67
limonene	64.28	68.38	68.23	64.63	64.15	68.59	71.80
Total aldehydes	3.61	4.02	3.09	3.73	4.11	3.18	3.44
citral	1.68	1.91	1.36	1.80	1.88	1.38	1.02
Total alcohols	1.88	1.63	1.38	2.02	1.50	1.55	0.10
linalool	0.56	0.56	0.44	0.56	0.58	0.58	0.07
Total esters	2.97	3.07	2.97	2.92	3.19	2.78	3.02
linalylacetate	0.42	0.35	0.27	0.34	0.42	0.46	0.09
Oxygenated compounds	8.57	8.82	7.53	8.76	8.91	7.60	6.58

Now, it can be said that even though the total hydrocarbon contents exhibit very similar value for all oils, CO_2 extracts contain less limonene than the cold-pressed oil which agrees with a previous report by Calame and Steiner (1982). Regarding alcohols, any CO_2 extract exhibits more than ten times higher content than the cold-pressed oil which also agrees with the result of the report.

Although the total amount of aldehydes are similar in all oils, including the cold-pressed oil, any of the CO_2 extracted oils contains more citral by a factor of 1.3–1.9 than the cold-pressed oil which does not agree with the previous report. However, these differences can be regarded as within the seasonal and regional variation of citral content because we used an ordinary commercial cold-pressed oil and a lemon from different sources. According to Staroscik and Wilson (1982b), this variation was by a factor of greater than 2. Another possible reason for disagreement is that the scale of the experiment was different; they used a 4-l extraction vessel, whereas we used an extraction vessel having only a 60-μl volume. Therefore, only about 600 mg of *flavedo* portion of lemon peel was used as a material in our system. However, they might have used whole peel including fiber organism which was not clear in their report.

The total amount of esters is also similar to each other. However, its major component linalylacetate is richer in the CO_2 extracts by a factor of 3–5 than in the cold-pressed oil. It is unknown whether this is due to the different extraction method or to the different sources of original material.

The oil obtained at 200 kg/cm² and 58°C that gave a slight off-flavor did not show any special difference in constituents that can be differentiated by the chromatogram. It is assumed that the reason why the oil exhibited off-flavor may be attributed to the higher solubility of the fluid than that under lower temperature and pressure, thus additional undesirable aromatic compounds were extracted as well as necessary compounds. An increase in the total yield reinforces this assumption.

Comparison of appearances of peel surface before and after extraction. Figure 7 shows photo micrographs of the lemon peels before and after extraction. In the photograph of the peel before extraction shown on the left-hand-side, oil cells are seen through the skin. On the other hand, in the

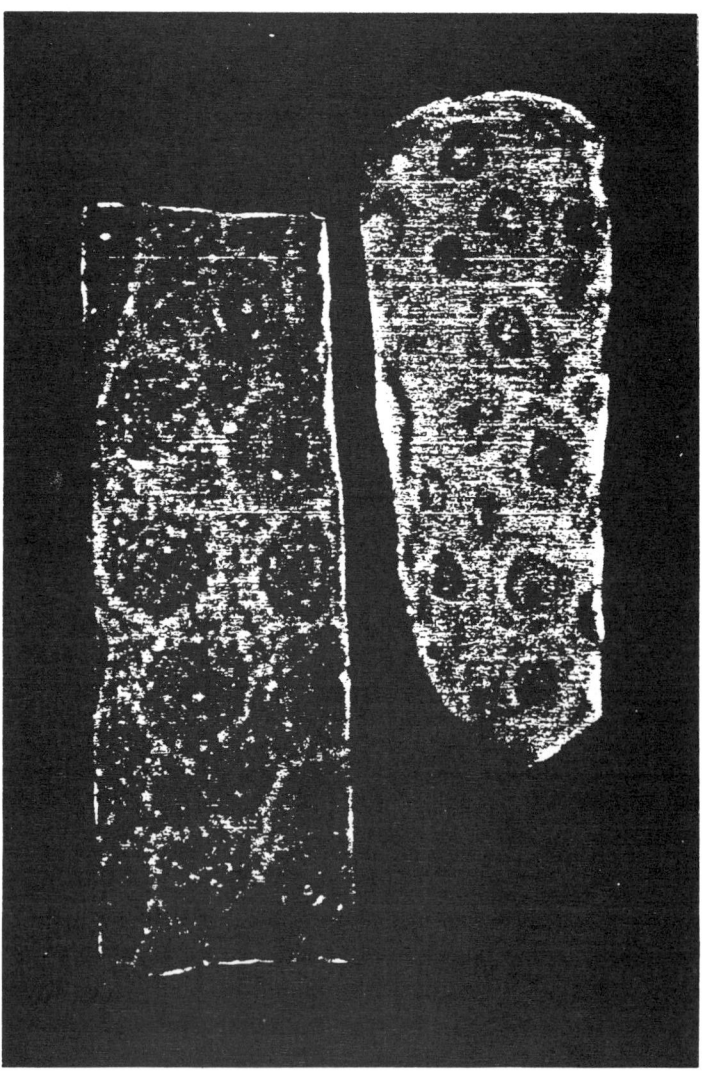

Figure 7. Microscopic photographs of lemon peels before and after extraction. The peel before extraction shown on the left-hand-side, oil cells are seen through the skin. On the other hand, after extraction shown on the right-hand-side, oil was drawn out of cells and they look like craters. Reproduced from Sugiyama and Saito (1988) with permission × 12.

photograph of the peel after extraction, oil was drawn out of cells and they look like craters. The skin itself looks less transparent due to a lack of oil. These photographs prove that the oil was not simply squeezed out by the pressure of carbon dioxide, but carbon dioxide diffused into oil cells and dissolved and drew the oil out from the cells, i.e., the oil was extracted.

Conclusion. Micro-SFE is a very simple and easy extraction technique. It is much easier to perform, as we have demonstrated, than one might consider. In addition, by changing the extraction conditions of temperature and pressure of carbon dioxide, extracts of different components can be obtained as different solvents are used in solvent extraction. It should be noted that an extract is obtained as frozen in dry ice, and separation of the extract from carbon dioxide can be carried out by simply leaving it at room temperature. There is no need to elevate the temperature of the extract, but carbon dioxide will sublimate automatically which is very desirable for extraction of thermally labile substances from natural products.

DIRECTLY COUPLED MICROSCALE SUPERCRITICAL FLUID EXTRACTION–SUPERCRITICAL FLUID CHROMATOGRAPHY

Introduction

Traditionally the analysis of SFE extract has been performed off-line using GC, HPLC, TLC, etc., and there are only a few reports on direct coupling of SFE with chromatographic techniques. Stahl and Schiltz (1976, 1977) developed an extraction system which was combined with TLC. Unger and Roumeliotis reported (1983) a coupling device which allows on-line HPLC analysis of extracts.

Although SFC seems to be closest to SFE among the different types (GC, LC and TLC) of chromatography, SFE and SFC have little to do with each other, as reviewed by Saito and Hondo (1986), and direct coupling of SFE with SFC had not yet been attempted, until the authors reported (Saito *et al.*, 1985; Sugiyama *et al.*, 1985) the new method.

In this section the instrumentation of the directly coupled micro-SFE/SFC system, and its application to caffeine extract from roasted coffee beans, are described.

Instrumentation

A directly coupled micro-SFE/SFC system can be regarded as a new double-stage separation analysis method which includes micro-SFE directly combined with SFC. In this new method SFE is utilized as the first separation step in a similar way to a sample pretreatment in HPLC; SFC is used as the second separation step. This configuration allows an analyst to place a raw and/or solid sample in the system in order to obtain a chromatogram of the sample extract. A multiwavelength UV detector, equipped with a high-pressure cell, is utilized as an extraction and/or chromatographic monitor. Three-dimensional spectrometric data, namely absorbance, wavelength and time, graphically presented in various fashions by computer-aided techniques, are very effective in the close examination of components in the SFE extract. Furthermore, application of peak deconvolution, which was reported by Hoshino *et al.* (1984, 1985), allows further investigation of chromatographic peak components of the extract.

In order to operate the micro-SFE/SFC system successfully;

(1) the volume of the extraction vessel should be appropriate for the sample size for SFC;
(2) the pressure decrease of the supercritical carbon dioxide should be kept to a minimum during transfer of the extract from the extraction cartridge to the sample loop of the SFC system;
(3) the SFC system should be pre-pressurized and equilibrated at the SFC analysis pressure before introducing the extract.

The hydraulics of the micro-SFE/SFC system we designed are shown in Fig. 8. The system allows several modes of operation:

(1) ordinary SFC with syringe injection;
(2) real-time extraction profile monitoring;
(3) micro-SFE with an extract trap column, which can be followed by off-line GC, HPLC, etc. analyses;
(4) directly coupled SFE/SFC, i.e. batch SFE with a trap loop, followed by SFC analysis with direct sample introduction.

Liquefied carbon dioxide from the cylinder (1) is fed to the pump (2) whose pump heads are cooled with a coolant jacket at $-5°C$ (BIP-1 modified for liquefied carbon dioxide delivery, Jasco, Tokyo, Japan). Coolant is supplied by a circulating cooling bath (3) (Model LC-101, Scinics, Tokyo, Japan). An entrainer or modifier solvent is delivered by the pump (4) (Jasco BIP-1) and mixed with liquefied carbon dioxide before entering the oven (21). The six-way switching valves (7), (9) and (11) (Model 7000, Rheodyne, CA, USA) are switched in accordance with the desired mode.

Hydraulics of micro-SFE with extract trap column
In micro-SFE with extract trap column mode the fluid flows through the extraction cartridge (8), the detector (14), bypassing the extract trap loop (10), the injector (12) and the separation column (13), via the extract trap column (16), the back-pressure gauge (17), and t   back-pressure regulator (18), and is then vented to atmosphere. After SFE the extract trap column (16) is disconnected from the system, and the extract is eluted with solvent. Then the extract is applied to other analytical instruments, such as GC and HPLC systems.

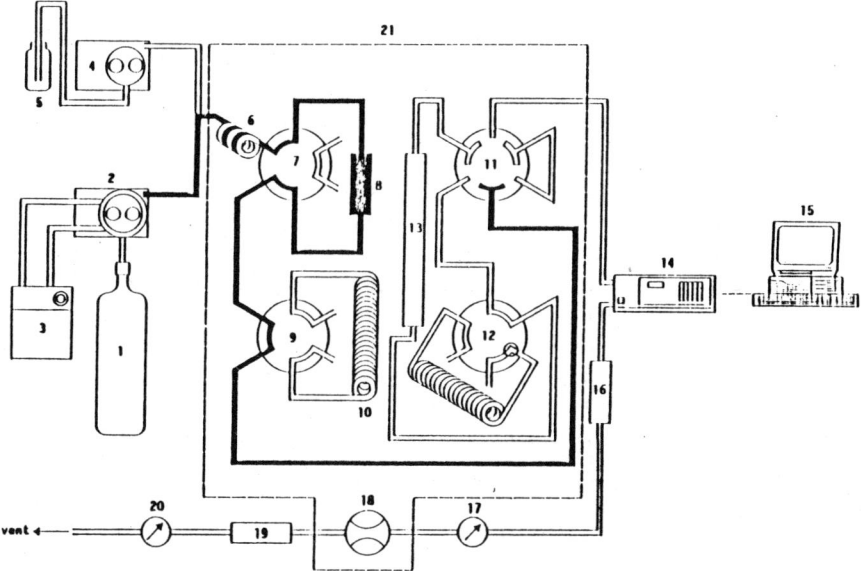

Figure 8. Hydraulics of directly coupled micro-SFE/SFC for extraction. Components: 1 = carbon dioxide cylinder, 2 = liquefied carbon dioxide delivery pump, 3 = coolant circulating bath, 4 = modifier delivery pump, 5 = modifier solvent reservoir, 6 = heat exchanger coil, 7 = six-way valve for bypassing extraction cartridge, 8 = extraction cartridge, 9 = six-way valve for bypassing trap loop, 10 = extract trap loop, 11 = six-way valve for bypassing injector and separation column, 12 = injector, 13 = separation column, 14 = multiwavelength detector, 15 = detector data processor, 16 = extract trap column, 17 = back-pressure gauge, 18 = back-pressure regulator, 19 = trap for mass flowmeter, 20 = mass flowmeter, 21 = oven. After SFE the valve (9) is switched to load the extract trap loop (10) with the extract. The valve is then switched back to bypass the loop for pre-pressurization and equilibration of the separation column (13), while the loop holds the extract.

Hydraulics of directly coupled micro-SFE/SFC

The flow line for the directly coupled micro-SFE/SFC mode is as indicated by the heavy line in Fig. 8. Carbon dioxide is delivered to the extraction cartridge (8), where extraction takes place, then to the six-way valve (9) with the extract trap loop (10), which is purged with carbon dioxide gas at atmospheric pressure prior to the extraction. The valve (11) is set in the non-connecting position to make a blocked end for the extraction line, and at the same time the valve (11) maintains the pressure of the column, which has been pre-pressurized and is to be equilibrated at the SFC pressure. The extraction cartridge (8), the separation column (13) and the extract trap column (16) are thermostated in the oven (21) (Jasco TU-100). When SFC is performed at a different temperature, a separate oven is used.

At the beginning of the extraction the pump (2) delivers liquefied carbon dioxide at its maximum flow rate to pressurize the extraction cartridge (8) quickly. As the pressure approaches the preset extraction pressure the flow rate gradually decreases, and finally the flow will automatically stop when the pressure reaches the preset value. Then the pressure will be maintained throughout extraction period. On completion of the extraction the six-way valve (9) is switched, in the position shown in Fig. 9, to load the trap loop (10) with the extract, and the pump (2) automatically starts flowing to compensate the pressure decrease due to the transfer of carbon dioxide and the extract in the extraction cartridge (8) to the trap loop (10) which has been purged with carbon dioxide gas at atmospheric pressure. When the transfer is completed, and the pressure is restored, the pump (2) stops. Then the six-way valve (7) is switched back again, in the position shown in Fig. 8, so that the loop (10) is bypassed, and the extract dissolved in the super-critical carbon dioxide is held in the trap loop (10) until the injection is made. Then valves (7) and (11) are switched to the SFC separation line, as indicated by the heavy line in Fig. 9. The system is now operated in the chromatography mode for equilibration of the separation column (13).

Finally, the six-way valve (9) is switched, in the position shown in Fig. 9, to inject the extract into the separation column (13). The chromatography mode can be easily converted from SFC to ordinary HPLC by using ordinary solvent without any hardware modification. A highly sensitive multiwave-length UV detector (14) (Jasco Multi-320, modified for high-pressure appli-cation) together with its personal computer-based (Oki iF-800, Oki Electric, Tokyo, Japan) data processor (15) are used as an extraction and/or chroma-tographic monitor. The flow cell, whose volume is` 4 μl, is modified to

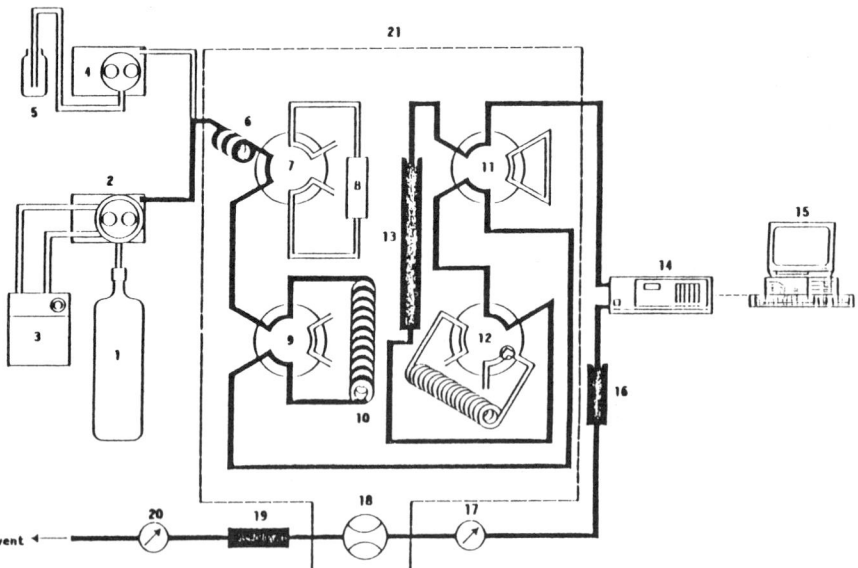

Figure 9. Hydraulics of directly coupled micro-SFE/SFC for chromatography. After equilibra-tion of the separation column under the chromatographic conditions the valve (10) is swtiched to inject the extract held in the trap loop into the column. The valve (10) in this figure is shown in the position for injection. Note that injector (13) is used only for ordinary syringe injection in the simple SFC mode.

withstand 300 bar pressure to meet pressure requirements in micro-SFE and SFC. The flow line from the pump through the separation column to the detector cell is kept under necessary pressure for SFE and SFC by a back-pressure regulator (18), Tescom model 26-3200-24 (Tescom, MN, USA), where the main pressure drop takes place. The back pressure is monitored by the back-pressure gauge (17) Jasco model PG-350, and the column effluent is vented to atmosphere.

Application of microscale supercritical fluid extraction–supercritical fluid chromatography to the optimization of caffeine extraction from roasted coffee beans

For our preliminary work we started with caffeine extraction from coffee beans, which is one of the classical applications of SFE, in micro-SFE with an extract trap column mode. The coffee extract, which was eluted out from the extract trap column, has been applied to an HPLC system (off-line micro-SFE/HPLC). Micro-SFE was performed under various conditions, and the contribution of each extraction parameter to the extracted amount of caffeine was examined. After examining the extraction conditions, directly coupled micro-SFE/SFC was performed successfully, and three-dimensional SFC chromatographic data of the coffee extract were obtained by placing the coffee powder in the system. The data were represented, by the data processor, as three-dimensional plots, chromatograms at 250 and 270 nm and a spectrum at 9.60 min, which showed clear characteristics of the caffeine spectrum.

Experimental

(a) *Micro-SFE/HPLC analysis of coffee beans.* In industrial applications, decaffeination or caffeine extraction is usually performed on green coffee beans having a certain water content. In our experiment roasted coffee beans were used instead of green beans. Roasted coffee beans, obtained from a grocery store, were ground and sieved to 30–60 mesh. They had 1–2% water content; i.e. significantly lower than that in green beans. In order to give a different water content a different amount of water was added to about 20 g of the coffee powder kept in glass vessels (100-ml capacity) with air-tight stoppers, and mixed by shaking, then equilibrated for at least 24 hours. Then about 350 mg of the moistened powder was packed into an extraction cartridge of 4.6 mm i.d. × 50 mm length, by tapping. Extraction was performed with a trap column of the same dimensions, packed with activated carbon (30–60 mesh) (Gasukuro Kogyo, Tokyo, Japan). After SFE the column was disconnected from the system, and the extract was eluted with 25 ml of methanol/water (45/55) mixture. Then 20 µl of the solution was injected into the HPLC system, consisting of a Tri-Rotar-V pump, a VL-614 injector, a Fine Pak SIL C18 column and Uvidec-

100-V UV detector (all from Jasco). Carbon dioxide of food additive grade (Toyoko Kagaku, Kawasaki, Japan) was used as an extraction medium.

(b) Directly coupled micro-SFE/SFC analysis of coffee beans. About 100 mg of the same coffee powder was prepared and placed in the extraction cartridge by the same procedure in off-line micro-SFE/HPLC. Then batch micro-SFE was performed with a trap loop of 500 μl instead of the extract trap column. After SFE the carbon dioxide containing the extract was transfered to the·trap loop, and directly introduced into the separation column by switching the injector valve, as described in 'Hydraulics of directly coupled micro-SFE/SFC' above.

Results and discussion

(a) Amount of caffeine extracted from coffee beans under various coditions obtained by off-line micro-SFE/HPLC analysis. Figure 10 shows an HPLC chromatogram of the coffee extract, obtained by the procedure described above. The coffee powder contained the added water amounting to 20% of the coffee weight, besides the original water content. The extraction pressure was 200 bar, the temperature was 48°C, and the time was 60 min.

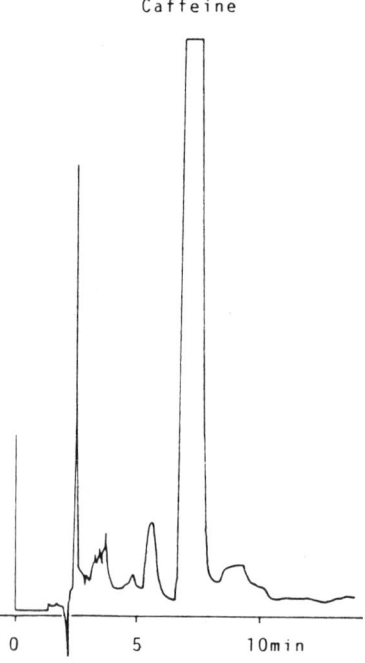

Figure 10. HPLC chromatogram of coffee extract by off-line micro-SFE/HPLC analysis. SFE conditions: pressure, 200 bar; temperature, 48°C; added water, 20%; time, 60 min. HPLC conditions; Jasco Fine Pak SIL C_{18}; eluent, methanol/water (55/45); flow rate, 1.2 ml/min; UV monitored at 272 nm and 0.64 AUFS. Reproduced from Sugiyama *et al.* (1985) with permission.

The amount of caffeine extracted from coffee beans was examined under various conditions of pressure, extraction time, added water and temperature, by the off-line micro-SFE/HPLC method. The results are shown in Fig. 11, where the amounts of caffeine extracted are represented as percentages of the amount extracted with hot water, i.e. as percentages of the caffeine level in ordinary drinking coffee. The amounts increased with increasing extraction pressure and time, shown by heavy lines. However, the amounts rapidly decreased with increasing temperature, and in the range above 60°C only small amount of caffeine were extracted. This decrease is considered to be due to the decrease of caffeine solubility in carbon dioxide, resulting from the density reduction. As the amount of added water decreased, the amount of caffeine also decreased. This suggests that the water content of

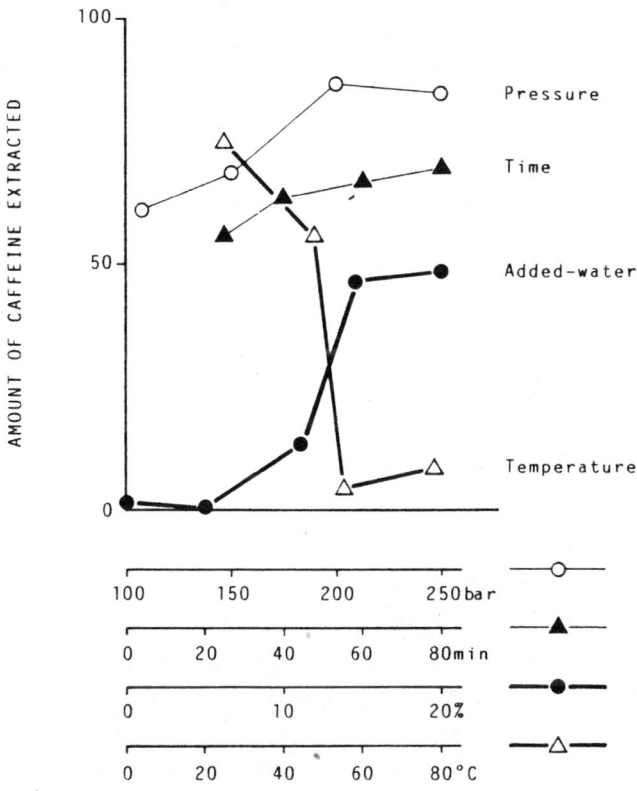

Figure 11. Percentage amounts of caffeine extracted from roasted coffee beans under various conditions against amount of caffeine extracted with hot water. Symbols: ○; various pressures with the added water, temperature, and extraction time constant at 20%, 48°C, and 60 min, respectively. ▲; various extraction times with other parameters constant, at 150 bar, 20% and 48°C. ●; various amounts of water added to coffee powder with other parameters constant, at 150 bar, 48°C and 60 min. △; various temperatures with other parameters constant, at 150 bar, 60 min and 20%. The temperature and the amount of added water have significant affect on the extraction efficiency, as shown by heavy lines. Reproduced from Sugiyama *et al.* (1985) with permission.

coffee plays the role of an entrainer solvent in extraction. Therefore, in order to extract caffeine from the roasted coffee beans efficiently, the extraction temperature should be below 50°C and an amount, at least 15% of the coffee weight, of water should be added.

The variation of the extracted amounts of caffeine in five successive experiments was calculated to be ±8% under conditions: pressure, 200 bar; temperature, 48°C; added water, 20%; time, 60 min.

(b) Identification of caffeine peak in directly coupled micro-SFE/SFC analysis of coffee beans by multiwavelength UV detector. The three-dimensional chromatogram, obtained by the micro-SFE/SFC method is shown in Fig. 12. A large peak of caffeine is clearly seen at 9.2 min in the chromatogram. Ordinary chromatograms monitored at 250 and 270 nm are shown in Fig. 13A. These chromatograms are not very informative without spectral data. In order to identify the caffeine peak chromatographically one might subject the caffeine standard to SFC. In SFC, however, the solvent in which caffeine is dissolved acts as modifier and influences the retention behavior significantly. This results in identification difficulties. Therefore spectral data are very useful and necessary for efficient identification of peak components in SFC. The bottom graph in Fig. 13B shows the spectrum taken at 9.60 min. The curve shows the clear characteristics of the caffeine spectrum.

So far we have discussed the directly coupled micro-SFE/SFC method from the viewpoint of qualitative analysis. The quantitative accuracy of the method has not been closely examined, partly because the amount of coffee powder was so excessive that the chromatographic peak gave absorbances too high for quantitation, and partly because the volume of the extraction

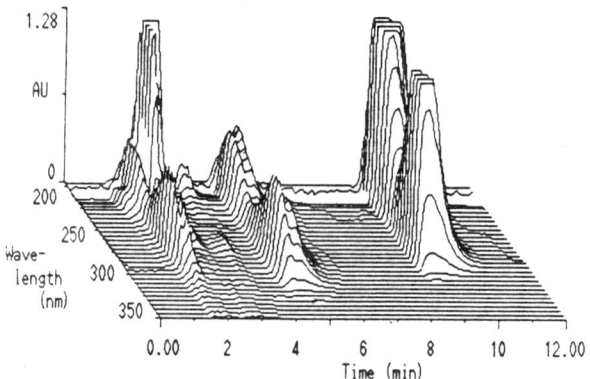

Figure 12. Three-dimensional SFC chromatogram obtained by directly coupled micro-SFE/ SFC. SFE conditions: pressure, 200 bar; temperature, 40°C; added water, 20%; time, 15 min. SFC conditions; mobile phase, supercritical carbon dioxide (*ca.* 5 ml/min); methanol (100 μl/ min); pressure, 150 bar, column, Jasco Fine Pak SIL C_{18} (6 mm i.d. × 150 mm length); temperature, 40°C. Reproduced from Sugiyama *et al.* (1985) with permission.

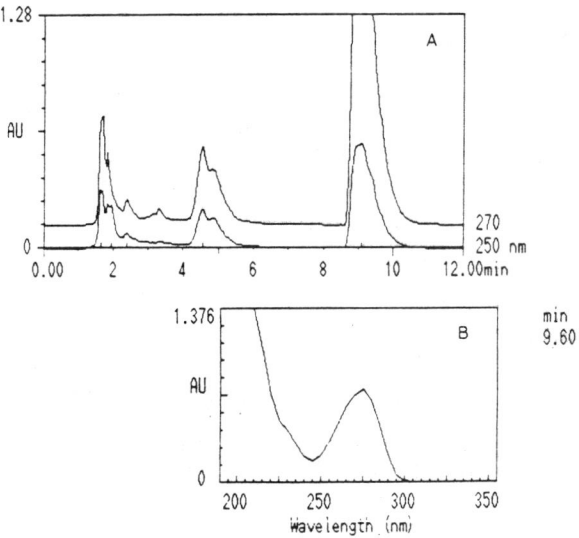

Figure 13. Chromatograms monitored at 250 and 270 nm (A) and UV spectrum taken at 9.60 min. The chromatograms were produced from the three-dimensional chromatogram shown in Fig. 12. The UV spectrum shows that the chromatographic peak eluted at 9.2 min is that of caffeine. Reproduced from Sugiyama *et al.* (1985) with permission.

cartridge did not properly match the volume of the trap loop for quantitative analysis. A study of quantitative analysis by this method is currently under way.

CONCLUSION

As we have demonstrated, the directly coupled micro-SFE/SFC system allows the analyst to apply raw and/or solid samples to the system to obtain chromatograms of sample extracts. This could be a powerful technique, extending the application of SFC to natural products, biological compounds, petrochemical products, etc., where extraction is necessary before chromatographic analysis.

In addition, a highly sensitive multiwavelength detector permits on-line UV spectrum monitoring of the extraction process, which has not been possible in a large-scale extraction system heretofore. Therefore one can easily investigate optimal extraction parameters at a lower cost, without operating a pilot-plant extraction system, which requires large amounts of sample and extraction medium.

REFERENCES

Analytical Methods Committee (1984). Application of gas–liquid chromatography to the analysis of essential oils. *A 109*, 1343–1360.

Baaliouamer, A., Meklati, Y. B., Fraisse, D., and Scharff, C. (1985). Qualitative and quantitative analysis of petitgrain eureka lemon essential oil by fused silica capillary column gas chromatography mass spectrometry. *J. Sci. Food. Agric. 36*, 1145−1154.

Bartmann, D. and Schneider, G. M. (1973). Experimental results and physico-chemical aspects of supercritical fluid chromatography with carbon dioxide as the mobile phase. *J. Chromatogr., 83*, 135−145.

Bott, T. R. (1982). Fundamental of carbon dioxide in solvent extraction. *Chem. Ind. (London)*, 19 June 1982.

Brogle, H. (1982). CO_2 as a solvent: its properties and applications. *Chem. Ind. (London)*, 19 June 1982.

Brunner, G. and Peter, S. (1982). State of art of extraction with compressed gases (gas extraction). *Ger. Chem. Eng.*, 5, 181−195.

Calame, J. P. and Steiner, R. (1982). CO_2 Extraction in the flavour and perfumery industries. *Chem. Ind. (London)*, 19 June 1982.

Chester, T. L. (1984). Capillary supercritical-fluid chromatography with flame-ionization detection: reduction of detection artifacts and extension of detectable molecular weight range. *J. Chromatogr., 299*, 424−431.

Coenen, H. and Rinza, P. (1981). Extraktion mit uberkritischen Gasen-eine Problemlosung zeu Aufarbeitung von Altolen. *Tech. Mitt. Krupp-Werksberichte, 39* (1981) H1, Z1.

Coppella, S. J. and Barton, P. (1987) Supercritical carbon dioxide extraction of lemon oil. *ACS Sym. Ser. 329*, 202−212.

Cotroneo, A., Verzera, A., Lamonica, G., Giovanni, D., and Licandro, G. (1986a). On the genuineness of citrus essential oils. Part X. Research on the composition of essential oils produced from sicilian lemons using 'pelatrice' and 'sfumatrice' extractors during the entire 1983/84 production season. *Flavour fragrance J. 1*, 69−86.

Cotroneo, A., Giovanni, D., Licandro, G., Ragonese, C., and Di Giacomo, G. (1986b). On the genuineness of citrus essential oils. Part XII. Characteristics of sicilian lemon essential oils produced with the FMC extractors. *Flavour fragrance J. 1*, 125−134.

Crowther, J. B. and Henion, J. D. (1985). Packed column supercritical fluid chromatography with simultaneous UV/MS detection. *Pitts. Conf. Abs.* No. 539.

Eggers, R. (1980). Large-scale industrial plant for extraction with supercritical gases. In: Schneider, G. M., Stahl, E., and Wilke, G. (Eds), *Extraction with Supercritical Gases*. Verlag Chemie, Weinheim, pp. 155−164.

Eggers, R. and Tschiersch, R. (1980). Development and design of plant for high-pressure extraction of natural products. In: Schneider, G. M., Stahl, E., and Wilke, G. (Eds), *Extraction with Supercritical Gases*. Verlag Chemie, Weinheim, pp. 165−189.

Filippi, de, R. P. (1982). CO_2 as a solvent: application to fats, oils and other materials. *Chem. Ind. (London)*, 19 June 1982.

Fujimoto, C., Hirata, Y., and Jinno, K. (1985). Supercritical fluid chromatography-infrared spectroscopy of oligomers: use of buffer-memory technique. *J. Chromatogr. 332*, 47−56.

Gardner, D. S. (1982). Industrial scale hop extraction with liquid CO_2. *Chem. Ind. (London)*, 19 June 1982.

Gere, D. R., Board, R., and McManigill, D. (1982). Supercritical fluid chromatography with small particle diameter packed columns. *Anal. Chem. 54*, 736−740.

Greibrokk, T., Blilie, L. A., Johansen, J. E., and Lundanes, E. (1984). New system for delivery of the mobile phase in supercritical fluid chromatography. *Anal. Chem. 56*, 2681−2684.

Hannay, J. B. and Hogarth, J. (1879). On the solubility of solids in gases. *Proc. R. Soc. Lond. 29*, 324−326.

Hirata, Y. and Nakata, F. (1984). Supercritical fluid chromatography with fused-silica packed columns. *J. Chromatogr., 295*, 315−322.

Hoshino, T., Senda, M., Hondo, T., Saito, M., Tohei, S. (1984). Application of photodiode array ultraviolet detector to unresolved peak analysis. *J. Chromatogr. 316*, 473−486.

Hoshino, T., Hondo, T., Senda, M., Saito, M., and Tohei, S. (1985). Quantitative deconvolution of heavily fused chromatographic peaks of biological components using a multiwavelength UV detector. *J. Chromatogr. 332*, 139−146.

Hubert, P. and Vitzthum, O. G. (1980). Fluid extraction of hops, spices, and tobacco with supercritical gases. In: Schneider, G. M., Stahl, E., and Wilke, G. (Eds), *Extraction with Supercritical Gases*. Verlag Chemie, Weinheim, pp. 25–44.

Jentoft, R. E. and Gouw, T. H. (1970). Pressure-programmed supercritical fluid chromatography of wide molecular weight range mixtures. *J. Chromatogr. Sci. 8*, 138–142.

Jentoft, R. E. and Gouw, T. H. (1972). Apparatus for supercritical fluid chromatography with carbon dioxide as the mobile phase. *Anal. Chem. 44*, 681–686.

Jinno, K., Saito, M., Hondo, T., and Senda, M. (1986). Correlation between retention data of polycyclic aromatic hydrocarbons and several descriptors in supercritical-fluid chromatography. *Chromatographia 21*, 219–222.

Jinno, K., Hoshino, T., Hondo, T., Saito, M., and Senda, M. (1986). Computer enhanced spectroscopic separation of coeluted components in supercritical fluid chromatography coupled with an ultraviolet multichannel detector. *Anal. Lett. 19*, 1001–1018.

Johnson, C. C., Jordan, J. W., Skelton, R. J., and Taylor, L. T. (1985). Supercritical fluid chromatography using packed columns with Fourier transform infrared detection. *Pitts. Conf. Abs.*, No. 538.

Kimura, K., Nishimura, H., Iwata, I., and Mitzutani, J. (1983). Deterioration mechanism of lemon flavor. 2. Formation mechanism of off-odor substances arising from citral. *J. Agric. Food Chem. 31*, 801–804.

Klavons, J. A. and Bennet, R. D. (1985). The nature of the protein constituent of commercial lemon juice cloud. *J. Agric. Food Chem. 33*, 708–712.

Klesper, E., Corwin, A. H., and Turner, D. A. (1962). High pressure gas chromatography above critical temperatures. *J. Org. Chem. 27*, 700–701.

Myers, N. M. and Giddings, J. C. (1965). High column efficiency in gas liquid chromatography at inlet pressures to 2500 p.s.i. *Anal. Chem. 37*, 1453–1457.

Myers, N. M. and Giddings, J. C. (1966). High inlet pressure micro column system for use in gas chromatography. *Anal. Chem. 38*, 294–297.

Norris, T. A. and Rawdon, M. G. (1984). Determination of hydrocarbon types in petroleum liquids by supercritical fluid chromatography with flame ionization detection. *Anal. Chem. 56*, 1767–1769.

Novotny, M., Bertsch W., and Zlatkis, A. (1971). Temperature and pressure effects in supercritical-fluid chromatography. *J. Chromatogr. 61*, 17–28.

Novotny, M., Springston, S. R., Peaden, P. A., Fjeldsted, J. C., and Lee, M. L. (1981). Capillary supercritical fluid chromatography. *Anal. Chem. 53*, 407A–408A.

Olesik, S. V., French, S. B., and Novotny, M. (1984). Development of capillary supercritical fluid chromatography/Fourier transform infrared spectrometry. *Chromatographia 18*, 489–495.

Peaden, P. A. and Lee, M. L. (1982). Supercritical fluid chromatography: methods and principles. *J. Liq. Chromatogr. 5*, 179–221.

Peaden, P. A. and Lee, M. L. (1983). Theoretical treatment of resolving power in open tubular column supercritical fluid chromatography. *J. Chromatogr. 259*, 1–16.

Peaden, P. A., Fjeldsted, J. C. Lee, M. L. Springston, S. R., and Novotny M. (1982). Instrumental aspects of capillary supercritical fluid chromatography. *Anal. Chem. 54*, 1090–1093.

Peter, S. and Brunner, G. (1980). The separation of nonvolatile substances by means of compressed gases in countercurrent process. In: Schneider, G. M., Stahl, E., and Wilke, G. (Eds), *Extraction with Supercritical Gases*. Verlag Chemie, Weinheim, pp. 141–154.

Rawdon, M. G. (1984). Modified flame ionization detector for supercritical fluid chromatography. *Anal. Chem. 56*, 831–832.

Saito, M., Sugiyama, K., Hondo, T., Senda, M., and Tohei, S. (1985). Laboratory-scale supercritical fluid extraction and chromatography using multiwavelength UV detector. *International Symposium Kyoto, Jan., 1985, Abstract*, pp. 84–86.

Saito, M. and Hondo, T. (1986). Directly coupled supercritical fluid extraction — supercritical fluid chromatography and its applications. *Yukagaku 35*, 273–280.

Shafer, K. H. and Griffiths, P. R. (1983). On-line supercritical fluid chromatography/Fourier transform infrared spectrometry. *Anal. Chem. 55*, 1939–1942.

Shaw, P. E., and Wilson III, C. W. (1982). Volatile sulfides in headspace gases of fresh and processed citrus juices. *J. Agric. Food Chem. 30*, 685–688.

Smith, R. D., Felix, W. D., Fjeldsted, J. C., and Lee, M. L. (1982). Capillary column supercritical fluid chromatography/mass spectrometry. *Anal. Chem. 54*, 1883–1885.

Smith, R. D., Kalinoski, H. T., Udseth, H. R., and Wright, B. W. (1984). Rapid and efficient capillary column supercritical fluid chromatography with mass spectrometric detection. *Anal. Chem. 56*, 2476–2480.

Stahl, E. (1977). Coupling of extraction with supercritical gases and thin-layer chromatography. *J. Chromatogr. 142*, 15–21.

Stahl, E. and Schiltz, W. (1976). Extraction with supercritical Gases in coupling with thin-layer chromatography. *Z. Anal. Chem. 280*, 99–104.

Stahl, E., Schutz, E., and Mangold, H. (1980). Extraction of seed oils with liquid and supercritical carbon dioxide. *J. Agr. Food Chem. 28*, 1153–1157.

Staroscik, A. J. and Wilson, A. A. (1982a). Quantitative analysis of cold-pressed oil by glass capillary gas chromatography. *J. Agric. Food Chem. 30*, 507–509.

Staroscik, A. J. and Wilson, A. A. (1982b). Seasonal and regional variation in the quantitative composition of cold-pressed lemon oil from California and Arizona. *J. Agric. Food Chem. 30*, 835–837.

Sugiyama, K., Saito, M., Hondo, T., and Senda, M. (1985). New double stage separation analysis method directly coupled laboratory-scale supercritical fluid extraction — supercritical fluid chromatography, monitored with a multiwavelength ultraviolet detector. *J. Chromatogr. 332*, 107–116.

Sugiyama, K. and Saito, M. (1988). Simple home-made microscale supercritical fluid extraction system and its application to gas chromatography – mass spectrometry of lemon peel oil. *J. Chromatogr. 442*, 121–131.

Takeoka, G. R., Guentert, M., Macku, C., and Jennings W. (1986). Advances in the separation of biological aromas. *ACS Symp. Ser. 317*, 53–64.

Takeuchi, T., Ishii, D., Saito, M., and Hibi, K. (1984). Supercritical fluid chromatography with micro packed columns and carbon dioxide as a mobile phase. *J. Chromatogr. 295*, 323–331.

Unger, K. K. and Roumeliotis, P. (1983). On-line high-pressure extraction-high-performance liquid chromatography. *J. Chromatogr. 282*, 519–526.

Vollbrecht, R. (1982). Extraction of hops with supercritical CO_2. *Chem. Ind. (London)*, 19 June 1982.

Williams, D. F. (1981). Extraction with supercritical gases. *Chem. Engng. Sci. 36*, 1769–1788.

Zosel, K. (1980). Separation with supercritical gases; practical applications. In: Schneider, G. M., Stahl, E., and Wilke, G. (Eds), *Extraction with Supercritical Gases*. Verlag Chemie, Weinheim, pp. 1–24.

Progress in HPLC, Vol. 4, pp. 111–155.
Yoshioka *et al.* (Eds)
© 1989 VSP.

Supercritical fluid chromatography–mass spectrometry (SFC–MS)

RICHARD D. SMITH, BOB W. WRIGHT and HENRY T. KALINOSKI

Chemical Methods and Separations Group, Chemical Sciences Department, Pacific Northwest Laboratory, Richland, WA 99352, USA

INTRODUCTION

Interest in supercritical fluid chromatography (SFC) and its combination with mass spectrometry (SFC–MS) has been growing for the efficient separation and characterization of mixtures not amenable to gas chromatography (Fjeldsted *et al.*, 1983; Fjeldsted and Lee, 1984; Chester, 1984; Fields *et al.*, 1984; Guthrie and Schwartz, 1986; Novotny *et al.*, 1981; Yonker *et al.*, 1984; Randall and Wahrhaftig, 1981; Lauer *et al.*, 1983; Levy and Ritchey, 1986, and Sie and Rijnders, 1967). This interest is furthered due to limitations perceived in liquid chromatography–mass spectrometry (LC–MS) interfacing methods, application of SFC–MS to thermal labile compounds, higher chromatographic efficiency (compared to LC) of SFC, and the availability of fused silica capillary columns with nonextractable stationary phases for SFC (Peaden *et al.*, 1982; Wright *et al.*, 1982). The use of mass spectrometry in conjunction with SFC provides detection with both high sensitivity and selectivity regardless of the mobile phase composition (Smith *et al.*, 1982b; Smith and Udseth, 1983a; Wright *et al.*, 1984; Smith *et al.*, 1984a, b, 1985a). The role of SFC and SFC–MS is likely to initially center around the characterization of materials for which the more mature GC and HPLC methods are inappropriate, establishing SFC as an excellent complement to both techniques. However, the potential exists for SFC–MS to replace HPLC–MS for a substantial fraction of normal phase applications if appropriate fluid phases can be developed..

We have previously described capillary SFC–MS interface designs for chemical ionization (Smith *et al.*, 1982b; Smith and Udseth, 1983a; Wright *et al.*, 1984) and electron impact ionization (Smith *et al.*, 1984b), with quadrupole mass spectrometry; high-speed separations using rapid pressure programming (Smith *et al.*, 1984a, 1985a); the introduction of solvent mixtures (Wright *et al.*, 1985); detailed evaluation of capillary restrictor performance (Smith *et al.*, 1986a); and potential analytical application to several problems where advantages exist compared to alternative GC or HPLC methods (Smith et al., 1985d; Kalinoski et al., 1986b; Smith et al., 1986b; Kalinoski et al., 1986a; Smith et al., 1985c). Most recently, approaches for the analysis of less volatile analytes with increased sample capacity (Smith and Udseth, 1987) and the interface with high-resolution mass spectrometry (Kalinoski et al., 1987) have been described.

In this review we first summarize the properties of supercritical fluids relevant to SFC−MS, as the advantages and range of applicability of the technique are essentially defined by the physical properties and solvating characteristics of the fluid phase. Following this we describe the practice and application of SFC−MS and give a detailed description of current instrumentation and interfacing methods. The scope of the present review includes both capillary and packed column SFC methods as well as a brief comparison of the relative merits and required compromises of each approach. Finally, the range of applicability of the SFC−MS technique will be illustrated, with examples of a variety of analytical problems and related applications. Attempts to define current limitations of SFC−MS, and the potential for circumventing these restrictions, will also be made.

PROPERTIES OF SUPERCRITICAL FLUIDS

The combination of physical properties (viscosities and diffusion rates) with variable solvent properties is the basis for the advantages of supercritical fluids for use in chromatography. The physical properties of a supercritical fluid are variable, roughly between the limits of a gas and that of a liquid at ambient conditions, through control of pressure. The solvent characteristics of a supercritical fluid are a function of density and the chemical nature of the fluid. Typically, supercritical fluids in SFC are used at densities from 0.1 to 0.8 of their liquid density. Under these conditions their diffusion coefficients are substantially greater than liquids. Similarly, the viscosity of supercritical fluids mirrors the diffusivity, and is typically $10-10^2$ times less than that for liquids (McHugh and Krukonis, 1986). These more favorable physical properties define the advantages of supercritical fluids in extraction, chromatography, and related applications.

The solvation properties of a fluid have been demonstrated to be roughly proportional to density (Giddings et al., 1968). The density of the supercritical fluid will be typically 10^2 to 10^3 times greater than that of the gas at ambient pressures. Consequently, molecular interactions are greater due to the shorter intermolecular distances. The 'liquid-like' density of a supercritical fluid results in greatly enhanced solubilizing capabilities. Table 1 gives the critical parameters for a number of common and potential supercritical fluid solvents for SFC. Solute−solvent interactions are expected to be sensitive to changes in both temperature and pressure. Since changes in the fluid density may impact both the quantity and qualitative nature of solute−solvent intermolecular interactions, the general solvent power of a supercritical fluid can be easily altered. For a pure supercritical fluid and,

Table 1.
Common SFC solvents

Compound	Boiling point (°C)	Critical temperature (°C)	Critical pressure (bar)	Critical density (g/cm^3)
CO_2	−78.5 (sublimes)	31.3	72.9	0.448
NH_3	−33.4	132.4	112.5	0.235
H_2O	100	374.2	218.3	0.315
N_2O	−88.6	36.5	71.7	0.45
Ethane	−88.6	32.3	48.1	0.203
Ethylene	−102.7	9.2	49.7	0.218
Propane	−42.1	96.7	41.9	0.217
Pentane	36.1	196.6	33.3	0.232
Benzene	80.1	288.9	48.3	0.302
Methanol	64.7	240.5	78.9	0.276
Isopropanol	82.5	235.3	47.0	0.273

indeed, fluid solutions at concentrations relevant to SFC—MS (i.e. essentially infinite dilution), the relationships between pressure, temperature and density are easily estimated with reasonable precision from equations of state (except near the critical point).

The range of solvating power of practical supercritical fluids for SFC—MS is of primary importance and ultimately determines the limits of application. The meaning of the terms 'volatility' and 'solubility' can become ambiguous under certain conditions in supercritical fluids. The solubility of a component in a supercritical fluid generally reflects its density (pressure) behavior with the greatest increase in solubility at the fluid's critical point (Smith and Udseth, 1983a). The general dependence of solubility for a solid in a supercritical fluid as a function of temperature and pressure is illustrated in a simplified manner in Fig. 1. The solute typically exhibits a pressure above which solubility increases significantly; the region of maximum increase in solubility as a function of pressure is near the critical pressure, where the change in density with pressure is greatest. This results from the fact that there is often a nearly linear relationship between log[solubility] and fluid density for dilute solutions of nonvolatile compounds (up to concentrations where solute—solute interactions become important). In contrast, at densities less than or near the critical density, increasing temperature will typically decrease solubility when volatility is low (Smith and Udseth, 1983a). However, 'solubility' may increase at sufficiently high temperatures, where the solute vapor pressure can also become significant. Thus, while the highest supercritical fluid densities at a given pressure are obtained near the critical temperature, the greatest solubilities (within given experimental pressure limitations) will often be obtained at somewhat lower densities but higher temperatures.

As with liquids, polar solutes are most soluble in polar supercritical fluids, although nominally nonpolar fluids can be remarkably good solvents for

SOLUBILITIES IN SUPERCRITICAL FLUIDS

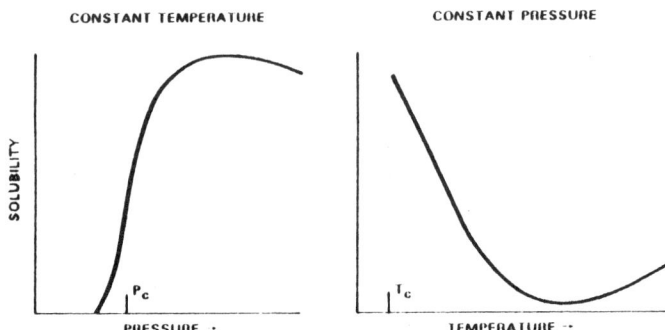

Figure 1. Typical trends for the solubilities for solid solutes in supercritical fluids as a function of pressure or temperature (see text).

many moderately polar compounds (McHugh and Krukonis, 1986). Carbon dioxide, for example, can exhibit solvating properties intermediate between pentane and methylene chloride at higher pressures, and solvatochromic studies of the solvent environment around solute probe molecules confirm their variable solvent properties (Yonker *et al.*, 1986). At normal operating pressures, half to several times the critical pressure, solubility typically increases with pressure under isothermal conditions. Under conditions of constant density, solubility generally increases with temperature. However, a temperature increase under isobaric conditions will generally result in decreased solubility at pressures less than several times the critical pressure; at higher pressures increased solubility will typically be obtained for an isobaric temperature increase.

Highly polar fluids typically have critical temperatures which are too high to allow operation with currently available columns or with labile biological analytes (see Table 1). Therefore, it is often advantageous in SFC to use a fluid mixture to obtain enhanced solvating power, alter the critical temperature of the mobile phase, or change chromatographic selectivities. The phase behavior of binary fluid systems is highly varied and much more complex than single-component systems and has been well described for selected systems (see for example McHugh and Krukonis (1986) and references therein). It is vital that fluid mixtures for SFC be selected so that they can be mixed and pumped as a single phase (which is most conveniently done at ambient temperatures, imposing additional constraints). Particular attention is required when operating over a range of pressures, as typical in SFC, so as to avoid producing a two-phase system. Comparison with simple predictions in the absence of actual phase equilibria (e.g. compare results from Crowther and Henion, 1985) shows that considerable error can result, which can lead to inadvertent operation in the vapor−liquid region of the phase diagram. These considerations are much more important when pressure programming methods are used to vary density, and can be relatively unimportant for packed column separations when isobaric methods are used (where a constant average density is acceptable). In fact, a considerable fraction of the work reported with packed columns using large fluid modifier concentrations, >5%, has actually involved a subcritical, or 'near critical', liquid mobile phase. An additional concern relevant to SFC−MS is that the fluid be maintained as a single phase until the point of injection into the mass spectrometer, which often occurs at a higher temperature than used for the chromatographic separation.

It is not possible to understate the role of supercritical fluid solubilities in SFC−MS. It has been shown that retention in nonselective stationary phases used in capillary SFC mirrors fluid phase solubilities (unpublished work, C. R. Yonker and R. D. Smith). Fluid phase solubilities and their dependence upon density will be shown to be important criteria for the application of specific analytical techniques involving the separation and identification of volatile and nonvolatile compounds, and for the transfer of nonvolatile compounds in the SFC−MS interface to the ionization region.

SUPERCRITICAL FLUID CHROMATOGRAPHY

Renewed interest in SFC is due in large part to the limitations in both chromatographic efficiency and detection methods for HPLC, the obvious applications to thermally labile components, and the availability of fused silica capillary columns with nonextractable stationary phases for SFC (Peaden *et al.*, 1982 and Wright *et al.*, 1982). The ability to interface SFC with mass spectrometric detection has provided a powerful analytical tool which extends the well-recognized capabilities of GC−MS to less volatile compounds. A schematic representation of capillary SFC instrumentation interfaced with quadrupole mass spectrometry typical of that currently used

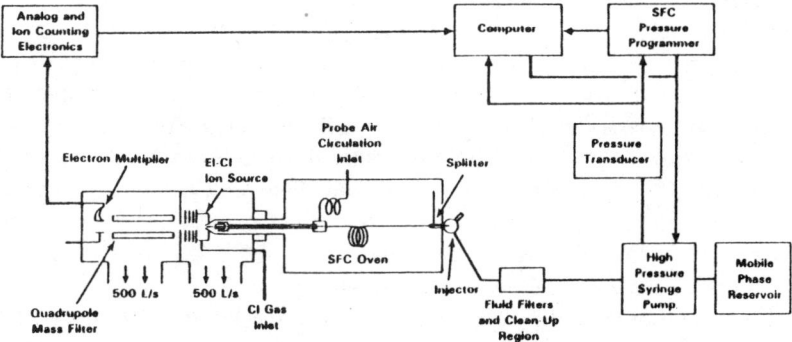

Figure 2. Schematic illustration of typical instrumentation for capillary SFC−MS.

is shown in Fig. 2. The lower viscosities and high diffusion coefficients relative to liquids result in significantly enhanced chromatographic efficiency per unit time compared to HPLC. Studies under isobaric conditions have demonstrated more than 3000 and 12 000 theoretical plates/m can be obtained with 50 μm and 25 μm (i.d.) columns, respectively (Wright and Smith, 1986). In SFC the mobile phase is maintained at a temperature somewhat above its critical point (at reduced temperatures of 1.02 to 1.4°C), and it is generally advantageous to utilize the highest temperature compatible with the SFC system and the material being analyzed. The density of the super-critical phase is usually several hundred times greater than that of the gas, but less than that of the liquid at typical SFC pressures (25−500 bar). The mild thermal conditions (determined by the choice of supercritical mobile phase) allow application to many labile compounds which cannot be ad-dressed by GC without derivatization. Similarly, the use of derivatization techniques with SFC extends the effective molecular weight range well beyond that possible with GC. The use of open tubular capillary columns results in negligible pressure drops for typical linear velocities (<10 cm/s) and column dimensions (50 μm i.d., <20 m length). This is an important consideration if the pressure programming capability of SFC for manipula-tion of mobile phase solvating power is to be fully exploited. Pressure programming in SFC provides many of the advantages of gradient elution in HPLC, including very rapid pressure programming to affect high-speed capillary SFC separations (Smith et al., 1985a).

Nearly all current SFC research utilizes either columns packed with 3−10 μm surface-modified silica particles (as prepared for HPLC) or wall-coated (0.1−1 μm film thickness) open tubular fused silica capillaries of 25 to 100 μm i.d. Packed columns of typical diameter for HPLC (1−5 mm i.d.) provide much greater flow rates than capillary columns, and generally allow relatively large sample loadings and improved detection limits, but have obvious disadvantages for mass spectrometric interfacing. Capillary columns have demonstrated greater numbers of effective plates than packed columns since they have much greater permeabilities. Thus, considerably longer columns can be used before an excessive pressure drop occurs through the column. Capillary columns allow programming over a wider range of pres-sure than packed columns (with the precise comparison depending upon column length, packing, linear velocity, and fluid viscosity). The small on-column injection volumes (<50 nl) typically required with capillary columns are most readily obtained by flow splitting. The flow rates with capillary columns provide easier interfacing to mass spectrometry but can present a practical barrier to the range of applicability for SFC−MS due to constraints upon interface operation (see later discussion).

Capillary columns coated with bonded and cross-linked stationary phases provide significantly greater flexibility for regulating retention through varia-tion of either pressure or temperature than is feasible with conventional packed columns. In general, pressure programming at the highest tempera-ture compatible with the separation offers the advantages of simplicity, improved reproducibility, and maximum chromatographic efficiency (since efficiencies are improved by operation at higher temperatures and lower densities due to more favorable diffusion rates). However, simultaneous pressure−temperature programming to either increase or maintain constant density can offer some advantages (Fields and Lee, 1984), and the availability of improved commercial instrumentation is resulting in better programming options.

Other considerations relating to relative merits of the two column technologies are derived from the large pressure drops incurred with packed columns and the generally greater importance of active surface sites for packed columns, as opposed to the relative inertness of deactivated capillary columns. Pressure drops for 50 μm i.d. capillary columns are negligible for linear velocities <20 cm/s and columns <10 m in length. In addition, small volumetric flow rates with capillary columns provide for rapid thermal equilibration. Columns packed with 3−5 μm particles can generate $10-10^2$ times as many plates per second as 50 μm i.d. capillary columns (Guiochon and Colin, 1984), but are more limited in the range of flow rates or column length due to the large pressure drop. The pressure drop for 5 μm packings at optimum linear velocities is ~3 × 10^4 greater per plate than for 50 μm i.d. open tubular capillary columns (Guiochon and Colin, 1984). This difference in column permeability limits effective high-speed application to very short packed columns. Separations requiring large numbers of plates using packed columns are also ultimately limited due to the rapid increase in plate height at lower than optimum linear velocities. The maximum efficiency obtainable in such separations depends upon the particular chromatographic conditions and the maximum pressure drop which can be tolerated during a separation (which is directly proportional to linear velocity). Capillary columns are less subject to this limitation and maximum possible efficiencies (given the same pressure drop) will be about 10^2 to 10^3 greater than for packed columns.

The reduced pressure drop in capillary columns is also of practical importance in pressure programming, where much greater flexibility exists than for packed columns. While 5 μm micro-particle packed columns generate more plates/second than 50 μm i.d. capillary columns, the lower pressure drop with capillaries allows greater numbers of plates to be obtained in conjunction with pressure programming. A pressure drop does not intrinsically prevent pressure programming, but it does result in more subtle difficulties. For example, the low pressure drop in capillaries allows the programmed pressure rise to take place across the entire column length almost instantaneously (since the pressure rise will be transmitted at near the speed of sound). As the pressure drop increases, the time to reach steady-state conditions also increases; thus changes in k' with pressure will be much more difficult to predict for packed columns. The pressure drop also limits the minimum pressure necessary at the column entrance (required to prevent analyte precipitation in the column), and thus the effective range of pressure and density which can be used to vary retention.

The different retention characteristics and phase ratios of commercially available packed columns and current capillary SFC columns is also an important consideration. For example, typical capacity ratios for phenanthrene in carbon dioxide with a 5 μm C_{18} microbore column are approximately 10^2 times greater than with a SE-54 capillary column (0.25 μm film thickness) at the same density. Greater capacity ratios are observed for packed columns with carbon dioxide in all cases. Since one of the areas of interest in SFC is for analysis of more polar and higher molecular weight compounds, which often show greater retention and can require relatively high fluid densities, capillary columns are usually more appropriate. Similar limitations arise from the greater adsorptive activity of packed column stationary phases. The above considerations combine to make pressure programming in packed columns far less attractive than for capillaries. These factors have resulted in the greater interest for packed column applications in using solvent modifers and solvent gradient methods to reduce retention than for capillary SFC. (Another reason is that the popular FID detector for capillary SFC is precluded for most mixed fluid systems.)

The practical advantages of pressure programming with capillary SFC are demonstrated in Fig. 3, which shows the separation of C_{12} to C_{40} n-alkanes obtained using a short 25 μm i.d. capillary column and a relatively fast pressure programming rate of 160 bar/min. The theory and practical considerations relevant to pressure programmed capillary SFC, with particular emphasis on rapid separations, has been described in detail elsewhere (Smith et al., 1985a). In practice, higher linear velocities and faster pressure ramp rates can be provide better separations when measured as separation number or Trennzahl number per minute (Wright et al., 1986). These performance characteristics allow capillary columns to often be competitive with packed columns for fast separations.

With all the advantages of capillary columns there are situations where the use of packed columns for SFC is favored. These applications include: (a) very high-speed separations where only very few effective plates are required, (b) alteration of selectivity using the wider range of stationary phases currently available, (c) situations where high flow rates are required, (d) obtain improved detection limits, and (e) situations where large sample loadings are necessary. Higher fluid flow rates can be useful in the transport of less volatile analytes through the SFC−MS interface to the ionization region. However, capillary columns can also be used successfully with high loading and high flow rates, and yield separations comparable to packed column isobaric separations (Smith and Udseth, 1987).

The retention process for capillary columns most often involves the partitioning of a solute between the bonded polymeric phase and the supercritical mobile phase. Solute retention in a chromatographic separation system is a dynamic equilibrium between the two phases, and will be dependent on the solubility of the solute in the fluid mobile phase and in the bonded organic stationary phase. Recent work with highly deactivated and nonselective stationary phases in capillary SFC has suggested that, in the absence of adsorption and specific chemical interactions with the stationary phase, fluid phase solubility is the major determinant of retention (Smith et al., 1985c). The functional relationship between retention and pressure at constant temperature has been described by Van Wasen and Schneider (1975). The trend

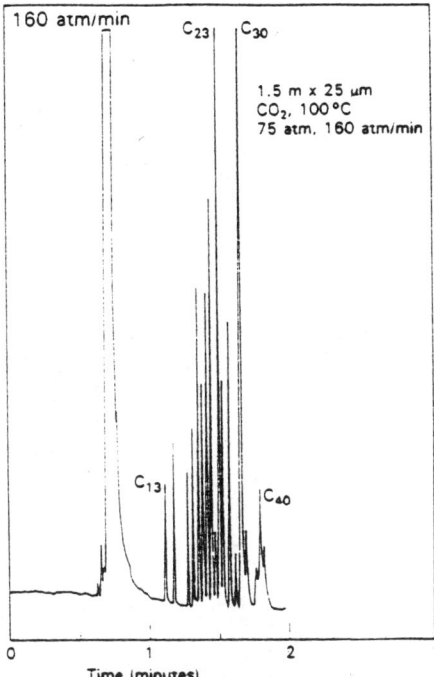

Figure 3. Fast SFC chromatogram for the *n*-alkane mixture using a 1.5 m × 25 μm column with a carbon dioxide mobile phase at 100°C with a 160 bar/min pressure program rate.

in retention was shown to depend on the partial molar volume of the solute in the mobile and stationary phases, coupled with the isothermal compressibility of the fluid mobile phase. A simple relationship between solubility and chromatographic retention based upon the thermodynamics of solute retention behavior has been examined (Yonker *et al.*, 1985; Smith *et al.*, 1986c). Utilizing the solubility of a pure, incompressible solute in a fluid, given by Gitterman and Procaccia (1983), this thermodynamic relationship has been shown to describe the features of solute retention as a function of pressure at constant temperature for SFC (Yonker *et al.*, 1985; Smith *et al.*, 1986c; Yonker and Smith, 1988).

The ability to modify solute−mobile phase intermolecular interactions in SFC with pure fluids is limited. Chemically tailored mobile phase solvents composed of binary or ternary mixtures with organic modifiers have an important advantage in the separation of complex sample mixtures by producing increased selectivity and overall decreased retention times. The addition of a solvent modifier to the supercritical fluid can produce enhanced selectivity, altered retention, and improved peak shape (Gitterman and Procaccia, 1983; Hirata and Nakata, 1984; Yonker and Smith, 1986). The ability to modify the solute−fluid mobile phase intermolecular interactions, while maintaining the critical conditions for the binary solution comparable to those of the pure fluid, can be important in the separation of polar and thermally labile compounds.

Potential binary fluid mixtures for SFC must be carefully considered to ensure they remain as a single phase at the temperature and pressure used during the separation. For the simplest class (Type 1) of binary fluid systems (methanol−carbon dioxide is one such system), an increase in the mole fraction of a polar modifier increases the critical pressure and critical temperature of the binary system. There exists between the two pure fluid critical points a continuous critical locus of temperatures and pressures as a function of the modifier composition. This critical locus for methanol−carbon dioxide is shown in Fig.4, which contains experimental data from Brunner (1985) and McHugh and Seckner (1987) along with the critical locus (solid line) calculated from a method outlined by Chuch and Prausnitz (1967) and Reid *et al.* (1977). The fit between the calculated critical locus and the experimental data is quite reasonable. Generally, the critical locus for a Type 1 system can be estimated based on the critical parameters of the pure fluid components and experimental conditions chosen so as to obtain the desired range of densities in the supercritical region of the $P-T-X$ phase diagram.

For SFC with binary fluids using methanol, acetonitrile, and 2-propanol as polar modifiers for carbon dioxide, the retention values of five probe molecules all decreases compared to pure CO_2, as shown in Fig. 5. Although the decrease in retention for 0.2 mole fraction of the various modifiers was substantial in all cases, significant changes in the selectivity are also observed dependent upon both the solute and the modifier.

The selectivity of the separation for the probe molecules relative to decylbenzene as a function of isopropanol modifier concentration is shown in Fig. 6. The selectivity (α) of the separation process changes with increasing concentration of 2-propanol in the supercritical mobile phase. This can be explained by increased solvent strength of the binary fluid phase as the concentration of 2-propanol increases for the polar probe solutes as compared to the nonpolar compounds. The greater intermolecular interactions

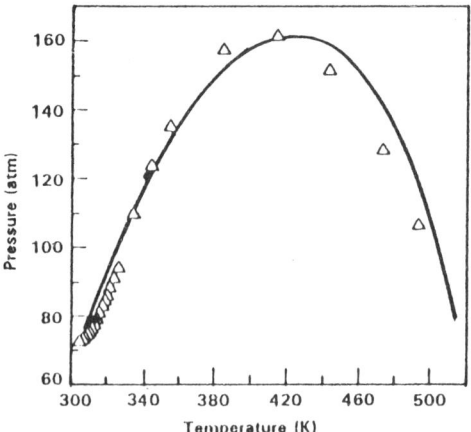

Figure 4. Critical loci for CO_2-methanol as a function of temperature and pressure. Experimental data of Brunner ($\triangle$) and of McHugh and Seckner ($\bullet$) plotted against predicted critical loci.

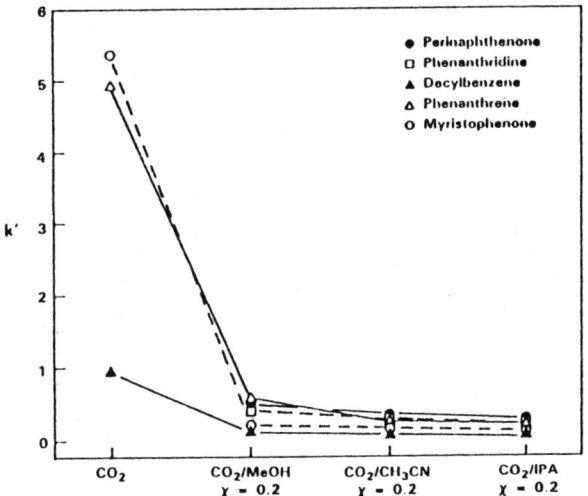

Figure 5. Plot of solute retention k') as a function of various organic modifiers, methanol (MeOH), acetonitrile (CH_3-CN) and 2-propanol (IPA) at constant mole fraction (20%).

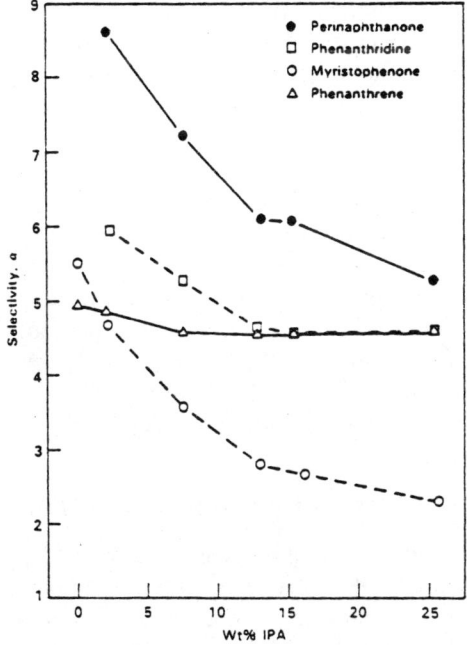

Figure 6. Plot of selectivity (α) versus weight percent of 2-propanol in CO_2 (wt% IPA) for selected probe molecules.

between the polar analytes of *peri*-naphthenone, phenanthridine and myris-
tophenone contributed to the change in selectivity with modifier concentration. In contrast, the relative retention (α) of phenanthrene compared to
decylbenzene appeared independent of 2-propanol concentration in carbon
dioxide. Figure 6 demonstrates the ability to chemically tailor binary supercritical phases to alter retention or selectivity. The same properties are
exploited in the use of entrainers in extraction for potential industrial processes, and SFC offers a mechanism to evaluate potential entrainers and to
decouple their role from effects related to those in condensed phases.

The transition from a two-phase vapor−liquid equilibrium region (where
only one phase will usually serve as the mobile phase) to that of a single
supercritical phase for binary fluids will obviously impact both retention and
selectivity of the separation process. Figure 7 shows an SFC−MS separation
obtained for a coal tar extract using a 2-propanol−propane binary fluid
mixture. At 120°C the binary mobile phase was estimated by the Prausnitz
method to be in the two-phase region of the $P-T-X$ phase envelope. One
can see from Fig. 7 the decrease in selectivity and retention of the sample
components of the coal tar mixture in the two-phase subcritical region
(120°C), where a liquid mobile phase enriched in 2-propanol is evident. As
one exceeds the critical temperature (somewhere between 120°C and 130°C)
the resolution between sample components increases with a concomitant
increase in retention time while pressure programming at a constant rate. At
even higher temperatures the change in resolution and retention time is not
as dramatic, being more removed from the critical point. These observations
suggest illustrate the basis of concurrent pressure and temperature programming with binary supercritical fluid solvents to decrease separation
times while maintaining adequate resolution of complex sample mixtures.

Supercritical fluids are flexible solvents for use in separation processes
where chemically tailored solvents for a specific application are required.
Supercritical fluid chromatography and extraction provide an efficient means
of separation and transport of nonvolatile or thermally labile compounds in
a wide range of applications. An important application of supercritical fluids
and SFC lies in their coupling with mass spectrometry as a selective detector. In this case supercritical fluids serve as a transport mechanism for solute
introduction into the ion source of the mass spectrometer. These advantages
and applications related to direct fluid injection mass spectrometry and
SFC−MS will be discussed in the following sections.

DIRECT FLUID INJECTION (DFI) MASS SPECTROMETRY

Since supercritical fluids are simply dense pressurized gases, considerable
flexibility exists for SFC−MS interfacing. The direct fluid injection (DFI) approach is currently the simplest and most widely used approach to SFC−MS
and is based upon the rapid expansion of the supercritical fluid mobile phase
into a low pressure region where ionization can occur. Since proper conditions
produce a gas after expansion, conventional mass spectrometric ionization

methods are readily adapted. Alternatively, the supercritical fluid can, in principle, be cooled to form a liquid and addressed with LC−MS interfaces. The fluid can also be decompressed and the analytes deposited on a surface (e.g. a moving ribbon) for subsequent desorption and ionization using a

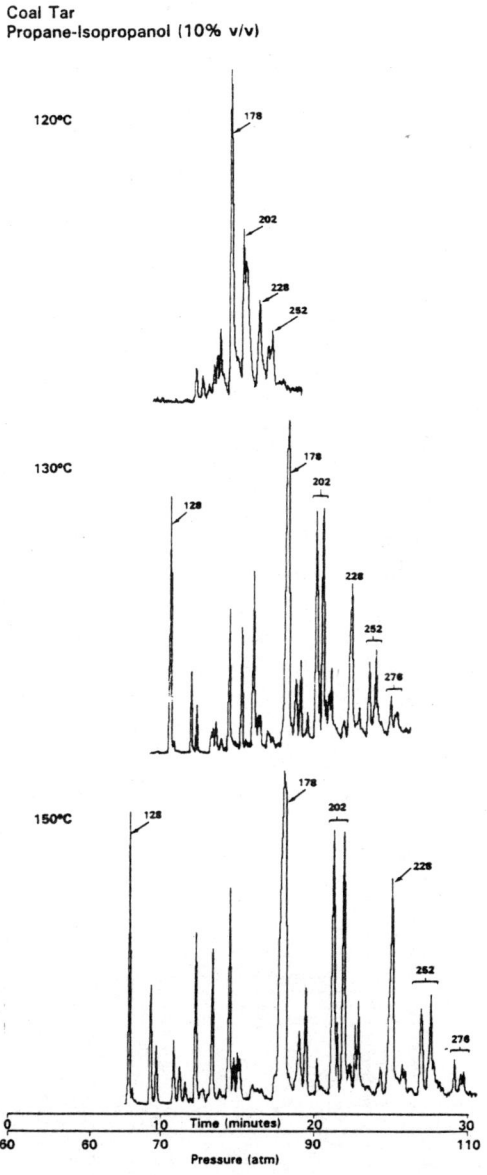

Figure 7. SFC−MS separations of a coal tar extract using a 2-propanol−propane mixture as the mobile phase as a function of temperature.

variety of mass spectrometric methods (e.g. thermal desorption, fast atom bombardment (FAB), secondary ion mass spectrometry (SIMS), laser desorption, etc.).

The fluid expansion process through the pressure restrictor is a vital step (and primary problem area) of the technique which defines both the capabilities and ultimate limitations. Restrictor performance constitutes the most vital component of an SFC–MS interface. The restrictor can be a short open nozzle with a relatively small inner diameter, or longer with a larger i.d., providing a fast expansion. Alternatively, the restrictor can have a complex structure providing a relatively slow expansion through a tortuous path. At present, most SFC or SFC–MS utilizes either some form of capillary or porous frit restrictor with the dimensions empirically selected to give the desired SFC linear velocity for the chromatographic temperature and pressure range.

The characteristics of an SFC–MS restrictor, and to a significant degree the efficiency of transfer for less volatile solutes to the gas phase during the expansion, depend upon the state of the fluid prior to expansion, the dimensions of the restrictor, restrictor heating (and heat transfer properties of the restrictor), and the state of the fluid before and after the expansion. Figure 8 schematically illustrates an adiabatic process for a restrictor where the enthalpy (H) of the fluid is the same before and after the expansion process. Expansions relevant to SFC occur under conditions where the Joule–Thomson coefficient $(\partial T/\partial P)_H$, is generally positive and net cooling results (Cambel and Jennings, 1958). If the restrictor is very short the process may be considered adiabatic. In this case the equilibrium state of the fluid after expansion (if also isolated from the surroundings) can be predicted directly from thermodynamic data. However, the state of the fluid during and shortly after expansion depends upon the physical processes related to the expansion which are kinetically controlled and inherently nonequilibrium.

Detailed descriptions of the thermodynamic considerations for expansion of supercritical fluids, estimation of fluid flow through SFC–MS restrictors

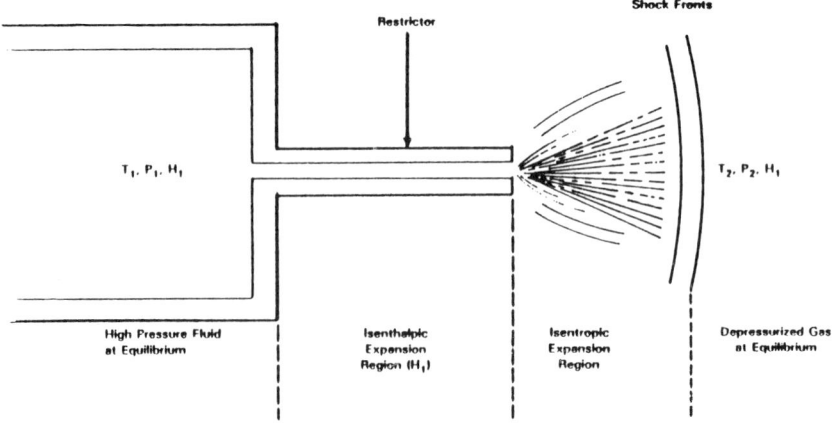

Figure 8. Schematic illustration of the supercritical fluid expansion for an adiabatic process.

and the performance of capillary restrictors in SFC and SFC—MS have been published (Smith *et al.*, 1986a, 1987) and will not be addressed here. However, some considerations in the transport of less volatile materials, a primary area of application for supercritical fluids, through capillary restrictors, will be described.

A vital property of SFC and SFC—MS restrictors is their effectiveness for transporting 'nonvolatile' compounds to the ion source. Success depends on the restrictor design, fluid pressure and temperature, the particular demands imposed by the ionization method and, under some conditions, the volatility or melting point of the 'nonvolatile' compounds. For truly nonvolatile compounds two distinct modes of operation appear feasible, the first leading to detection 'spiking' by a precipitation process, and the second resulting in a 'nucleation regime' where spiking is unlikely. In the precipitation mode the nonvolatile analyte tends to collect on the walls of capillary restrictors and (if the local temperature is above the analyte melting point) flow towards the end of the restrictor. The analyte liquid has been observed using optical microscopy to collect at the end of the restrictor, and be periodically entrained in the expanding fluid. This is a complex process with details which likely depend on the liquid (analyte) viscosity and restrictor geometry. Solid particle agglomerates also appear to form by a related mechanism.

The fluid expansion process has been directly observed, and various materials expanded from supercritical fluid solutions through 5 to 75 μm i.d. fused silica capillaries have been collected and analyzed (Petersen *et al.*, 1987; Matson *et al.*, 1986a; Smith *et al.*, 1986a). In one set of experiments, polycarbosilane, which has a reported mean molecular weight of ~1430, a melting point of ~240°C, and decomposes to yield refractory silicon carbide at >900°C (Yajima *et al.*, 1978), was studied. The molten polycarbosilane (Dow Corning) solute was observed to collect at the fused silica capillary exit and be periodically entrained in the gas flow (Smith *et al.*, 1986a). For polycarbosilane in pentane at 350°C and 100 bar, short fibers were formed (Fig. 9b). The high shear forces can apparently also elongate polymer droplets to yield fibers. The particle size (either fibers or spheres, depending upon the solute, fluid, and restrictor conditions) was found to increase with capillary diameter. These observations are consistent with observed FID spiking phenomena, and suggest a process involving the restrictor walls and solute melting point.

The second mode of operation producing much smaller particles occurs when analyte nucleation is delayed to near the end of the restrictor. Figure 9a shows polycarbosilane collected under such a set of conditions (30 ppm pentane solution at 240 bar, 250°C) with a 6 mm × 25 μm i.d. restrictor. In this case large quantities of nearly monodisperse particles with an average diameter of ~0.03 μm were collected (Smith *et al.*, 1986a). One nanogram of analyte will produce on the order of 10^8 particles of this size.

The major requirement for operation in the 'nucleation mode' is to maintain solvating conditions to near the end of the restrictor, which is facilitated by the use of very short restrictors and conditions enhancing analyte solubil-

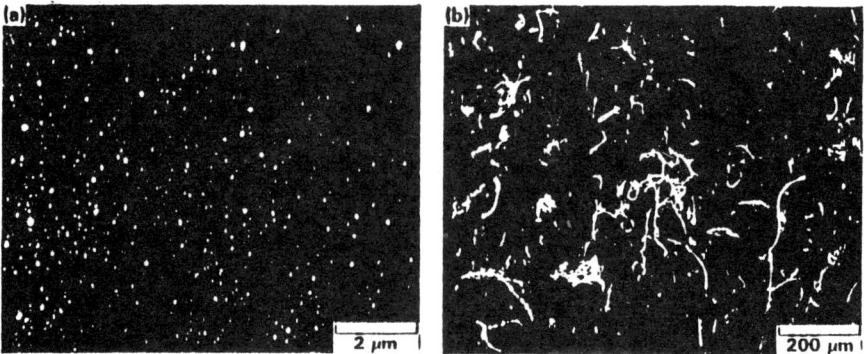

Figure 9. Photographs showing products obtained from the expansion of supercritical polycarbosilanepentane solutions. (a) Ultrafine polycarbosilane powder formed by expansion from 250°C and 240 bar; (b) fibers collected from a solution expanded from 350°C and 100 bar.

ity. This does not necessarily correspond to elevated restrictor temperatures. For example, the polycarbosilane particles shown in Fig. 9b were formed at 350°C and 100 bar, while the very fine, nearly monodisperse, particles were produced at 240 bar and 250°C. The solvating power of pentane for polycarbosilane is much greater under the latter condition. In addition, the particles were formed at gas temperatures below their melting point (due to cooling upon expansion), and where adhesion to the capillary walls was apparently insignificant. Under these conditions a very long restrictor ($L/D \gg 10\,000$) would plug rapidly, but typically would produce large particles (as in Fig. 9b) if heated to >300°C.

Increased fluid pressure serves to enhance the transport of nonvolatile analytes, but in practice the maximum pressure is limited by either the SFC conditions needed to yield sufficient retention or the maximum pressure for the system. Heating of the fluid just prior to expansion, and heating of the restrictor itself, have been shown to improve and extend detection for compounds having low volatilities. It has been shown that volatility plays a major role in detection of such analytes, both directly and indirectly, since fluid phase solubilities (at a given density) generally mirror analyte vapor pressures.

For nonvolatile compounds, heating of the fluid before or during expansion can be counterproductive due to lower fluid densities and the resulting lower fluid phase solubilities. For less volatile analytes improved detection results with shorter restrictors and higher flow rates. Larger solute concentrations and less soluble analytes will also lead to more rapid nucleation and particle growth. These small solute particles, typically well under 0.1 μm diameter for solute concentrations under 100 ppm, form rapidly (10^{-7}–10^{-5} s) and apparently pose no difficulty for FID. The particle sizes formed from concentrations more indicative of SFC for individual analytes at the restrictor (<1 ppm), however, are too small to be resolved by conventional electron microscopy (i.e., likely <200 Å diameter).

Figure 10 shows the wide range of restrictor geometries which have been used for SFC or SFC—MS. Drawn and tapered fused silica restrictors (Fig. 10B) are readily produced. They are made by a relatively slow controlled drawing of a heated region of fused silica tubing from which the polyimide coating (shown as dark lines in Fig. 10) has been removed. Drawn capillaries or other variations resulting in larger aspect ratios, typically in the range of 100—10000, can be easily fabricated and are less likely to plug than straightwalled capillaries (Fig. 10A) (Smith *et al.*, 1986a; Jentoft and Gouw, 1969). The fluid expansion through such restrictors can be approximated using the relationships described elsewhere (Smith *et al.*, 1986a) by using an exit orifice diameter slightly larger than the actual diameter. An advantage of the drawn restrictor, compared to a similar length of small-diameter capillary, results from the fact that the undrawn portion of the capillary allows effective heating of the fluid prior to expansion. This eliminates the need for an additional zero dead volume union in a heated region just prior to expansion. A major disadvantage of the drawn restrictor is its fragility, which results from removal of the polyimide coating.

Other designs yielding shorter restrictors with more favorable (i.e. lower) aspect ratios are shown in Fig. 10, C—E). The polished 'integral' restrictor

SFC Restrictors (not to scale)

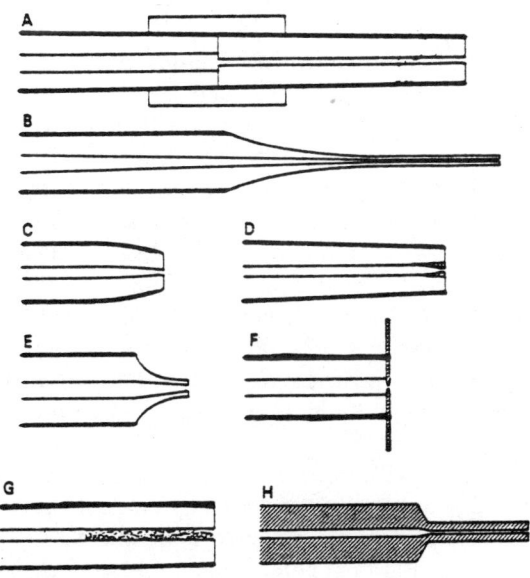

Figure 10. Cross-section diagram (not to scale) of various restrictor designs which have been explored for SFC. All restrictors, except the pinhole laser-drilled orifice (F) and the pinched restrictor (H), are fabricated from fused silica capillary tubing coated with polyimide (dark lines). See Table 2.

(C) is widely used and fabricated by carefully polishing the closed end of a piece of fused silica capillary (obtained by careful heating) until a small orifice giving the desired flow rate is obtained (Guthrie and Schwartz, 1986). Another approach being investigated in several laboratories involves internal deposition of a material to largely close the end of a capillary. These techniques produce relatively robust restrictors; however, their production can be a somewhat tedious process. Fast drawing techniques have also been used in our laboratory with fused silica to produce restrictors of similar dimensions (E). These restrictors can be fabricated in a relatively reproducible fashion, but have the drawback of the drawn region being relatively fragile. The diaphragm, or laser-drilled 'pinhole' orifice (F), was explored in earlier SFC−MS work and has severe practical limitations due to the alignment and low dead volume requirements (Smith *et al.*, 1982). Properly used this approach provides a nearly optimum expansion process and is quite feasible for use with the higher flow rates or conventional diameter packed columns. The thin diaphragm, however, must be carefully supported to withstand high SFC pressures. The porous frit restrictor (G) is quite different from the other restrictors since it provides numerous flow paths (Markides *et al.*, 1986). The frit resembles a column packed with submicron particles, resulting in multiple fluid paths with aspect ratios $\gg 10^5$. The frit restrictor is relatively rugged and will not be easily plugged by particles entrained in the fluid. Fluid velocity in the porous frit restrictor is always subsonic and the tortuous flow path, combined with the relatively long residence time, should provide for more efficient heat transfer to the expanding gas. This results in improved transport due to enhanced volatility, but prevents detection of compounds with insufficient vapor pressures at the restrictor temperature. Truly nonvolatile materials will not be transmitted through this restrictor and will ultimately result in plugging. The 'pinched' restrictor (H) is fabricated by carefully crimping ~1 mm length of the termination of a segment of capillary platinum−iridium tubing (50−100 μm i.d.) so as to obtain the desired flow rate (Smith, *et al.*, 1982). This approach has the disadvantage of being irreproducible and giving unstable behavior, particularly at low flow rates. An alternative method of producing restrictors from platinum−iridium tubing which have dimensions similar to (C), has also been described, but appears quite difficult (Grob, 1983). The advantage associated with these metal restrictors is their compatibility with higher temperatures and more polar fluids (e.g. water and ammonia) than fused silica restrictors.

Table 2 summarizes observations relevant to the various restrictor designs, and provides an estimate of their effective aspect ratios. Clearly, the ideal restrictor has not yet been developed and, due to its crucial role in SFC−MS, and SFC detection in general, remains an area of active research. The major consideration in selection of a restrictor should be the volatility of the analytes; less volatile analytes are expected to benefit from a restrictor with a low aspect ratio and operation which maximizes fluid density at the restrictor exit.

Table 2.
Restrictors used for SFC and SFC–MS

Type[a]	Effective aspect ratio[b]	Fabrication	Reproducibility	Primary advantage	Primary disadvantage
A. Straight-walled capillary	~2000–20 000	—	Excellent	Easily assembled	Requires union to capillary, excessive precipitation of nonvolatiles.
B. Long drawn capillary	100–10 000	Easy	Very good	Easily fabricated, moderate aspect ratio	Fragile
C. Polished	100–2000	Tedious	Fair–good	Low–moderate aspect ratio, rugged	Difficulty in fabrication
D. Internal deposition	100–2000	Tedious	?	Low–moderate aspect ratio, rugged	Difficulty in fabrication
E. Fast (short) drawn	50–1000	Easy	Good	Low aspect ratio	Fragile
F. Pinholed (laser-drilled)	5–25	Difficult	Poor–fair	Near ideal expansion	Difficult alignment and elimination of dead volume
G. Porous (multipath) frit	$\gg 10^5$	[c]	Good	Commerically available rugged	Loss of nonvolatiles
H. Pinched	?	Easy	Poor	Applicable with nearly any fluid	Unstable and irreproducible

[a] All fabricated from fused silica tubing except for the pinhole orifice substrate (F) and pinched platinum–iridium restrictor (H).
[b] Appropriate range for comparison with straight-walled capillary (A).
[c] Limited information available on fabrication.

In the direct fluid injection (DFI) capillary SFC–MS interface, supercritical conditions are maintained in the column to a region immediately adjacent to the ion source, where the fluid is rapidly decompressed by expansion through an orifice or restrictor (Sie and Rijnders, 1967; Peaden *et al.*, 1982; Smith *et al.*, 1982a). The lower fluid flow rates with capillary columns avoid complex multi-stage pumping systems and utilize shock fronts and rapid collisional processes in the expanding jet to help disrupt solvent clusters.

When an expansion occurs in a chamber with a finite background pressure, the expanding gas will interact with the background gas producing a shock wave system. This includes barrel and reflected shock waves as well as a shock wave perpendicular to the jet axis (the Mach disk), (Randall and Wahrhaftig, 1981). Recent laser studies have indicated that solute–solvent clusters do not persist beyond the Mach disk (Goates *et al.*, 1987). The distance from the restrictor orifice to the Mach disk may be crudely estimated as $0.67D(P_1/P_2)^{1/2}$, where D is the orifice diameter, P_1 is the fluid pressure and P_2 is the expansion region pressure. For a capillary restrictor the fluid pressure is more correctly described by the fluid pressure at the end of the capillary (P_{exit}), which can be estimated (Smith *et al.*, 1986c), and represents typically 5–20% of the SFC column pressure. Thus, if $P_1 = 60$ bar, $P_2 = 1$ torr, and $D = 5$ μm, the distance to the Mach disk is ~0.7 mm. The extent of cluster formation is related to the fluid pressure, temperature, and the orifice dimensions. Since initial cluster formation involves volatile solvent molecules, heat applied in the later stage of this expansion reduces solvent clustering but, as discussed earlier, does not necessarily facilitate transfer of solute molecules. It is expected that, as the capillary restriction becomes longer, clustering will become more significant.

A simplified schematic illustration of one SFC–MS configuration, with quadrupole mass spectrometry, developed at the authors' laboratory, was shown in Fig. 2. A high-pressure, programmable, syringe pump is used to generate a pulse-free flow of a high-purity fluid. Sample injection utilizes a commercial HPLC high-pressure valve with volumes from 0.06 to 0.2 μl and flow splitting at (typically) ambient temperatures. The sample splitter should be as close to the injector as possible, and the temperature is maintained sufficiently low for the split to occur as a subcritical liquid, such that discrimination between sample components is expected to be negligible. Split ratios range from as little as 1:3 for conventional separations on long (>15 m) 50 μm i.d. columns, to as high as 1:80 for fast separations on 25 μm i.d. columns. Splitless injection techniques allowing larger sample volumes, often at a cost of chromatographic efficiency, are also becoming widely used. The column is mounted in a constant-temperature oven (typically a slightly modified GC oven) which also serves to heat the air circulated through the DFI transfer probe. Careful control of temperature along the SFC column length is crucial. The restrictor and probe tip region are typically heated independently from the SFC column to result in a single-phase expansion and optimize analyte transport to the ionization region. An electrically heated stainless-steel capillary is utilized for restrictor temperature control, which can be operated at over 600°C. The mass spectrometer ion source and required pumping speeds are similar to those utilized for GC−MS.

Both electron impact and chemical ionization have been explored with SFC interfacing (Yonker, *et al.*, 1984; Randall and Wahrhaftig, 1981; Smith and Udseth, 1983b; Wright *et al.*, 1984; Smith *et al.*, 1984a, b; Wright *et al.*, 1985; Smith *et al.*, 1985b, d; Kalinoski *et al.*, 1986b; Smith *et al.*, 1986b; Kalinoski *et al.*, 1986a). An advantage of SFC−MS with small-diameter capillary columns is that the flow rates are sufficiently low that any CI reagent gas may be used. Methane, isobutane or ammonia are the most frequently used chemical ionization reagent gases. Typical detection limits range from 0.1 to 10 pg depending upon the compound, analysis time, separation efficiency, and CI reagent gas. While sensitivity was somewhat less than that obtained by CI, good EI mass spectra can also be obtained with 100 pg injections, under conditions of true EI, *not* CO_2 charge exchange CI. (Smith *et al.*, 1984b). The flexibility in selection of the ionization method, and the ability to use the existing EI spectral libraries, provides an additional advantage for SFC−MS relative to most LC−MS methods.

Figure 11 gives a schematic illustration of the SFC−MS interface and ion source originally developed for EI (Smith *et al.*, 1984b). The fused silica capillary restrictor allowed injection of the fluid into a heated expansion region (1 cm × 0.14 cm i.d. with a 0.1 cm orifice to the ionization volume), which provided the higher pressure (0.1−1 torr) necessary for solvent and solvent−solute cluster break-up prior to the ionization region. The temperature of this region was typically 50−150°C higher than the mobile phase critical temperature to avoid a two-phase region (formation of a condensed solvent phase) during the expansion. When the two-phase region was avoided there was no evidence of solvent clustering. However, at temperatures which resulted in a two-phase system during expansion, extensive solvent clustering was observed (Smith *et al.*, 1984b). The expansion region of the EI interface directed the SFC effluent into the ionization volume of a high-efficiency ionizer. Ion source chamber pressure was in the range of $1-5 \times 10^{-5}$ torr and the mass spectrometer chamber pressure was $<10^{-6}$ torr, allowing conventional 70 eV EI.

Truly useful SFC−MS instrumentation must also address problems related to sample injection methods and quantitation. Nearly all earlier capillary SFC work with small-diameter (≤100 μm i.d.) capillaries has used split injection techniques, which can have low reproducibility and result in constraints on sample size due to the sample volume and the often limited solubility in the sample solvent. This is further complicated by the relatively small sample size typically required to avoid overloading such columns (typically *ca.* 1−10 ng per component for 50 μm i.d. columns with a 0.25 μm film thickness), explaining the increasing application of packed columns for SFC due to their higher phase ratios.

Application of capillary restrictors is limited by solute precipitation for nonvolatile analytes, particularly at low flow rates (Smith *et al.*, 1986a). Thus, the development of practical SFC−MS applications would be greatly

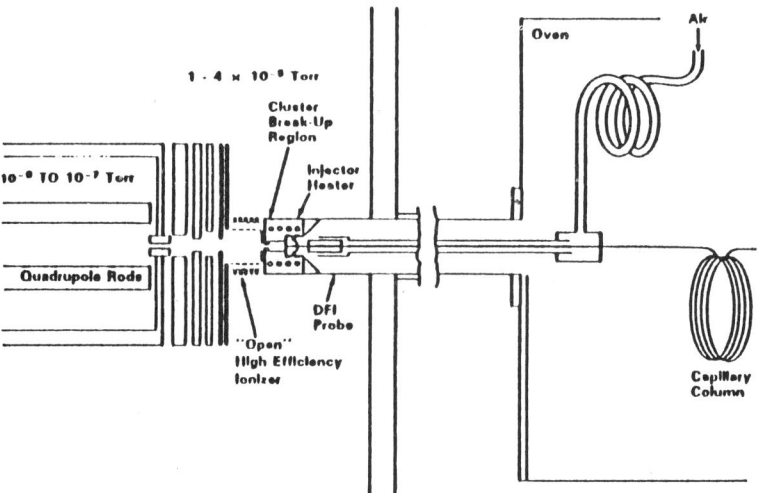

Figure 11. Schematic of a capillary SFC–MS interface developed for electron impact ionization.

facilitated by improvements in SFC injection methods, significantly increased sample capacity, greater detector sensitivity and improved interface operation with less volatile analytes. As an attempted solution to these limitations, an alternative high flow rate (HFR) SFC–MS interface was developed (Smith and Udseth, 1987) with the aim of: (a) operation at high fluid flow rates, (b) improved control of restrictor temperature, (c) enhanced sensitivity for less volatile compounds, and (d) allowing more rapid restrictor replacement and operational convenience. Similar but less ideal variations upon this approach can be implemented using instruments designed for thermospray LC–MS.

The SFC–MS interface probe was constructed from a 1.27 cm o.d. stainless steel tube which was temperature-regulated using air flow heated by the chromatograhic oven, similar to previous capillary interfaces. Figure 12 gives a schematic illustration of the interface and probe. The capillary restrictors used for the HFR interface consisted of a 5–10 cm length of 25–50 μm i.d. fused silica capillary tubing with a drawn region, typically 3–10 mm in length at its termination and an exit diameter of 8–10 μm. A vacuum seal was made to the probe body via a 'half-union' which was fabricated using a capillary stainless-steel sheath which extended approximately 0.7 cm beyond the end of the probe. The diameter of the CI source entrance aperture (in the repeller) was selected to give a 1:25 'split' between flow to the source and the direct pumping port. Mass flow rates of 150–200 mg/min for carbon dioxide were obtained without any degradation of source performance with this interface for easily fabricated tapered restrictors. Higher flow rates can be readily accommodated by decreasing the diameter of the orifice between the expansion region and the CI source. This approach provides relatively high fluid densities at the capillary exit, which was anticipated to delay solute nucleation and make transfer to the mass spectrometer ionization region more efficient.

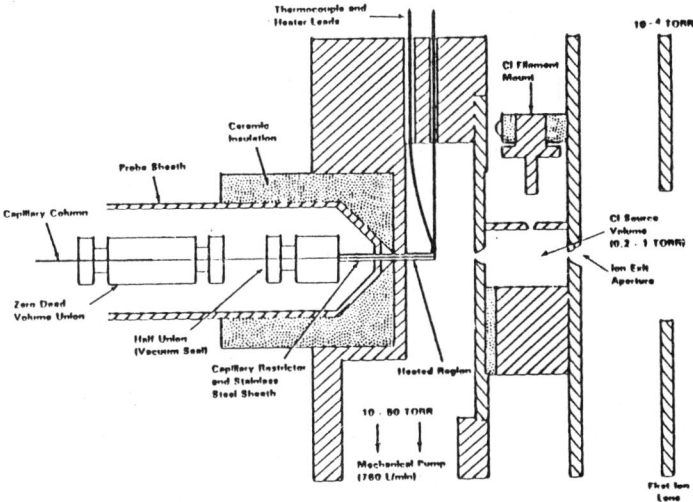

Figure 12. Schematic illustration of the high flow rate (HFR) SFC−MS interface.

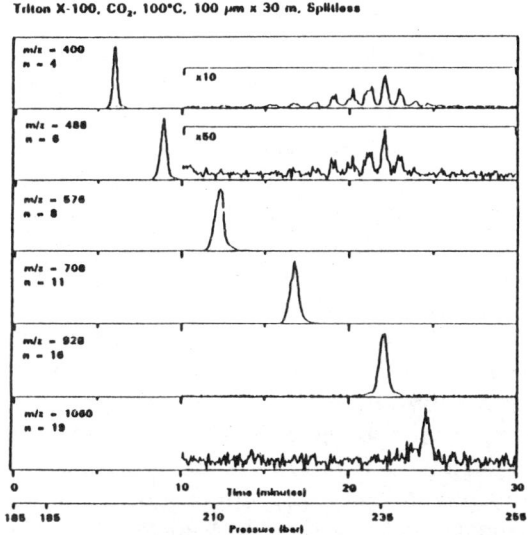

Figure 13. Selected ion chromatograms for the oligomers from a Triton X-100 separation using the HFR interface with splitless injection (0.2 µl) and a 2.5 bar/min pressure increase.

The improved performance for higher molecular weight components is demonstrated in Fig. 13, which shows capillary SFC−MS selected ion chromatograms for an 0.2 µl injection of a solution of the nonionic surfactant Triton X-100. The separation was obtained using supercritical carbon dioxide at 100°C as the mobile phase and a 100 µm i.d. × 30 m fused silica capillary column. The separation utilized a pressure program of 2.5 bar/min from 185 bar starting 2 min after injection. Mass spectrometric detection utilized ammonia chemical ionization, but good results were also obtained using methane CI (Smith and Udseth, 1987). Capillary SFC−MS with previous interface designs and similar sample sizes was limited to $n \sim 16$ (Smith et al., 1987). The selected ion chromatograms show that the $n = 19$ oligomer was readily observed, although it has a concentration at least an order of magnitude smaller than $n = 16$. It should be noted that, even for ideal interface operation, the mass spectrometer TIC will show discrimination against the heavier components compared with chromatograms obtained with the FID. This is due to the decreased transmission efficiency of quadrupole mass filters at high m/z, and the fact the FID response is approximately proportional to analyte mass flux, whereas MS ion current is proportional to molar flux.

Figure 14 compares SFC−MS separations obtained with microbore and capillary columns using the HFR interface. The separation of a polarity test mixture with CO_2 at 50°C and 400 bar inlet pressure with a 1 mm × 25 cm C_{18} microbore column is shown in Fig. 14A. The separation shows significant tailing for the polar components (e.g. N-ethylaniline) and excessive retention for several components (p-chlorophenol and decanol, not shown.) Since the separation was conducted at 400 bar the only options for significantly reducing retention with this column would be to increase temperature or alter the mobile phase. Increasing the temperature is limited by the thermal stability of the stationary phase and would not be universally advantageous since the maximum fluid density (and thus solvating power) obtainable would decrease. Figure 14B shows the improved SFC−MS separation obtained for the same polarity test mixture upon addition of 1% methanol modifier, primarily due to interaction of the modifier with the stationary phase. Peak tailing is largely eliminated, retention reduced, and substantial selectivity changes observed. As expected, the nonpolar components showed no significant changes in retention upon addition of the modifier. Figure 14C shows a separation of the same polarity test mixture obtained with a 100 μm i.d. × 30 m capillary column at 50°C using a pressure program rate of 12.5 bar/min and an average linear velocity of approximately 19 cm/s. The sample size (0.2 μl) and concentration ($\sim$0.5 μg/μl) were identical, so that approximately 100 ng/component was injected (splitless operation) on column in each case. Even though the separation was substantially better than obtained with the microbore column, the mass spectrometer was needed to resolve two sets of overlapping peaks. It should be noted that the order of elution using the capillary column was similar to the microbore column for the CO_2−methanol mobile phase.

Experience with capillary restrictors for both FID and MS detection has shown that heating of the fluid just prior to (or during) the expansion through the restrictor can facilitate transfer of less volatile compounds (Smith et al., 1986a; Chester et al., 1985; Richter, 1985; Smith et al., 1984b, 1986b). The improved transport to the detector is primarily associated with enhanced vapor pressure or delayed precipitation of less volatile analytes (Smith et al., 1985d). Materials with very low volatility still present a problem for mass spectrometric detection. The origin of this difficulty is the nucleation process which occurs for nonvolatile compounds under the fluid expansion conditions. Studies of the expansion process and the role of the capillary restrictor

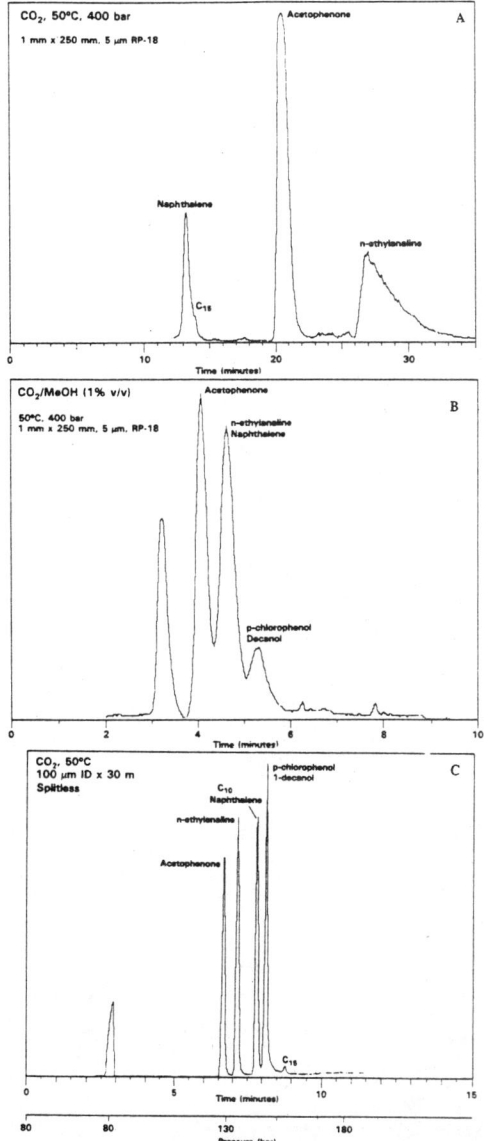

Figure 14. Comparison of SFC−MS separations obtained for a polarity test mixture using CO_2 (A) and CO_2−1% CH_3OH (B) mobile phases with a C_{18} microbore column and CO_2 with a capillary column at high flow rates (C).

conducted in this laboratory (Smith *et al.*, 1986a), indicate heating of the capillary restrictor can lead to early precipitation of nonvolatile analytes (due to warming of the fluid and loss of solvating power). The analyte need not be lost but, under the proper conditions, can be transported as large aggregates (Matson *et al.*, 1986b). These aggregates are not directly detectable by mass spectrometry but their subsequent pyrolysis would provide one practical approach to obtaining useful mass spectra.

Pyrolysis—mass spectrometry has been used extensively for the characterization of nonvolatile biological and other materials (Meuzelaar *et al.*, 1973; Voorhees, 1984; Meuzelaar *et al.*, 1982; Irvin, 1982), and should provide an additional tool for the characterization of nonvolatile compounds following SFC. Evidence for this supercritical fluid chromatography pyrolysis—mass spectrometry (SFC—Py—MS) process has been found in the analysis of the nonionic surfactant Triton X-100. Figure 15 shows the TIC and selected ion chromatograms for m/z 268 (a common pyrolysis product ion for the Triton oligomers) and m/z 620 (an ion associated with the intact $n = 9$ oligomer) from the SFC—MS analysis of a Triton X-100 sample with a restrictor heater temperature of 650°C. The time delay associated with observation of the pyrolysis products indicates an origin not associated with the gas phase expansion, since any time lag would be much smaller. The products are almost certainly due to early precipitation of the analyte and flow to the restrictor exit where it could undergo decomposition on hot surfaces of the capillary heater. The addition of a pyrolysis step to the SFC—MS interface should constitute a relatively straightforward modification of present interface designs. Such an interface would also have the advantage of allowing more efficient detection of higher molecular weight analytes since numerous lower molecular weight fragments would be formed from each molecule. Additionally, it is well established that such information provides a characteristic 'fingerprint' for the identification of unknown compounds.

A recent advance in the evolution of the SFC—MS interface is the development of a capillary direct supercritical fluid injection interface for use with high-voltage, high-resolution magnetic sector mass spectrometry (Chess *et al.*, 1987). A key to the successful operation of this interface is the in-

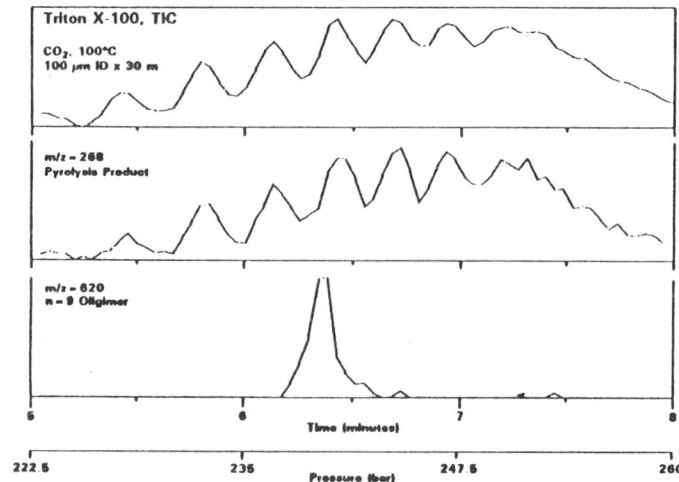

Figure 15. Capillary SFC—MS TIC and selected ion chromatograms for m/z 268 and m/z 620 from a sample of Triton X-100. The restrictor heater temperature was ~650°C resulting in substantial pyrolysis. Note the time delay between the $n = 9$ oligomer molecular ion (m/z 620) and the pyrolysis product at m/z 268.

corporation of features that permit the ion source and interface probe tip to remain electrically isolated from the chromatographic system. This design permits operation of the mass spectrometer at high accelerating potentials (up to 8 kV), essential for higher overall instrument sensitivities. The interface is also designed to permit careful control of temperatures, recently demonstrated to be vital in the transfer of less volatile analytes (Smith *et al.*, 1986a) to the mass spectrometer ion source. The interface was produced with modification to an existing ion source design (Chess *et al.*, 1987) and the total effluent from a capillary SFC column can be directly injected without additional ion source pumping other than that required for chemical ionization mass spectrometry. This technology permits access to the extended mass range (possible with such mass spectrometers), high resolution capability, sensitivity for higher masses, and the availability of tandem mass spectrometric (MS/MS) techniques available with double focusing, mass spectrometers.

The modifications made to the mass spectrometer (not including the ion source) were to replace the standard 190 l/min mechanical roughing pump attached to the source housing diffusion pump with a 400 l/min pump to allow the larger volume of gas associated with DFI to be handled, and the inclusion of a 1.27 cm i.d. vacuum ball valve centered on the vacuum housing end flange containing the feed-throughs for the ion source. The interface probe vacuum lock and end flange are dedicated to SFC–MS application allowing simple and rapid conversion to other operational modes. A schematic diagram of the interface probe is shown in Fig. 16 and a photograph of the disassembled probe is shown in Fig. 17.

The stainless-steel ion source for supercritical DFI–MS was based on the design of a standard VG electron impact/chemical ionization (EI/CI) source for use on the ZAB 2-F instrument. Changes to the standard source design included: modification to accommodate the entry of the 1.27 cm DFI probe through the back side of the source, co-axial with the center of the ion exit slit; removal of the electron trap (allowing CI operation only); reduction of the internal volume by one-third; elimination of all unnecessary re-entrant holes; and fabrication of a repeller plate with a hole in the center to allow direct entry of the effluent from the flow restrictor into the source volume. Further, the ion exit slit width was fixed at 0.13 mm to 'tighten' the ion source to maintain the relatively high pressure required for optimal CI operation. Copper spring contacts for a thermocouple and restrictor heater electrical connections are also mounted on the ion source support structure. The interface includes all the design features and advantages of the interfaces for quadrupole MS instruments (Smith *et al.*, 1986c, 1987), with characteristics which permit operation at the high accelerating potentials (source voltages) required for sector mass spectrometers. These details include the Vespel® (Dupont) barrel and cap (see Fig. 16) and zero-dead-volume (ZDV) seal, for a gas-tight connection between the probe interior and the ion source and probe tip (at source potential) from the stainless-steel probe shaft (at ground potential).

It was recently demonstrated (Kalinoski *et al.*, 1986a) that supercritical fluid extraction coupled with mass spectrometry (SFE−MS) is useful in the analysis for low levels of contaminants in complex natural matrices. The ability to obtain exact mass data using high-resolution MS further extends the application of the SFE−MS technique. As an example of the use of the interface, an oscillographic trace of the ammonium adduct ion (*m/z* 384) region for the trichothecene mycotoxin diacetoxyscirpenol (DAS, molecular weight 366) is shown in Fig. 18. With the mass spectrometer operating at a resolving power of 7400 (defined at 5% valley), and with ion matching (±3 milli-mass units), the calculated exact mass of the ammonium adduct ion,

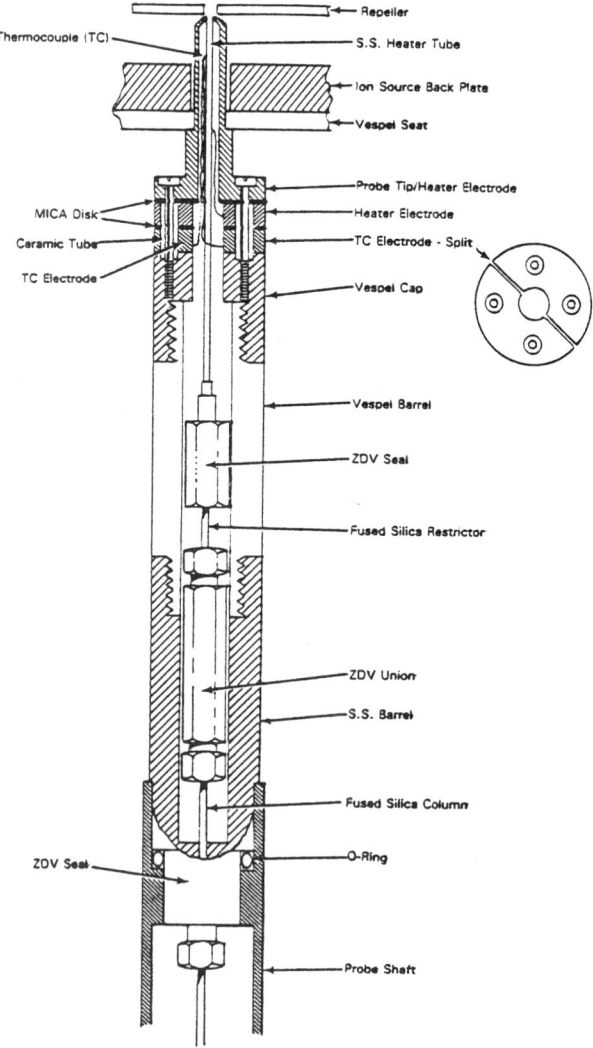

Figure 16. Schematic diagram of the interface probe for direct supercritical fluid injection−high resolution mass spectrometry.

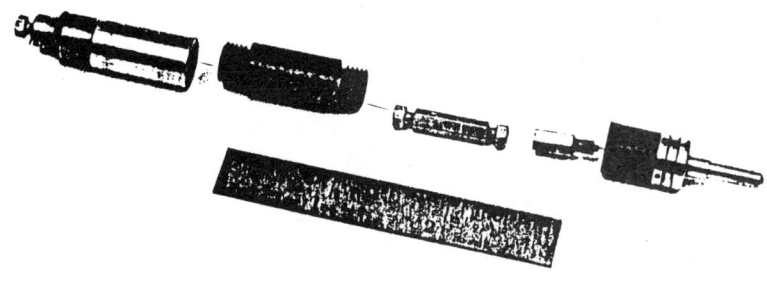

Figure 17. Photograph of the interface probe, for SFC with high resolution MS, disassembled.

$(M + NH_4)^+$, for DAS (m/z 384.2022) was detected in the presence of numerous isobaric interferences in the supercritical CO_2–isopropanol extract of a wheat sample containing DAS and the related trichothecene T-2 toxin. T-2 toxin was also identified based on exact molecular weight data (m/z 484.257 for the $M + NH_4^+$ ion) in the same extract. The extraction experiment has also been performed at a resolving power of 20 000 (defined at 10% valley), indicating that even higher specificity is achievable for this type of analysis, but at the anticipated cost of some sensitivity. Based on these and earlier studies (Kalinoski *et al.*, 1986a and references therein) detection limits of 10 ppb for the trichothecenes appear feasible.

APPLICATIONS

The greatest immediate areas of applicability of SFC with DFI–MS are for mixtures containing moderately polar compounds which are thermally labile or less polar compounds of relatively high molecular weight. The solvating powers available with supercritical fluids 'extend' the applicability of GC to allow analyses which are difficult by HPLC (e.g. moderate–high molecular weight polymers) or due to insufficient volatility, lability or (in the case of HPLC) insufficient chromatographic resolution or selectivity.

It must be recognized that SFC or SFC–MS does not provide a general alternative to LC–MS, but rather provides a basis for analysis of otherwise intractable mixtures and, in cases where problems are amenable to both LC

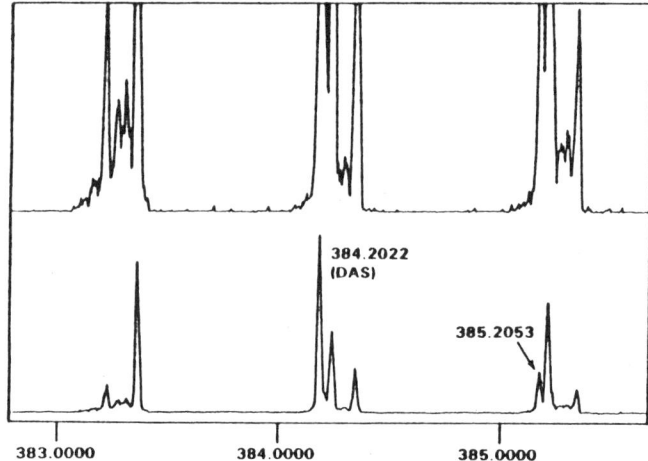

Figure 18. Oscillographic trace of the ammonium adduct ion (m/z 384) region of the trichothecene mycotoxin diacetoxyscirpenol (DAS) obtained during supercritical fluid extraction–direct fluid injection mass spectrometry (SFE–MS) of a wheat sample containing DAS and T-2 toxin. The resolving power of the ZAB mass spectrometer was 7400 (defined at 5% valley).

and SFC, significant advantages. On the basis of chromatographic resolution, GC, where applicable, should always be the method of choice. Similarly, given a choice between LC and SFC, SFC should generally provide superior separations. In terms of SFC–MS sensitivity, similar considerations are also relevant (given the same ionization process) since, to a reasonable approximation, analyte molar concentration in the mobile phase (liquid, supercritical fluid, or gas) will be inversely proportional to the density in the mobile phase and directly proportional to peak width. Thus, both chromatographic resolution and sensitivity are expected to be in the order GC > SFC > HPLC. Of course, either chromatographic resolution or detection sensitivity can be improved at the expense of the other.

The most important questions for chemical analysis, then, are what fraction of analyses which currently require HPLC are amenable to SFC, and how can one determine whether a specific application is feasible by SFC. As discussed earlier, the range of application is essentially defined by the solvating power of practical supercritical fluids. At present, operation with highly polar supercritical fluids (e.g. NH_3) appears feasible, but still presents difficulties for routine application due to the chemical stability of available stationary phases and other materials used in SFC systems (e.g. polymeric materials in valves and unions). It also appears that the specific chemical interactions (i.e. hydrogen bonding) which can be obtained using supercritical fluids over the *normal* range of fluid densities are reduced compared to the liquid phase. Specific highly polar solvents, such as water, are often impractical due to high critical temperatures (see Table 1) and useful fluid mixtures are defined and limited by the phase equilibria. The limited knowledge of supercritical fluid solvation phenomena and phase equilibria in many potentially attractive systems also constitutes a practical barrier.

As a practical matter, the current applicability of SFC—MS with less polar mobile phases (e.g. CO_2, pentane, propane, propane—isopropanol, etc.) can be estimated on the basis of suitable sample solvents. SFC appears generally applicable to materials soluble to any significant degree in hexane, and a large portion of low to moderate molecular weight compounds showing even marginal solubility in methylene chloride. On the other hand, SFC with less polar supercritical fluid mobile phases is unlikely to be successful if an aqueous sample solvent is required.

A few of the applications of SFC—MS which have been explored to date are briefly described below.

Polymeric materials

One particularly significant application of supercritical fluid separation techniques has been the characterization of polymeric mixtures (Nieman and Rogers, 1975; Klesper and Hartmann, 1976, 1977a—c; Klesper, 1978; White and Houck, 1986; Wright *et al.*, 1986; Schmitz and Hilgers, 1986). Both packed and capillary column SFC have been used for the analysis of polymeric mixtures consisting of oligomers of over 2200 molecular weight. These analyses were usually performed using flame ionization or ultraviolet absorption for detection. For complete characterization of polymers, information regarding the molecular weight of oligomers, as well as the relative abundance of each oligomer, is required. Figure 19 gives the total ion chromatogram (TIC) of a 400 average molecular weight polyethylene glycol (PEG) sample separated using a supercritical carbon dioxide—isopropanol (5% by volume) mixture as the mobile phase. Due to the terminal alcohol group, and presence of oxygen in the polymer chain, PEG is capable of significant hydrogen bonding. The separation was accomplished under relatively mild thermal conditions (125°C) and the pressure programming capability of the supercritical fluid mixture permitted elution of all oligomers using a 10 m × 50 μm i.d. capillary column in less than 30 min. Mass spectrometry for this sample was conducted using a quadrupole MS system. The ability to chromatographically separate oligomers of such moderately polar polymers prior to mass spectrometric analysis is of great benefit when compared to desorption ionization mass spectrometric techniques such as fast atom bombardment (FAB) and electrohydrodynamic ionization, where excessive fragmentation and complications associated with desorption can skew the apparent oligomer distribution.

The ammonia and methane chemical ionization mass spectra of the $n = 12$ oligomer of polyethylene glycol (nominal molecular weight 546), shown in Fig. 20, were obtained than a similar separation using the high-resolution MS system. The ammonia CI spectrum is characterized by an abundant ammonium adduct ion contribution $(M + 18)^+$, at m/z 564 and fragment ions produced by cleavage of carbon—oxygen bonds along the polymer chain. This fragmentation process is similar to that proposed for FAB ionization of polyethylene glycol samples (Lattimer, 1983). The major fragment ions are terminal diols of the structure $[H-(OCH_2CH_2)_n-OH +$

NH$_4^+$]. A small protonated molecule at m/z 547 is also present in the spectrum. Fragmentation may have been assisted by the input of thermal energy in the capillary restrictor heater region. For this analysis the stainless-steel restrictor heater was operated at approximately 450°C (which is much greater than, and does not correspond to, the temperature of the expanding fluid mobile phase — Smith et al., 1985d). Methane CI mass spectra of PEG oligomers (Fig. 20B) are characterized by a protonated molecular ion (M + H)$^+$, of low relative abundance and by significantly lower molecular weight fragment ions. The higher-energy methane chemical ionization can be used in conjunction with high-resolution mass spectrometry to obtain structural information on samples analyzed by SFC−MS.

Other polymeric materials have been analyzed using SFC−MS including polystyrenes, polybutadienes and the nonionic surfactants such as Triton X-100 (Smith et al., 1987). The use of the magnetic sector mass spectrometer, with greater mass range and better high mass sensitivity than the quadrupole instruments, and new interface designs, promise to extend the molecular weight range of polymeric materials that can be analyzed using SFC−MS.

Labile and less volatile compounds

The combination of supercritical fluid chromatography with mass spectrometry is particularly attractive for the analysis of thermally labile and less volatile compounds not amenable to gas chromatography, and for which liquid chromatographic and LC−MS techniques offer limited selectivity or sensitivity. The range of compounds that can be successfully analyzed using SFC−MS will be illustrated by selected examples. A brief list of other compound classes, not described herein, to which the various SFC−MS

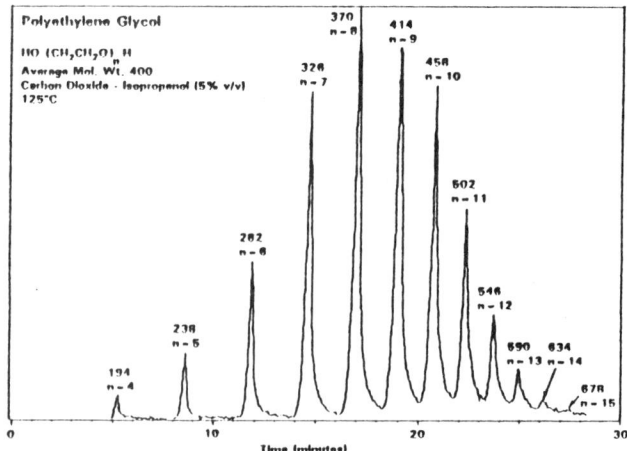

Figure 19. Total ion chromatogram (TIC) for the SFC−MS separation of a 400 average molecular weight polyethylene glycol (PEG) sample chromatographed on a 10 m × 50 μm SE-54 column using 5% isopropanol−carbon dioxide supercritical mobile phase at 125°C.

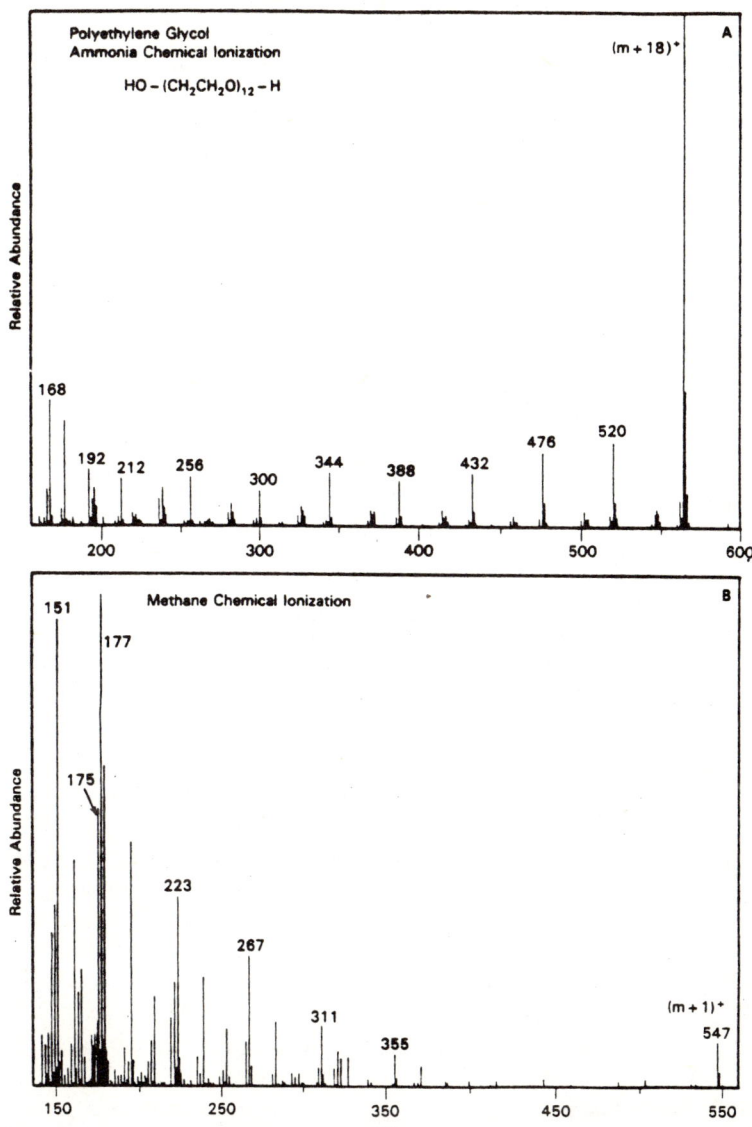

Figure 20. Ammonia (A) and methane (B) chemical ionization mass spectra of the $n = 12$ oligomer (nominal M.W. 546) of polyethylene glycol obtained following SFC—MS using a carbon dioxide—isopropanol mobile phase, obtained using the magnetic sector instrument.

methods have been applied includes: pesticides and rodenticides (Kalinoski *et al.*, 1986b; Wright *et al.*, 1986; Berry *et al.*, 1986); ionophores (Kalinoski *et al.*, 1988), higher molecular weight glycerol esters, alkaloids, sulfonamides, steroids and xanthines (Berry *et al.*, 1986; Lee and Henion, 1986), polar substituted biphenyls (Smith *et al.*, 1984a, b), and carotenoids (Frew and Johnson, 1986).

An example of the application of fast pressure programmed SFC to the analysis of a group of acid and carbamate pesticides is shown in Fig. 21.

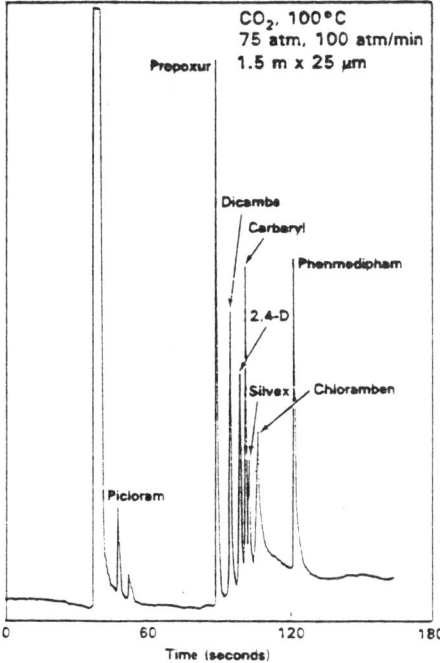

Figure 21. Rapid SFC with flame ionization detection of an acid and carbamate pesticide mixture.

Even more rapid separations should be feasible with continued development of these methods. The methane chemical mass spectra of three thermally labile herbicides obtained by capillary SFC—MS using a carbon dioxide mobile phase are shown in Fig. 22. All three compounds (linuron, diuron and alachlor) yield spectra characterized by the protonated molecule and structurally characteristic fragments. The urea herbicides, linuron and diuron, fragment to form a common dichloroisocyanate giving a protonated molecule at m/z 188. The temperature of the restrictor heater in the SFC—MS interface can affect the abundance of this fragment ion relative to the protonated molecule of the analyte. At high restrictor heater temperatures the relative abundance of the isocyanate ion is greater than that of the protonated molecule, indicating that the isocyanate may arise due to a thermal decomposition of the parent urea. Similar temperature sensitivity was not found for alachlor (Fig. 22 bottom), in which facile decomposition to an isocyanate is unlikely. The base peak for alachlor corresponds to the loss of methanol (32 dalton) from the protonated analyte molecule (m/z 270).

Mass spectra of the labile toxic compounds brucine (a rodenticide) and thiram (a seed disinfectant—fungicide), obtained following capillary SFC—MS with a carbon dioxide—isopropanol (5% v/v) supercritical fluid mixture, are shown in Fig. 23. For this analysis the fluid modifier, isopropanol, was also utilized as the chemical ionization reagent, which results in the formation of abundant protonated molecules $(M + H)^+$ for each sample, indica-tive of mild chemical ionization conditions. The ability to choose the CI reagent gas in SFC–MS adds an additional measure of selectivity to an analysis, and production of mass spectra with little or no fragmentation aids in the use of selected ion monitoring for detection to increase the sensitivity of an analysis. In such situations tandem mass spectrometry (SFC–MS/MS) is an ideal combination.

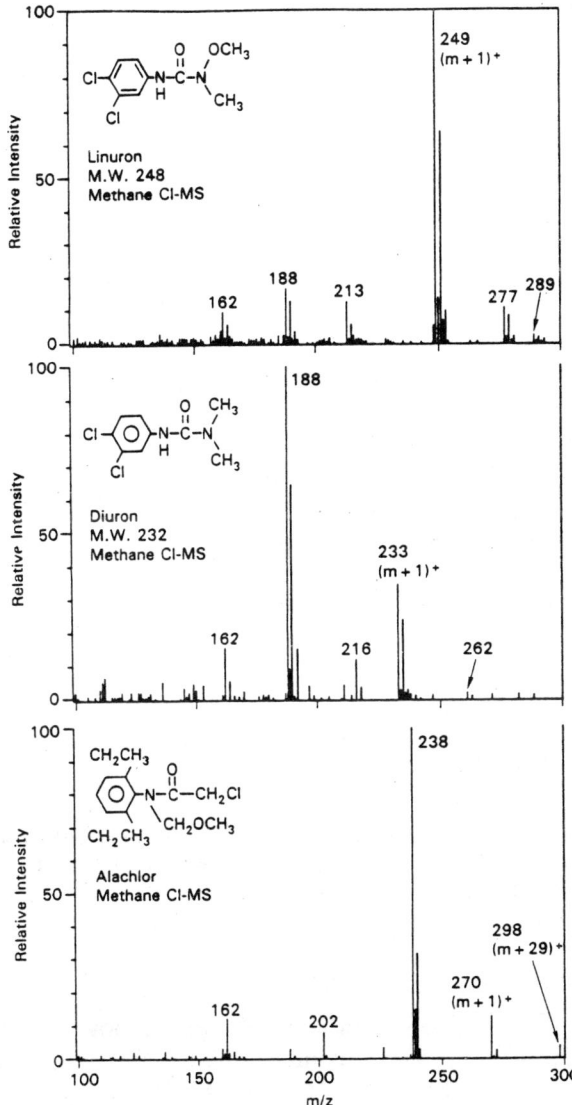

Figure 22. Methane chemical ionization mass spectra of the herbicides linuron (top), diuron (middle), and alachlor (bottom), obtained during SFC–MS separation using a supercritical carbon dioxide mobile phase.

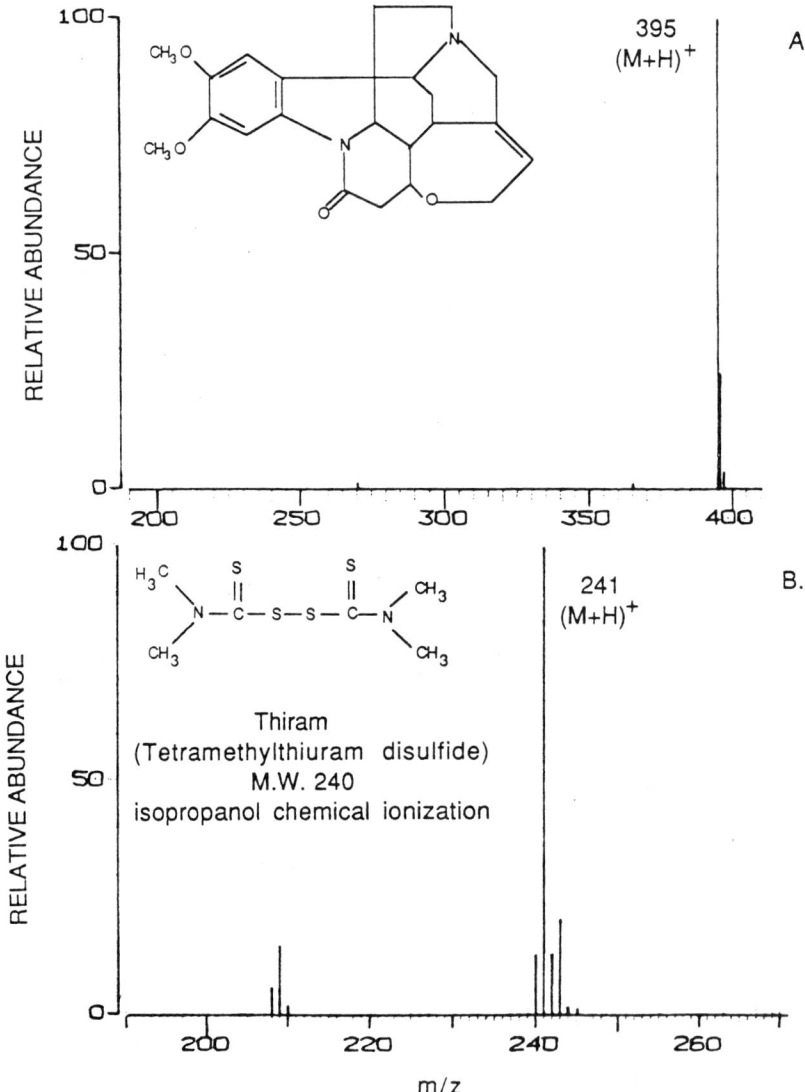

Figure 23. Isopropanol chemical ionization mass spectra of the anticoagulant rodenticide brucine (dimethoxystrychnine) (A) and the seed disinfectant-fungicide thiram (B) obtained following CO_2–isopropanol SFC–MS.

Complex mixture analysis

The capillary SFC–MS instrumentation provides a powerful new tool for
the characterization of complex mixtures not amenable to separation by gas
chromatography. Figure 24 shows a complex SFC–MS total ion chromatogram for a polar fraction of a diesel fuel marine material known to have
relatively large concentrations of nitrogen-containing compounds. Pressure
programming was used with a 6.5 mole% isopropanol–carbon dioxide fluid

mixture as the mobile phase in a 15 m × 50 μm i.d. capillary column at 100°C. The complexity of this fuel fraction is suggested by six typical single ion chromatograms from this separation shown in Fig. 25. For the ionization conditions used, little fragmentation resulted and most signals (after correction for isotopic contributions) can be attributed to fuel components. The signal at m/z 194, for example, suggests no less than twelve components which were tentatively ascribed to C_1-benzoquinolines and C_1-acridines. Over 1000 such contributions can be resolved from the data obtained during the separation shown in Fig. 24.

The use of selected ion monitoring for detection, coupled with rapid pressure programming SFC, can often provide a fast, specific method for the separation and determination of specific compounds in complex mixtures. Figure 26 shows the TIC and selected ion chromatograms of an extract of wheat containing three trichothecene mycotoxins. The extract was obtained from a wheat sample containing 10 ppm of diacetoxyscirpenol (DAS), and T-2 toxin that had been extracted using supercritical carbon dioxide (Kalinoski *et al.*, 1986a). Ammonia CI provided high detection selectivity although the SFC–MS chromatographic conditions were not optimized for separation of the mycotoxins, a relatively short (4 m × 50 μm) column and rapid-pressure ramp allowed detection of two of the compounds. The selected ion chromatograms of m/z 484 and m/z 384 indicated ample sensitivity and well-defined chromatographic peaks for T-2 toxin and DAS. Chromatographic retention times and the mass spectra obtained during elution of these components confirm the presence of these toxins.

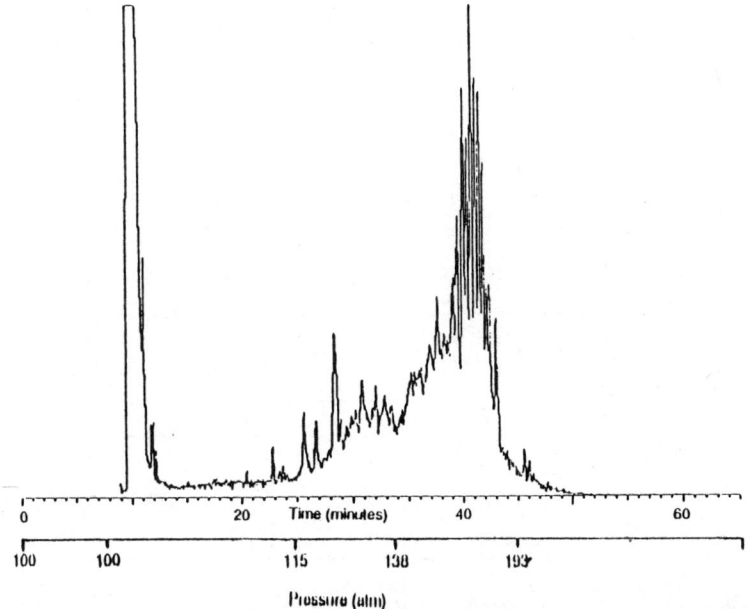

Figure 24. SFC–MS total ion chromatogram for a complex polar fraction of a diesel fuel marine. See Fig. 25.

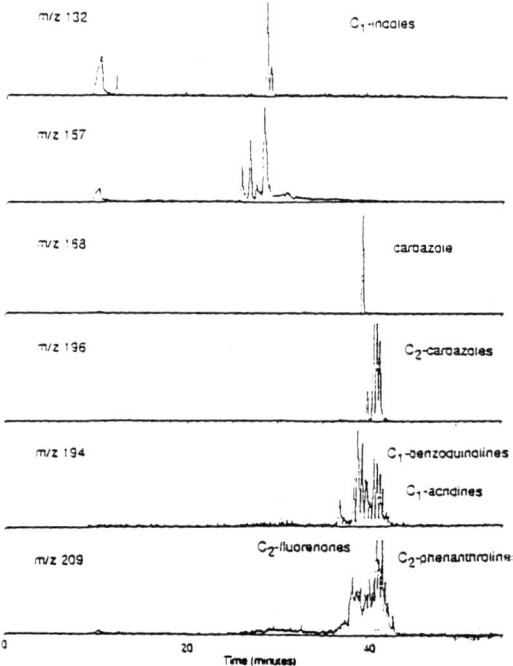

Figure 25. Selected ion chromatograms for the SFC−MS separation shown in Fig. 24. Tentative identifications are given.

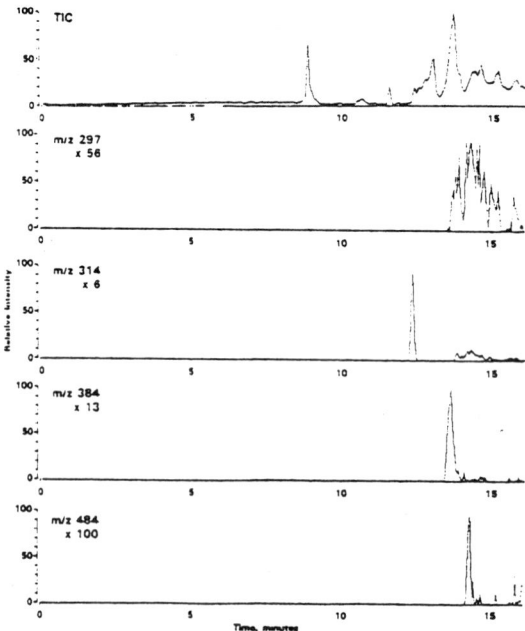

Figure 26. Total ion chromatogram (TIC) and selected ion chromatograms of an SFC−MS analysis using ammonia CI of an extract of wheat containing 10 ppm each of diacetoxyscirpenol, deoxynivalenol and T-2 toxin.

Direct supercritical fluid extraction—mass spectrometry

The direct supercritical fluid injection interface with mass spectrometry (DFI—MS) is also well suited for studies of supercritical fluid extractions or fractionation processes. On-line mass spectrometry of a supercritical fluid extract of a complex mixture, or less complex mixtures deposited on an adsorbent or other substrate, can be carried out using micro-scale extraction techniques (Smith *et al.*, 1985b, c; Kalinoski *et al.*, 1986a). This type of analysis is useful for overall characterization or chemical class fractionation of complex materials (Smith *et al.*, 1985b, c; Smith and Udseth, 1983c), as well as the determination of low levels of contaminants or impurities in some bulk materials as illustrated earlier using the high-resolution MS interface (Kalinoski *et al.*, 1986a).

Figure 27 shows three mass spectra obtained during supercritical ammonia extraction of an 'intractable' diesel fuel marine sediment sample at 150°C and 415 bar. Study of such highly polar complex mixtures with polar supercritical fluids, such as ammonia, is of interest due to the limited experimental data available from other methods. The effect of ammonia pressure upon the extraction is clearly evident, and higher molecular weight components are extracted at the higher pressures. At 415 bar sediment components are observed to at least m/z 850. Further, the higher molecular weight contributions from this sample likely extend well beyond the upper range of the quadrupole mass spectrometer. The interpretation of the spectrum in Fig. 27 is complicated by the complex, unknown composition of this material, and the possibility of reaction with ammonia.

Interpretation of such complex spectra obtained using supercritical fluid extraction is aided by the ability to perform additional separation steps on the analyte extract. Tandem mass spectrometric separation methods (MS/MS) can be effectively exploited to identify specific components. Data obtained by MS/MS in such studies have demonstrated that parts-per-million levels of contaminants in complex natural matrices (Kalinoski *et al.*, 1986a), can be detected. The mass spectra in Fig. 28 give MS/MS daughter ion spectra obtained during the supercritical ammonia extraction of the diesel fuel sediment described in Fig. 27. The identity of one set of components as C_1–C_3 alkyl carbazoles was clearly suggested by the MS/MS spectra. The possibility that reaction of ammonia with the sediment was the source of these components was eliminated, since largely comparable spectra for these compounds were also obtained at higher temperatures using supercritical pentane as the extraction solvent. The combination of tandem mass spectrometric techniques in SFC—MS applications is particularly favorable with the mild CI methods which have been demonstrated to date.

SUMMARY

Supercritical fluid chromatography interfaced with mass spectrometry continues to grow as an analytical method for the chemical characterization of

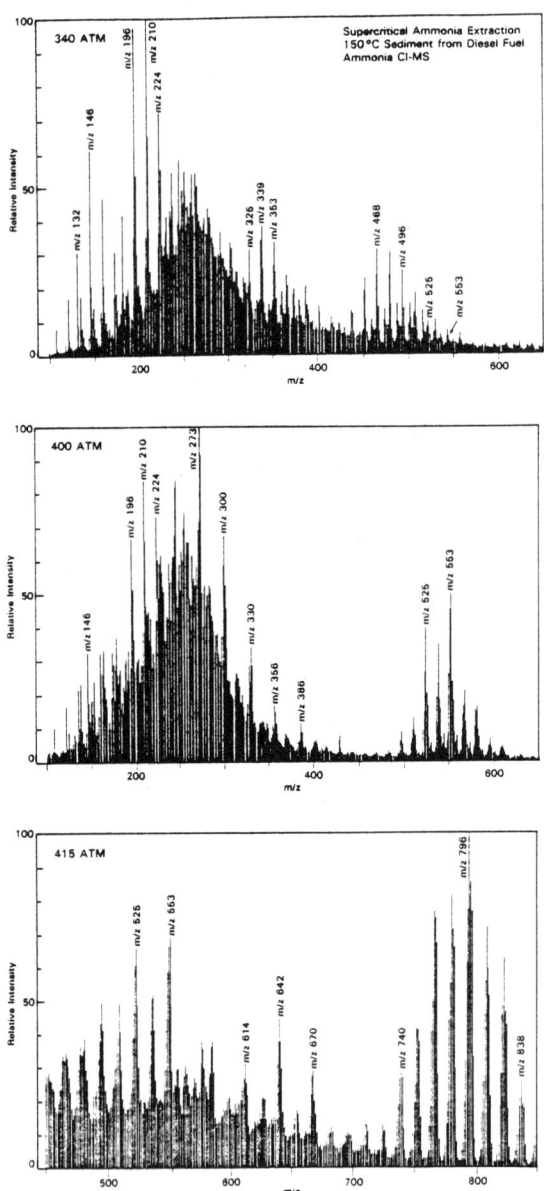

Figure 27. Mass spectra obtained during supercritical ammonia extraction of a diesel fuel marine sediment at 150°C and various pressures.

materials for which traditional gas and liquid chromatographic techniques are inadequate. Previously intractable samples and complex mixtures are able to be addressed, and new developments in instrumentation are pro-

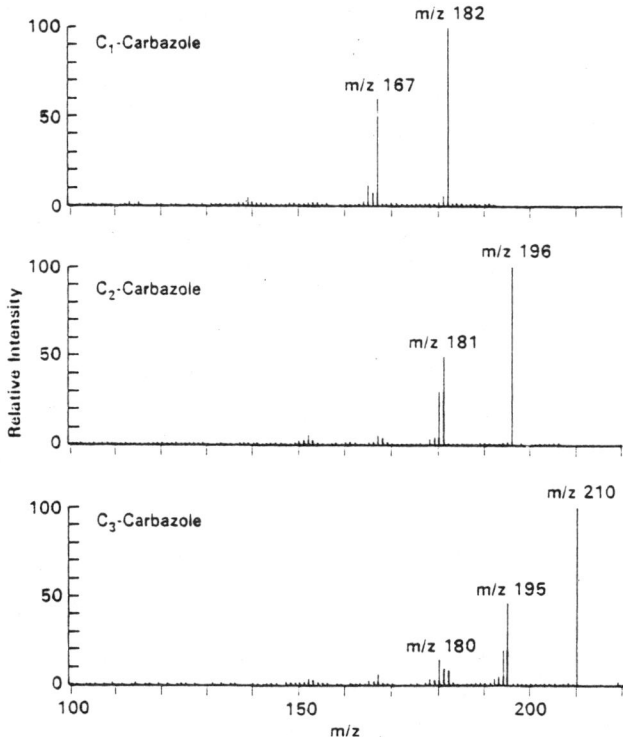

Figure 28. Collision-induced dissociation (CID) mass spectra of three alkyl carbazole samples obtained during supercritical ammonia extraction of an 'insoluble' diesel fuel sediment at 150°C and 400 bar.

viding for further application of the method. The use of supercritical fluids and mass spectrometry also provides for the study of fluid phase phenomena and fluid extraction processes. Newly available commercial SFC−MS instrumentation and new approaches for nonvolatile samples will serve to further broaden the application of this technique.

ACKNOWLEDGEMENT

This work has been supported by the US Department of Energy, Office of Health and Environmental Research, under Contract DE-AC06-76RLO 1830.

REFERENCES

Berry, A. J., Games, D. E., and Perkins, J. R. (1986). Supercritical fluid chromatographic and supercritical fluid chromatographic−mass spectrometric studies of some polar compounds. *J. Chromatogr. 363*, 147−158.
Brunner, E. J. (1985). Fluid mixtures at high pressures. I. Phase separation and critical phenomena of 10 binary mixtures (a gas + methanol). *Chem. Thermodyn. 17*, 671.

Cambel, A. B. and Jennings, B. H. (1958). *Gas Dynamics*. McGraw Hill, New York.

Chester, T. L. (1984). Capillary supercritical fluid chromatography with flame ionization detection: reduction of detection artifacts and extension of detectable molecular weight range. *J. Chromatogr. 299*, 424–431.

Chester, T. L., Innis, D. P., and Owens, G. D. (1985). Separation of sucrose polyesters by capillary supercritical-fluid chromatography/flame ionization detection with robot-pulled capillary restrictors. *Anal. Chem. 57*, 2243.

Chueh, P. L. and Prausnitz, J. M. (1967). Vapor liquid equilibria at high pressures: calculation of critical temperatures, volumes and pressures of non polar mixtures. *AIChE J. 13*, 1107.

Crowther, J. B. and Henion, J. D. (1985). Supercritical fluid chromatography of polar drugs using small particle packed columns with mass spectrometric detection. *Anal. Chem. 57*, 2711–2716.

Fields, S. M. and Lee, M. L. (1984). Effect of density and temperature on efficiency in capillary supercritical fluid chromatography. *J. Chromatogr. 349*, 305.

Fields, S. M., Kong, R. C., Fjeldsted, J. C., Lee, M. L., and Peaden, P. A. (1984). Effect of column diameter on efficiency in capillary supercritical fluid chromatography. *J. High Res. Chromatogr. Chromatogr. Comm. 7*, 312–318.

Fjeldsted, J. C. and Lee, M. L. (1984). Capillary supercritical fluids chromatography. *Anal. Chem. 56*, 619A–628A.

Fjeldsted, J. C., Kong, R. C., and Lee, M. L. (1983). Capillary supercritical fluid chromatography with conventional flame detectors. *J. Chromatogr. 279*, 449–455.

Frew, N. M. and Johnson, C. G. (1986). Use of combined supercritical fluid chromatography–mass spectrometry to characterize carotenoid pigments. Presented at the 34th Annual Conference on Mass Spectrometry and Allied Topics, Cincinnati, OH, 8–13 June.

Giddings, J. C., Myers, M. N., McLaren, L., and Keller, R. A. (1968). High pressure gas chromatography of nonvolatile species. *Science 162*, 67.

Gitterman, M. and Procaccia, I. (1983). Quantitative theory of solubility in supercritical fluids. *J. Chem. Phys. 78*, 2648.

Goates, S. R., Zabriskie, N. A., Simons, J. R., and Khocbehi, B. (1987). Detection of aerosol formation in the effluent of a supercritical fluid chromatography. *Anal. Chem. 59*, 2927–2930.

Grob, K. (1983). An attempt to extend ordinary capillary gas chromatography to supercritical fluid chromatography (SFC). *J. High Res. Chromatogr. Chromatogr. Comm. 6*, 178–184.

Guiochon, G. and Colin, H. (1984). Narrow bore and micro-bore columns in liquid chromatography. *J. Chromatogr. Lib. 28*, 1.

Guthrie, E. J. and Schwartz, H. E. (1986). Integral pressure restrictor for capillary SFC. *J. Chromatogr. Sci. 24*, 236–241.

Hirata, Y. and Nakata, F. (1984). Supercritical fluid chromatography with fused-silica packed columns. *J. Chromatogr. 295*, 315.

Irvin, W. J. (1982). *Analytical Pyrolysis, A Comprehensive Guide*. Marcel Dekker, New York.

Jentoft, R. and Gouw, T. H. (1969). Supercritical fluid chromatography of a 'monodisperse' polystyrene. *J. Polymer Sci., Polymer Letters Ed.*, 811–813.

Kalinoski, H. T., Udseth, H. R., Wright, B. W., and Smith, R. D. (1986a). Supercritical fluid extraction and direct fluid injection mass spectrometry for the determination of trichothecene mycotoxins in wheat samples. *Anal. Chem. 58*, 2421–2425.

Kalinoski, H. T., Wright, B. W., and Smith, R. D. (1986b). Ammonia and methane chemical ionization mass spectra of acid and carbamate pesticides using direct supercritical fluid injection. *Biomed. Environ. Mass Spectrom. 13*, 33–45.

Kalinoski, H. T., Udseth, H. R., Chess, E. K., and Smith, R. D. (1987). Capillary supercritical fluid chromatography–mass spectrometry. *J. Chromatogr. 384*, 3–14.

Kalinoski, H. T., Wright, B. W., and Smith, R. D. (1988). Chemical ionization mass spectra of high molecular weight, biologically active compounds following supercritical fluid chromatography. *Biochem. Environ. Mass Spectrom. 15*, 239–242.

Klesper, E. (1978). Supercritical fluid chromatography of styrene oligomers. *Angew. Chem. Int. Ed. Engl. 17*, 738–746.

Klesper, E. and Hartmann, W. (1976). Apparatus and separations in supercritical fluid chromatography. *J. Polymer Sci., Polymer Letters Ed. 14*, 77–88.

Klesper, E. and Hartmann, W. (1977a). Chromatography with supercritical fluids. *J. Polymer Sci., Polymer Letters Ed. 15*, 9–16.

Klesper, E. and Hartmann, W. (1977b). Parameters in supercritical fluid chromatography of styrene oligomers. *J. Polymer Sci., Polymer Letters Ed. 15*, 707–712.

Klesper, E. and Hartmann, W. (1977c). Preparative supercritical fluid chromatography of styrene oligomers. *J. Polymer Sci., Polymer Letters Ed. 15*, 713–719.

Lattimer, R. D. (1983). Fast atom bombardment mass spectrometry of polyglycols. I. *J. Mass Spec. Ion Processes 55*, 221–232.

Lauer, H. H., McManigill, D., and Board, R. D. (1983). Mobile phase transport properties of liquefied gases in near-critical and supercritical fluid chromatography. *Anal. Chem. 55*, 1370–1375.

Lee, E. D. and Henion, J. D. (1986). Open tubular column supercritical fluid chromatography/mass spectrometry on a benchtop mass spectrometer. *J. High Res. Chromatogr. Chromatogr. Comm. 9*, 172–174.

Levy, J. M. and Ritchey, W. M. (1986). Investigations of the uses of modifiers in supercritical fluid chromatography. *J. Chromatogr. Sci. 224*, 242–248.

Markides, K. E., Fields, S. M., and Lee, M. L. (1986). Capillary supercritical fluid chromatography of labile carboxylic acids. *J. Chromatogr. Sci. 24*, 254–257.

Matson, D. W., Petersen, R. C., and Smith, R. D. (1986a). Formation of silica powders from the rapid expansion of supercritical solutions. *Adv. Ceramic Mat. 1*, 242–246.

Matson, D. W., Petersen, R. C., and Smith, R. D. (1986b). The preparation of polycarbosilane powders and fibers during rapid expansion of supercritical fluid solutions. *Mat. Letters. 4*, 429.

McHugh, M. A. and Krukonis, V. J. (1986). *Supercritical Fluid Extraction*. Butterworths, Boston.

McHugh, M. A. and Seckner, A. J. (1987). Phase behavior of the carbon dioxide—methanol system. *J. Chem. Data*, submitted for publication.

Meuzelaar, H. L. C., Posthumus, M. A., Kistemaker, P. G., and Kistemaker, J. (1973). Curie point pyrolysis in direct combination with low voltage electron impact mass spectrometry. *Anal. Chem. 45*, 1546–1549.

Meuzelaar, H. L. C., Hauerkamp, J., and Hileman, F. D. (1982). *Pyrolysis of Recent and Fossil Biomaterials: Compendium and Atlas*. Elsevier, Amsterdam.

Nieman, J. A. and Rogers, L. B. (1975). Supercritical fluid chromatography applied to the characterization of a siloxane-based gas chromatographic stationary phase. *Sep. Sci. 10*, 517–545.

Novotny, M., Springston, S. R., Peaden, P. A., Fjeldsted, J. C., and Lee, M. L. (1981). Capillary supercritical fluid chromatography. *Anal. Chem. 53*, A407–A414.

Peaden, P. A., Wright, B. W., and Lee, M. L. (1982). Preparation of non-extractable methylphenylpolysiloxane stationary phases for capillary column gas chromatography. *Chromatographia 15*, 335.

Petersen, R. C., Matson, D. W., and Smith, R. D. (1987). Formation of polymer fibers from the rapid expansion of supercritical fluid solutions. *Polymer Engng. Sci. 27*, 1643–1697.

Randall, L. G. and Wahrhaftig, A. L. (1981). Direct coupling of a dense (supercritical) gas chromatograph to a mass spectrometer using a supersonic molecular beam interface. *Rev. Sci. Instrum. 52*, 1283–1295.

Reid, R. C., Prausnitz, J. M., and Shewood, T. K. (1977). *The Properties of Gases and Liquids*. McGraw-Hill, New York, 3rd edn, Chapters 5–7.

Richter, B. E. (1985). Modified flame ionization detector for the analysis of large molecular weight polar compounds by capillary supercritical fluid chromatography. *J. High Res. Chromatogr. Chromatogr. Comm. 8*, 297.

Schmitz, F. P. and Hilgers, H. (1986). Separation by means of supercritical fluid chromatography of 1-vinyl and 2-vinylnaphthalene oligomers prepared through radical and anionic initiation. *Makromol. Chem. Rapid Commun. 7*, 59–63.

Sie, S. T. and Rijnders, G. W. A. (1967). High pressure gas chromatography and chromatography with supercritical fluids. II. Permeability and efficiency of packed columns with high pressure gases as mobile fluids under conditions of incipient turbulence. *Sep. Sci. 2*, 699–727.

Smith, R. D. and Udseth, H. R. (1983a). Mass spectrometry with direct supercritical fluid injection. *Anal. Chem. 55*, 2266–2272.

Smith, R. D. and Udseth, H. R. (1983b). New method for direct analysis of supercritical fluid coal extraction and liquification. *Fuel 62*, 466–468.

Smith, R. D. and Udseth, H. R. (1983c). Direct mass spectrometric analysis of supercritical fluid extraction products. *Sep. Sci. Tech. 18*, 245–283.

Smith, R. D. and Udseth, H. R. (1987). Mass spectrometer interface for microbore and high flow rate capillary supercritical fluid chromatography with splitless injection. *Anal. Chem. 59*, 13–22.

Smith, R. D., Felix, W. D., Fjeldsted, J. C., and Lee, M. L. (1982a). Capillary column supercritical fluid chromatography/mass spectrometry. *Anal. Chem. 54*, 1883–1885.

Smith, R. D., Fjeldsted, J. C., and Lee, M. L. (1982b). Direct fluid injection interface for capillary supercritical fluid chromatography–mass spectrometry. *J. Chromatogr. 247*, 231–243.

Smith, R. D., Kalinoski, H. T., Udseth, H. R., and Wright, B. W. (1984a). Rapid and efficient capillary column supercritical fluid chromatography with mass spectrometric detection. *Anal. Chem. 56*, 2476–2480.

Smith, R. D., Udseth, H. R., and Kalinoski, H. T. (1984b). Capillary supercritical fluid chromatography/mass spectrometry with electron impact ionization. *Anal. Chem. 56*, 2971–2973.

Smith, R. D., Chapman, E. G., and Wright, B. W. (1985a). Pressure programming in supercritical fluid chromatography. *Anal. Chem. 57*, 2829–2836.

Smith, R. D., Udseth, H. R., and Hazlett, R. N. (1985b). Diesel fuel marine and sediment analysis. *Fuel 64*, 810–815.

Smith, R. D., Udseth, H. R., and Wright, B. W. (1985c). In: *Supercritical Fluid Technology*. J. M. L. Penninger, M. Radosz, M. A. McHugh, and V. A. Krukonis (Eds). Elsevier, Amsterdam, pp. 191–223.

Smith, R. D., Wright, B. W., and Udseth, H. R. (1985d). Rapid and high resolution capillary supercritical fluid chromatography and SFC/MS of trichothecene mycotoxins. *J. Chromatogr. Sci. 23*, 192–199.

Smith, R. D., Fulton, J. L., Petersen, R. C., Kopriva, A. J., and Wright, B. W. (1986a). Performance of capillary restrictors in supercritical fluid chromatography. *Anal. Chem. 58*, 2057–2064.

Smith, R. D., Wright, B. W., and Udseth, H. R. (1986b). In: *Chromatography and separation science*. S. Ahuja, Ed., ACS Symposium Series 297. American Chemical Society, Washington, DC, pp. 260–293.

Smith, R. D., Wright, B. W., and Yonker, C. R. (1987a). Characterization of supercritical fluid solutions using supercritical fluid chromatography and mass spectrometry. Theories and practice of supercritical gas extraction. NTS, Inc., 173–232.

Smith, R. D., Kalinoski, H. T., and Udseth, H. R. (1987b). Fundamentals and practice of supercritical fluid chromatography/mass spectrometry. *Mass Spectrometry Rev. 6*, 445–496.

Van Wasen, V. and Schneider, G. W. (1975). Pressure and density dependence of capacity ratios in supercritical fluid chromatography with carbon dioxide as mobile phase. *Chromatographia 8*, 274.

Voorhees, K. J. (Ed.) (1984). *Analytical Pyrolysis: Techniques and Applications*. Butterworths, London.

White, C. M. and Houck, R. K. (1986). Supercritical fluid chromatography and some of its applications: a review. *J. High Res. Chromatogr. Chromatogr. Comm. 9*, 4–17.

Wright, B. W. and Smith, R. D. (1986). Rapid analysis using capillary supercritical fluid chromatography. *J. High Res. Chromatogr. Chromatogr. Comm. 9*, 73.

Wright, B. W., Peaden, P. A., Lee, M. L., and Stark, T. J. (1982). Free-radical crosslinking in the preparation of non-extractable stationary phases for capillary gas chromatography. *J. Chromatogr. 248*, 17.

Wright, B. W., Udseth, H. R., Smith, R. D., and Hazlett, R. N. (1984). Supercritical fluid chromatography and supercritical fluid chromatography—mass spectrometry of marine diesel fuel. *J. Chromatogr. 314*, 253—262.

Wright, B. W., Kalinoski, H. T., and Smith, R. D. (1985). Investigation of retention and selectivity effects using various mobile phases in capillary supercritical fluid chromatography. *Anal. Chem. 57*, 2823—2829.

Wright, B. W., Kalinoski, H. R., Udseth, H. R., and Smith, R. D. (1986). Capillary supercritical fluid chromatography-mass spectrometry. *J. High Res. Chromatogr. Chromatogr. Comm. 9*, 145—153.

Yajima, S., Hagegawa, Y., Hayashi, J., and Iimura, J. (1978). Synthesis of continuous silicon carbide fibre with high tensile strength and high Young's modulus. *J. Mat. Sci. 13*, 2569.

Yonker, C. R. and Smith, R. D. (1986). Study of retention processes in capillary supercritical fluid chromatography with binary fluid mobile phases. *J. Chromatogr. 361*, 25.

Yonker, C. R. and Smith, R. D. (1988). Retention in supercritical fluid chromatography: Influence of the partial molar volume of the solute in the stationary phase. *J. Phys. Chem. 92*, 1664—1667.

Yonker, C. R., Wright, B. W., Udseth, H. R., and Smith, R. D. (1984). New methods for characterization of supercritical fluid solutions. *Berg. Bunseges. Phys. Chem. 88*, 908—911.

Yonker, C. R., Wright, B. W., Petersen, R. C., and Smith, R. D. (1985). Temperature dependence of retention in supercritical fluid chromatography. *J. Phys. Chem. 89*, 5526.

Yonker, C. R., Frye, S. L., Kalkwarf, D. R., and Smith, R. D. (1986). Characterization of supercritical fluid solvents using solvatochromic shifts. *J. Phys. Chem. 90*, 3022—3026.

Progress in HPLC, Vol. 4, pp. 157—178.
Yoshioka *et al.* (Eds)
© 1989 VSP.

Computer-assisted separation system in supercritical fluid chromatography

KIYOKATSU JINNO

School of Materials Science, Toyohashi University of Technology,
Toyohashi 440, Japan

INTRODUCTION

Although liquid chromatography (LC) is one of the techniques generally used for the analysis of mixtures, which cannot be separated by gas chromatography (GC) because of its limitation in volatility and thermal stability, efficient separations can only be obtained at the expense of a long analysis time. An alternative to solve this LC problem is supercritical fluid chromatography (SFC).

Since the first description of SFC in the separation of metal porphorins was reported in 1962 (Klesper *et al.*, 1962) much research has been done to explore the potential advantages of this analytical technique. The unique feature of SFC is that the mobile phase is subjected to pressures and temperatures near its critical point. Under these conditions its density approaches that of a liquid, and then solute diffusion coefficients are approximately 2 orders of magnitude greater than those found in liquids. A supercritical fluid possesses solvating properties similar to a liquid, and solute diffusivities intermediate between a gas and a liquid. Therefore, comparable efficiencies to LC can be obtained in shorter analysis times. This is the main reason why SFC has recently attracted much attention.

With the development of SFC and its ancillary techniques, complex mixtures can now be readily separated and components analyzed qualitatively and quantitatively. The rapid ascent of this technique is one of the most increasing phenomena in the history of modern chromatography. Furthermore, the utility of SFC has been catalyzed by the development of high-efficiency columns, various unique stationary phases, sensitive and specific on-line detection systems such as SFC—mass spectrometry (MS), and microprocessors for instrument operations, as well as data handling.

However, each step in SFC analysis still raises problems in automation and systemization, as with other chromatographic techniques. For example, the selection of satisfactory separation conditions is one of the major problems in this separation technique. In order to set the best separation condition in chromatography, analysts generally first consult other relevant literature, even though this is tedious and time-consuming. After finding a good combination of stationary phase, mobile phase and experimental conditions for his particular application, the analyst may then need to run several experiments to optimize the separation condition with the available chromatographic system. These tasks have traditionally been carried out by 'trial-and-error experiments' in chromatography, based on personal experience and intuition. In addition, these experiments require standard substances which are the same as, or similar to, compounds that one wants to analyze, even if those substances are highly toxic or not easily available. This is another weak point of general chromatography.

Recent developments in semiconductor devices, computer technology and information sciences now offer very valuable and useful tools, i.e. approaches such as chemometrics, which can change the situation relating to chromatography to be convenient and time-saving.

If one would prefer to use optimization whilst utilizing modern chemometric techniques, two approaches may be considered. One is the Simplex method and the other is retention prediction. In the Simplex approach, some experiments are required to be optimized and generate Simplex, using actual analytical samples. There are a number of reviews on this approach in optimization (for example, Berridge, 1986). The second approach is the concept of retention prediction. Studies on quantitative structure−retention relationships (QSRR) should be investigated as a basis of retention prediction of components in a mixture. The author has proposed a system for LC separations assisted by the retention prediction concept (Jinno and Kawasaki, 1984a, b, 1986). After optimizing separation conditions using those chemometric techniques and actual sample injection, identification of the components in the mixture that appear as peaks in the measured chromatogram can be easily accomplished by the use of a computerized multichannel detection system such as a photodiode array UV/VIS detector or a fourier-transform UV detector. Other computerized combined instrumentation, such as SFC−FT infrared (FTIR) (Shafer and Griffiths, 1983; Olesik et al., 1984; Fujimoto et al., 1985; French and Novotny, 1986) and SFC−MS (Smith et al., 1986a, b) are also useful at this stage for component identification. In this identification process the retention prediction approach for optimization can also assist in identifying approximate compound candidates automatically.

One of the typical applications for this concept in practical analytical problems in SFC is the separation of polycyclic aromatic hydrocarbons (PAHs) in various environmental samples. PAHs comprise the largest class of known chemical carcinogens (Harvey, 1985), and their identification and determination is an important analytical problem. Although LC is one of the techniques generally used for the analysis of samples containing PAHs, which have limited volatility and thermal stability, efficient separations can only be obtained at the expense of a long analysis time. SFC seems to be an alternative to solving the problem in PAH analysis. However, even though SFC could be a powerful technique for PAH analysis, identification of the components in complex samples is still difficult. From this point of view a computer-assisted system should be considered. The coupling of SFC with a UV multichannel detector will have a high potential in PAH analysis, because identification of the components by UV spectral matching is enhanced by microcomputer-assisted procedures. The potential of this SFC−UV multichannel detector system will be more enhanced by the microcomputer-assisted retention prediction system. The concept of the separation system assisted by computer is schematically demonstrated in Fig. 1.

This contribution will describe the computer-assisted separation system for PAHs in SFC (retention prediction system coupled with UV multichannel detector), in order to clarify the potential of above-mentioned concept for practical analytical problems.

CONSTRUCTION OF THE RETENTION PREDICTION SYSTEM

The first investigation required for constructing the retention prediction system of PAHs in SFC is to clarify the most promising descriptors which can describe PAH retention, i.e. a QSRR study on the relationships between PAH retention and various descriptors.

Capacity factors for a number of PAHs in SFC were collected from the literature, and are listed in Table 1. The basic conditions for the data listed in the table are described as follows:

(A) Column; 4.6 mm i.d. × 15 cm, packed with ODS-5 μm, mobile phase; carbon dioxide, pressure; 200 atm, temperature; 40°C, obtained by the author's system (Super-100, Jasco, Tokyo, Japan).

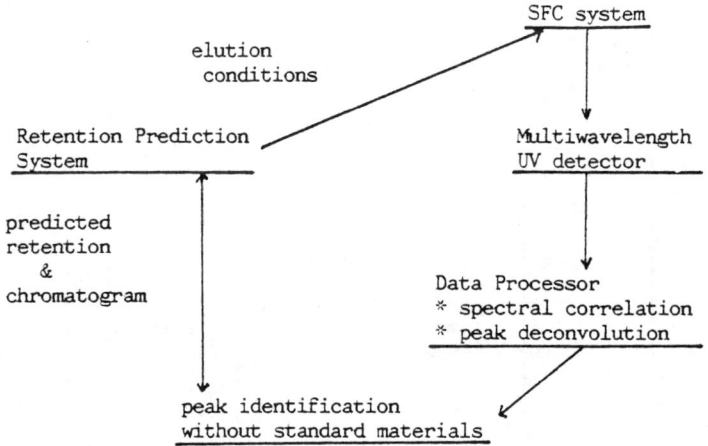

Figure 1. Block diagram of computer-assisted SFC system.

Table 1.
Capacity factors and various descriptors of PAHs on several SFC systems

Solute	Capacity factor, k'							Descriptor				
	Column A	Column B	Column C	Column D	Column E	Column F	Column G	F	L/B	χ	V_W	α
Naphthalene	1.17	0.23	—	0.20	0.20	0.27	0.48	5.0	1.24	3.405	73.96	17.48
Fluorene	1.97	0.23	—	0.80	—	—	—	6.5	1.57	4.612	93.67	21.69
Phenanthrene	3.07	—	—	—	0.60	—	1.45	7.0	1.46	4.815	99.56	24.70
Anthracene	3.07	0.81	0.20	1.00	0.65	0.60	1.50	7.0	1.57	4.809	99.56	25.93
Fluoranthene	5.07	1.25	0.24	2.60	—	—	—	8.0	1.22	5.565	109.04	28.35
Pyrene	6.52	1.35	—	3.36	1.00	0.75	2.60	8.0	1.27	5.559	109.04	29.34
Benz[a]anthracene	8.97	1.99	—	4.80	1.30	—	3.55	9.0	1.58	6.220	125.16	32.86
Chrysene	9.94	—	—	—	1.25	0.98	4.00	9.0	1.72	6.226	125.16	33.06
Naphthacene	—	—	—	—	1.50	1.02	4.10	9.0	1.89	6.214	125.16	32.27
Perylene	—	—	—	14.08	2.20	1.15	5.20	10.0	1.27	6.976	134.64	38.84
Benzo[b]fluoranthene	17.1	—	0.32	—	—	—	—	10.0	1.40	6.976	134.64	37.78
Benzo[k]fluoranthene	17.1	2.80	0.35	—	—	—	—	10.0	1.48	6.970	134.64	37.78
Benzo[e]pyrene	—	—	—	—	—	1.35	6.10	10.0	1.12	6.976	134.64	38.84
Benzo[a]pyrene	24.2	3.03	—	15.88	—	1.20	6.13	10.0	1.50	6.976	134.64	38.84
Dibenz[a,c]anthracene	—	4.09	0.42	—	—	—	—	11.0	1.24	7.637	150.76	41.22
Dibenz[a,h]anthracene	29.2	—	—	—	—	—	—	11.0	1.72	7.631	150.76	41.31
Coronene	—	5.71	0.54	—	—	1.60	10.0	12.0	1.00	8.464	153.60	42.50

Reproduced by permission from Vieweg, 1986.

(B) Column; 200 μm i.d. × 58 cm, packed with 10 μm silica gel, mobile phase; 10% ethanol in *n*-hexane, pressure; 40 atm, temperature; 260°C. Data of Hirata and Nakata (1984).

(C) Column; 200 μm i.d. × 58 cm capillary, bonded with phenylmethyl-polysiloxane, mobile phase; *n*-pentane, pressure; 32 atm, temperature; 210°C. Data of Novotny *et al.* (1981).

(D) Column; 4.6 mm i.d. × 15 cm, packed with ODS-3 μm, mobile phase; carbon dioxide, pressure; 325 atm, temperature; 32°C. Data of Gere *et al.* (1982).

(E) Column; 6 mm i.d. × 2 m, packed with PEG 6000 (23 wt%) coated on Sil-O-Gel (120−140 mesh), mobile phase; *n*-pentane, pressure; 41.3 atm, temperature; 213°C. Data of Sie and Rijnders (1967a).

(F) Column; 6 mm i.d. × 1 m, packed with alumina (120−140 mesh), mobile phase; isopropanol, pressure; 51.75 atm, temperature; 245°C. Data of Sie and Rijnders (1967b).

(G) Column; 6 mm i.d. × 1 m, packed with alumina (120−140 mesh), mobile phase; isopropanol, pressure; 46.5 atm, temperature; 245°C. Data of Sie and Rijnders (1967b).

Although Sie and Rijnders (1967a, b) determined the relationships between boiling point data of PAHs and their retention in SFC, this kind of physico-chemical parameter of some PAHs is still not accurately known because of the sublimating properties. Therefore, various descriptors were surveyed, which enables theoretical or experimental calculations to estimate for the correlation study. In Table 1 the values of F, L/B, Vw, χ and α, which are selected from many parameters and judged as promising descriptors, are also listed for each PAH solute.

The explanations of those parameters are as follows:

F: Correlation factor, F = (number of double bonds) + (number of primary and secondary carbons) − 0.5 × number of non-aromatic rings in the molecule. This indicates the size of the molecule (Schabron *et al.*, 1977).

χ: Connectivity index; this is topological parameter well organized by Kier and Hall (1976).

L/B: Breadth−length ratio of the molecule which shows the two-dimensional shape of the molecule, i.e. shape parameter (Wise *et al.*, 1981).

Vw: van der Waals volume, calculated by using van der Waals radii of the atoms which compose the molecule (Bondi, 1964).

α: If inductive interaction is mainly present in the solute−stationary phase interaction, the average molecular polarizability of the solute can be related to the retention (Lamparczyk and Radecki, 1983). This average molecular polarizability is calculated from the method proposed by Miller and Savchik (1979).

The correlation coefficients obtained by linear regression analysis for each data set are summarized in Table 2. The values in italics indicate the descriptor which gives the highest correlation coefficient for each combination between log k' and the individual descriptors.

The retention of PAHs shows good correlations between log k' and F, χ, α and Vw. This means that the size of the molecule is one of the dominant factors to control the retention of PAHs in SFC. As the regression analysis shows that the value of L/B does not correlate well with the retention, this observation in itself seems to indicate that the shape of the solute is not important in controlling the retention. However, close examination of the data implied that F, χ, α and Vw could not fully explain the order of elution of the structural isomers such as benzo[b]fluoranthene and benzo[a]pyrene. On the other hand, the L/B parameter can reasonably explain the order of elution of these isomers. These results imply that the shape of a solute is also a dominant factor in controlling PAH retention in SFC systems, and this information is very similar to that obtained in LC (Jinno et al., 1987a, b).

A more interesting feature is found in the correlation between log k' and molecular polarizability. As with GC and LC, the correlations between log k' and α are very high in all the cases investigated.

In order to construct a retention prediction system in SFC, the next step is to evaluate the possibility of predicting the retention of PAHs. Generally pressure is considered as the most important and effective way to control retention in SFC, as in mobile phase composition in LC and temperature in GC. Since there is a highly correlated relationship between log k' and various descriptors, the following N equations will be obtained by the linear regression analysis for N different pressures at fixed temperature conditions. This procedure is the same as that performed in the case of the retention prediction in LC:

$$
\begin{aligned}
P &= P_1 && \log k'_1 = A_1 *D + B_1 \\
P &= P_2 && \log k'_2 = A_2 *D + B_2 \\
&\ \ \vdots && \quad \vdots \qquad \vdots \\
&\ \ \vdots && \quad \vdots \qquad \vdots \\
P &= P_n && \log k'_n = A_n *D + B_n
\end{aligned}
\tag{1}
$$

Table 2.
Correlation coefficients between log k' and descriptors for PAHs on several SFC systems

SFC system	Correlation coefficient				
	F	L/B	χ	Vw	α
A	0.990	0.508	0.991	0.986	0.997
B	0.989	0.224	0.988	0.987	0.988
C	0.992	0.806	0.993	0.978	0.962
D	0.987	0.142	0.989	0.977	0.983
E	0.991	0.559	0.991	0.979	0.983
F	0.976	0.112	0.972	0.978	0.976
G	0.986	0.078	0.984	0.986	0.984

Table 3.
Retention data of five PAHs at various pressure conditions

Pressure (atm)	Capacity factor				
	Naphthalene	Fluorene	Anthracene	Pyrene	Chrysene
250	1.22	1.78	2.84	5.67	8.42
200	1.16	1.97	3.07	6.31	9.69
150	1.30	2.17	3.58	7.77	12.5
100	1.51	2.78	4.87	11.1	19.6

Column; 4.6 mm i.d. × 15 cm long, packed with ODS-5 µm; mobile phase: carbon dioxide; temperature: 40°C.
Reproduced by permission from Vieweg (1986).

where P is the pressure, D is the descriptor which shows the high correlation with $\log k'$, and A and B are constants.

In this contribution, α is selected as the most promising descriptor, since this gives the highest correlation coefficient in the authors' SFC system. Values of A and B listed in Table 5 were obtained by the use of the retention data of few standard substances as listed in Table 3. In this example we obtained those data by the following separation conditions: mobile phase; CO_2, column; 4.6 mm i.d. × 15 cm packed with ODS-5 µm, column temperature; 40°C, pressure; 100−250 atm, detection; UV multi-channel detector.

Good correlations were observed for all the data sets. Then, if A and B can be expressed as a function of pressure, P, the following two equations can be obtained by multiple regression analysis:

$$A = f_1(P) = \sum_{i=0}^{N} K_i P^i \tag{2}$$

$$B = f_2(P) = \sum_{i=0}^{N} L_i P^i \tag{3}$$

where N is a number less than n, and K and L are constants.

Multiple regression analyses have been carried out for the sets of $P-A$ and $P-B$ relations in Table 4. The equations derived are as follows:

$$A = -0.000109 \ P + 0.0824, \quad r = 0.987 \tag{4}$$

$$B = \quad 0.00121 \ P - 1.240, \quad r = 0.991 \tag{5}$$

then

$$\log k' = (-0.000109 \ P + 0.0824) \times \alpha + (0.00121 \ P - 1.240) \tag{6}$$

As shown in Fig. 2, one can obtain the surface of eqn (6) in the three-dimensional illustration. Equation (6) means that if P, the pressure, and α, molecular polarizability of a compound, are known, the logarithm of the capacity factor, $\log k'$, can be determined for the given chromatographic condition.

Table 4.
Results of linear regression analysis for the correlation between log k' and polarizability at different pressures: $\log k' = A\alpha + B$

Pressure (atm)	A	B	r*
250	0.0559	−0.932	0.991
200	0.0603	−1.011	0.994
150	0.0646	−1.048	0.993
100	0.0726	−1.121	0.995

* Correlation coefficient
Reproduced by permission from Vieweg (1986).

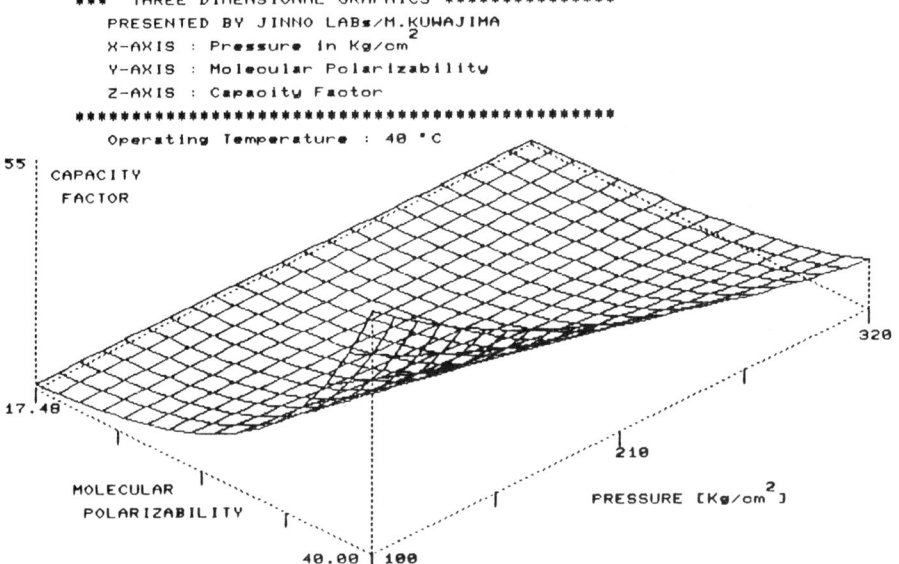

Figure 2. Three-dimensional surface of the function between capacity factor and molecular polarizability with pressure as a parameter.

In order to demonstrate the prediction ability of this concept, eqn (6) was used to predict the retention of 12 PAHs such as naphthalene, fluorene, phenanthrene, anthracene, fluoranthene, pyrene, benz[a]anthracene, chrysene, benzo[b]fluoranthene, benzo[k]fluoranthene, benzo[a]pyrene and dibenz[a,h]anthracene at 200 atm and 40°C with supercritical carbon dioxide as the mobile phase.

Comparison of predicted and measured capacity factors is demonstrated in Figure 3. It appears that prediction ability of the system was fairly good. It is also possible to predict the retention of PAHs under different temperature conditions if the procedure described above applies for a measured data

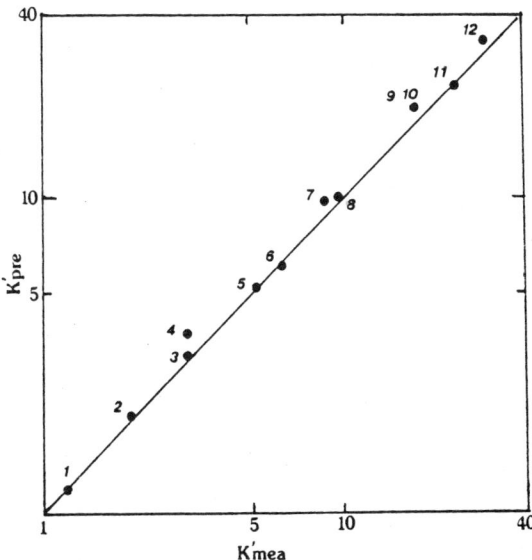

Figure 3. Comparison between predicted retention and measured retention for twelve PAHs: 1: naphthalene, 2: fluorene, 3: phenanthrene, 4: anthracene, 5: fluoranthene, 6: pyrene, 7: benz[a]anthracene, 8: chrysene, 9: benzo(b)fluoranthene, 10: benzo[k]fluoranthene, 11: benzo[a]pyrene, 12: dibenz[a,h]anthracene. (Reproduced by permission from Vieweg, 1986.)

set, where the parameter should be changed to T (temperature) from P (pressure).

This system can also offer an aid to the identification of PAHs from SFC retention data. Estimated polarizability values obtained from the retention value corresponding to the individual peak can be compared to those of standard compounds in the data files previously stored (using eqn (6)), and the approximate identification can be performed automatically. The algorithm of this function of the identification system is shown in Fig. 4. The system gives the compound name, predicted retention information and numerical possibility of the identification on the CRT or line-printer. Of course, this function can be performed for GC and LC separations using a similar concept.

The author believes that, in the near future, chromatographers will be able to obtain analytical columns for LC and SFC with software for the retention prediction of desired compound groups, which are readily prepared by the column suppliers, based on their routine quality control process. The work described herein, computer-enhanced chromatography, is intended to demonstrate the performance of the systems assuming the situation.

IDENTIFICATION OF PAHs IN DIESEL PARTICULATE MATTER EXTRACT

The identification of PAHs present in the fraction of the extract from diesel engine particulate matter will be described as a demonstration of the application (Jinno *et al.*, 1986c) using the concept described in the previous section.

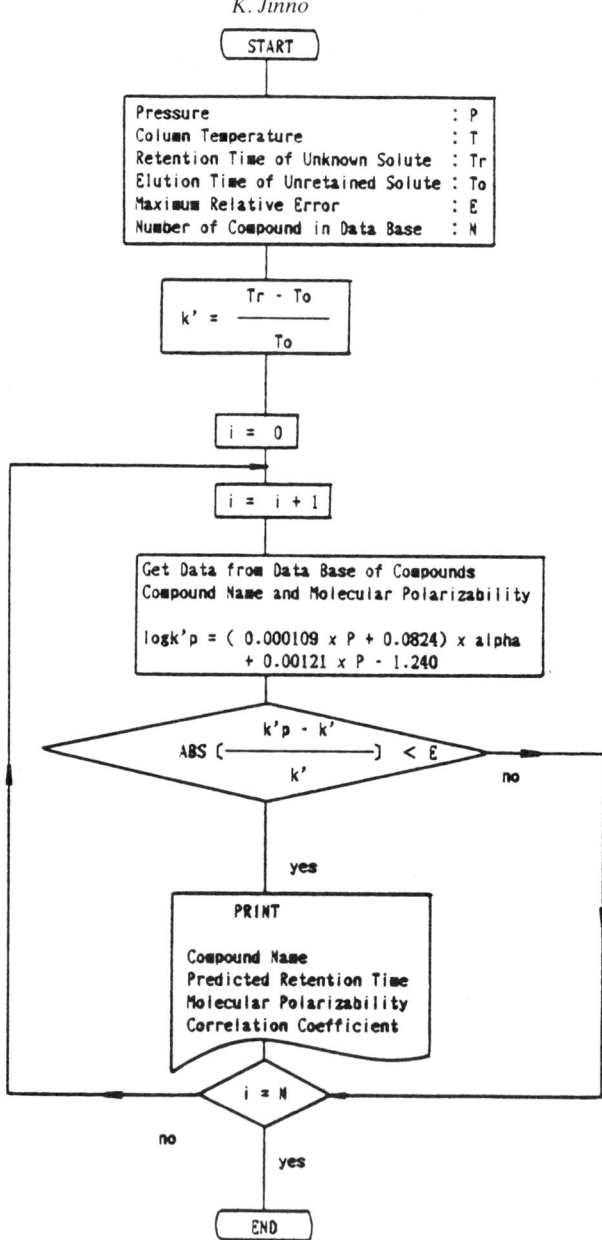

Figure 4. Flow-chart of the algorithm of automated identification system in SFC.

The sample was the extract from diesel engine particulate matter. The dichloromethane extract from particulate matter collected from the exhaust of a test-diesel engine was separated into seven fractions by a silica column with gradient elution mode from 100% *n*-hexane to 100% dichloromethane. As some fluorescence emission from the third fraction was observed by irradiation with an UV lamp, this fraction was considered as that containing PAHs. This PAH fraction of the concentrated extract was redissolved in acetonitrile and analyzed by the subsequent SFC separations.

For the column used, eqn (6) was obtained as described previously. This equation could be used for optimization of separation condition and approximate identification in SFC, since this means that, if P and α are known, the logarithm of capacity factor of any PAH solute can be determined for the given chromatographic condition.

Since most important carcinogenic or mutagenic compounds in PAHs are considered as middle-sized compounds, primary concern in this analysis is focused on these middle-sized PAHs. To establish the separation condition for these middle-sized PAHs, the retention prediction system was used, where the desired condition was that dibenz[a,h]anthracene (five-ring isomer) should elute in less than 10 min. The prediction system gave the most optimized condition as 200 atm and 40°C of carbon dioxide supercritical state. At this condition the predicted retention time of dibenz[a,h]anthracene is *ca.* 8.5 min. The synthesized chromatogram as a response of the retention prediction system is shown in Fig. 5A, in which retention of 12 PAHs such as naphthalene, fluorene, phenanthrene, anthracene, fluoranthene, pyrene, benz[a]anthracene, chrysene, benzo[b]fluoranthene, benzo[k]fluoranthene, benzo[a]pyrene and dibenz[a,h]anthracene is predicted. Figure 5B shows the actual chromatogram of the diesel extract obtained at the condition of 200 atm and 40°C. Since it is very difficult to identify some components eluted in less than 1.5 min because of the complexity of the chromatogram at this separation condition, the identification of the eight main peaks numbered in the measured chromatogram was first performed using a computerized identification system.

An approximate assignment for these peaks by the identification system has been accessed. One of the typical outputs of the system in this process is shown in Fig. 6 for peak No. 3. The system implies that this peak contains benz[a]anthracene with fairly high possibility. Similar procedures were performed with other peaks, and the results are summarized in Table 5. More precise identification of peaks in the measured chromatogram can be accomplished by the use of several functions of the computerized UV multichannel detector.

Since one of the functions offered by the multichannel detector in chromatography is to give the UV spectrum of the solute separated in one chromatographic run, this capability has been applied to assign these peaks. In addition, it is more powerful for that purpose to use the capability to search the standard UV spectrum from the stored data file which matches to measured one (Hoshino *et al.*, 1984).

The first step in the identification process concerned the component contained in peak No. 1. The automatic correlation study on matching between the observed UV spectrum and the stored UV spectra of PAHs yielded the computer output shown in Fig. 7. The results show that the UV spectrum of peak No. 1 is closely correlated to that of fluoranthene. The UV spectra show excellent agreement with each other in the wavelength range longer than 230 nm. Even though the measured UV spectrum is a little different from that of the standard UV spectrum of fluoranthene in the range of shorter wavelength, one can conclude peak No. 1 contains fluoranthene, for

<CHROMATOGRAM/MCASYST>

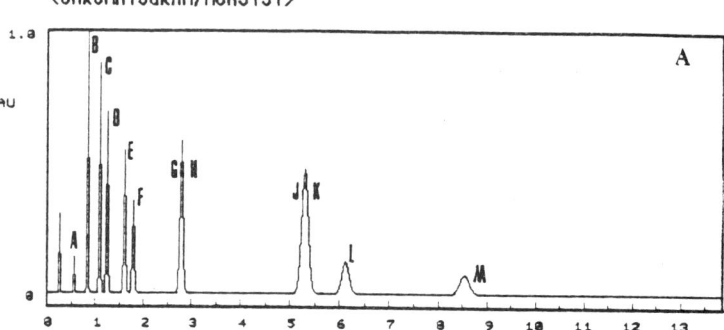

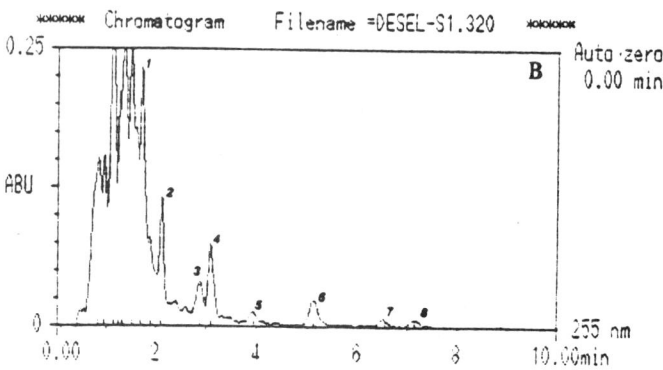

Figure 5. Comparison of predicted and measured chromatograms of PAHs with supercritical carbon dioxide as the mobile phase. **A**: Predicted chromatogram by the microcomputer-assisted retention prediction system. Mobile phase: carbon dioxide, pressure: 200 atm, temperature: 40°C; peak assignment: A: naphthalene, B: fluorene, C: phenanthrene, D: anthracene, E: fluoranthene, F: pyrene, G: benz[a]anthracene, H: chrysene, J: benzo[b]fluoranthene; K: benzo[k]fluoranthene, L: benzo[a]pyrene, M: dibenz[a,h]anthracene. **B**: Measured chromatogram of the extract from diesel engine particulate matter at the above SFC condition. (Reproduced by permission from the American Chemical Society, 1986.)

```
##################################################################

Pressure                             200   bar
Column Temperature                   40    °C
Retention Time of Unknown Solute     2.88  min
Maximum Relative Error                .1

-----------------------------------------------------------------

sample name                          alpha   log k'   coeff.

benz(a)anthracene                    32.86   0.993    0.554
chrysene                             33.06   1.005    0.426

##################################################################
```

Figure 6. Output of the automated identification system for peak No. 3 in the chromatogram of Fig. 5B.

Table 5.
Identification of PAHs in the extract from diesel engine particulate matter

Peak No.	Approximate assignment by retention prediction	Assignment UV multichannel detector
1	Fluoranthene	Fluoranthene
2	Pyrene	Pyrene
3	Benz(a)anthracene	Benz[a]anthracene
4	Chrysene	Chrysene
5	*	*
6	Benzo[b]fluoranthene	*f*: Benzo[b]fluoranthene
	Benzo[k]fluoranthene	*r*: Benzo[k]fluoranthene
7	Benzo[e]pyrene	Benzo[e]pyrene
8	Benzo[a]pyrene	Benzo[a]pyrene
9	*	*
10	*	Indeno[1,2,3-cd]pyrene
11	*	Benzo[ghi]perylene
12	*	*
13	Fluorene	Fluorene
14	*	*
15	Phenanthrene	Phenanthrene
16	*	*
17	Anthracene	Anthracene[a]
18	*	*

* Not identified.
f: Front part of the peak, *r*: rear part of the peak
[a] Correlation number of 0.80.
Reproduced by permission from the American Chemical Society, 1986.

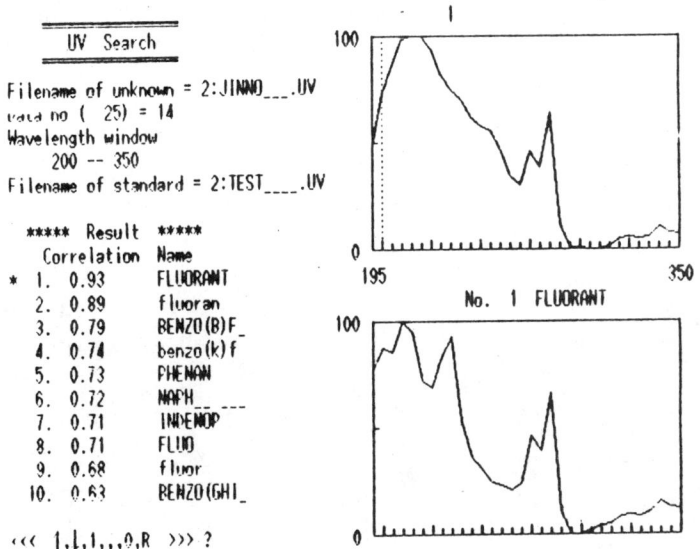

Figure 7. Output of the UV spectrum search process for peak No. 1 in the chromatogram of Fig. 5B. (Reproduced by permission from the American Chemical Society, 1986.)

the following two reasons: (1) that the retention time of peak No. 1 is almost consistent with the predicted retention of fluoranthene by the retention prediction system; (2) the range of shorter wavelength of the UV spectrum is sometimes critically affected from the calculation of background subtraction.

Similar procedures were then applied to the identification of components contained in peaks Nos. 2 to 8 in the measured chromatogram, and the results are listed in Table 5. The spectra of individual peaks and the standard spectrum assigned to them by the UV multichannel detector are also demonstrated in Fig. 8.

In the retention range longer than 10 min beyond the chromatogram shown in Fig. 5, three major peaks were apparent, as indicated in Fig. 9. With the same procedures, peaks Nos. 10 and 11 were identified as indeno-[1,2,3-cd]pyrene and benzo[ghi]perylene, respectively. Those UV spectra are compared with the matched standard spectra in Fig. 10. The reason why, is that in spite of the fact that benzo[ghi]perylene has a similar α value (41.22) to that of dibenz[a,h]anthracene (41.31), it eluted later than expected (*ca.* 8.4 min). This is the difference in the flatness of their molecular structures, and this concept has been proposed for LC separations (Jinno *et al.*, 1987a, b).

The identification of peaks Nos. 5 and 7 in Fig. 5, and No. 9 in Fig. 9, was difficult, even though their UV spectra were obtained clearly, and shown in Fig. 11, because no UV spectra stored in the data base closely matched those spectra. However, the component in peak No. 7 is tentatively identified as benzo[e]pyrene, because the UV spectrum is very similar to that of benzo[e]pyrene in the literature (Clar, 1964) and its predicted retention time (6.2 min) is good agreement with that of peak No. 7 (6.4 min).

As seen in Fig. 5, the separation condition of 200 atm and 40°C is not

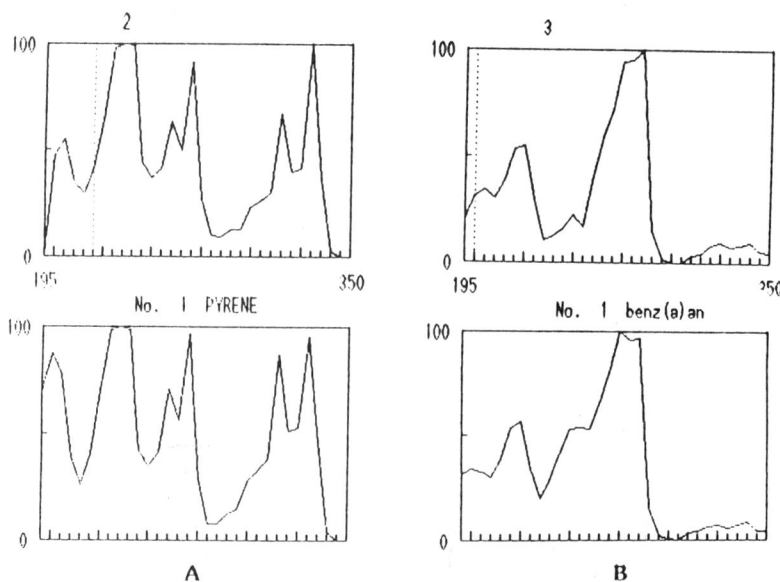

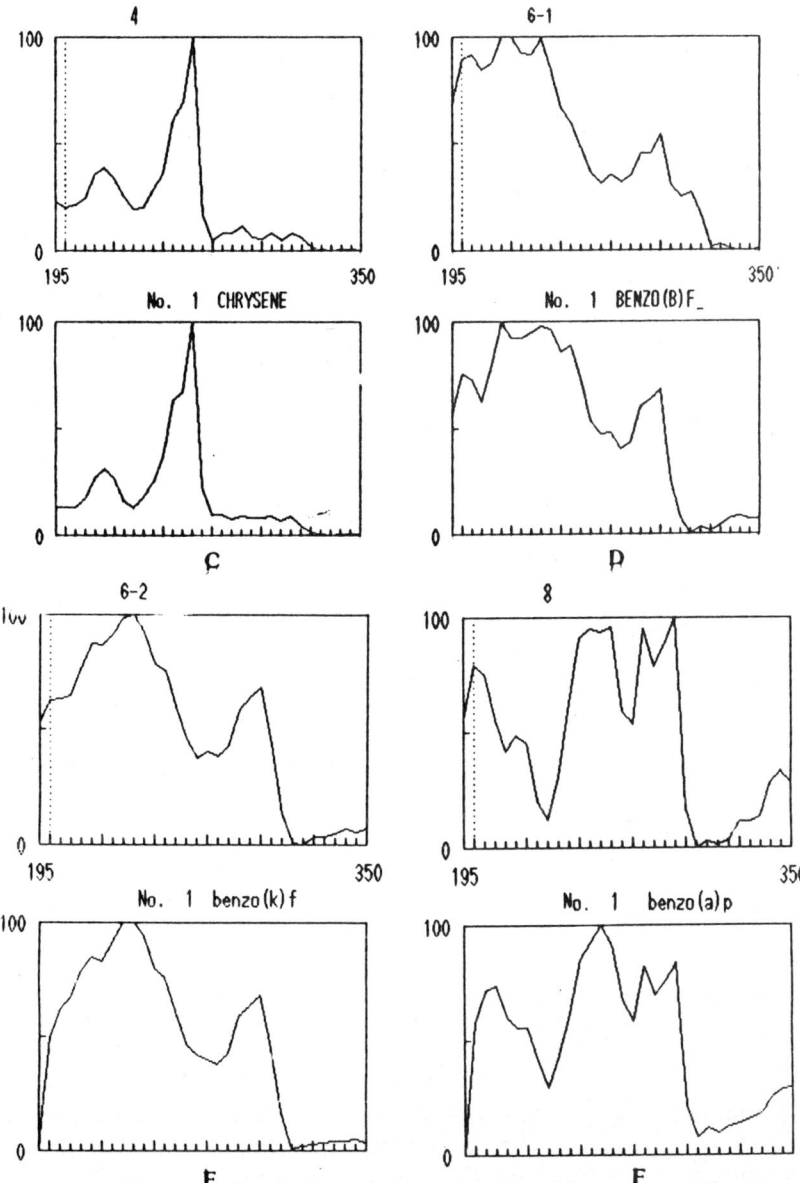

Figure 8. Comparison of the measured UV spectra of the separation components in Fig. 5B and the most correlated standard UV spectra. A: Peak No. 2, assigned to pyrene; B: peak No. 3, assigned to benz[a]anthracene; C: peak No. 4, assigned to chrysene; D: peak No. 6 (front part), assigned to benzo[b]fluoranthene; E: peak No. 6 (rear part), assigned to benzo[k]fluoranthene; F: peak No. 8, assigned to benzo[a]pyrene. (Reproduced by the permission from American Chemical Society, 1986.)

suitable to separate small-size PAHs (one to three rings). Therefore, to separate small PAHs in the extract, the separation condition was re-examined using the retention prediction system. The desired condition in this case is that anthracene elutes in less than 2 min. As a result of the retention

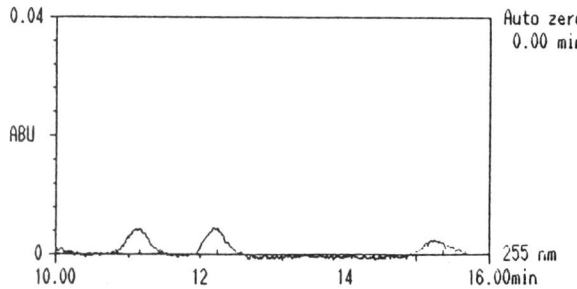

Figure 9. Chromatogram of diesel particulate extract at the retention time range after 10 min. Separation condition is the same as in Fig. 5.

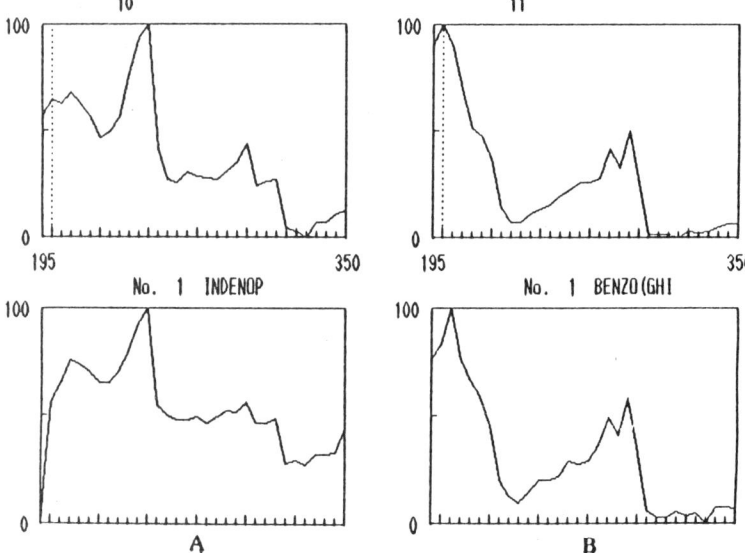

Figure 10. Comparison of UV spectra between measured and filed for peaks No. 10 to A and 11 to B in Fig. 9, respectively.

prediction, the condition of 100 atm and 40°C was shown on the CRT of the microcomputer. The predicted chromatogram at that condition is shown in Fig. 12A. The measured chromatograms at 255 nm, and the three-dimensional one, are shown in Figs 12B and 12C, respectively.

Seven prominent peaks are apparent in the measured chromatogram. Using the approximate identification tool of the retention prediction system, and the powerful functions of the multichannel UV detector such as the peak deconvolution technique, it has been confirmed that peaks Nos. 13 and 15 include fluorene and phenanthrene, respectively, although those peaks consist of multicomponents. For peak No. 13, two different UV spectra demonstrated in Fig. 13A are observed for retention times, of 0.94 min and 1.05 min. The spectrum of 1.05 min is almost the same as that of fluorene. The smoothing procedure was applied to that peak and the chromatograms

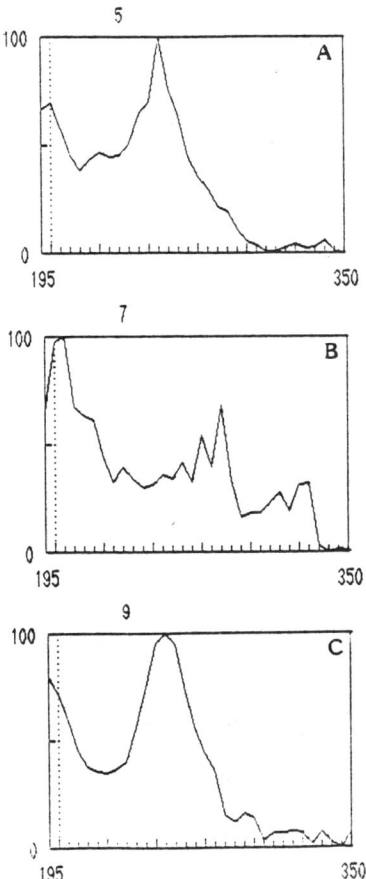

Figure 11. UV spectra of peaks No. 5, 7 and 9 in the chromatogram of Figure 5B. A: Peak No. 5, B: peak No. 7, C: peak No. 9.

obtained are shown in Fig. 13B (Jinno *et al.*, 1986a; Hoshino *et al.*, 1984). Obviously the peak at 1.05 min should be assigned to fluorene. A similar procedure has been used to the assignment of peak No. 15. The UV spectra at 1.47 min and 1.56 min, as shown in Fig. 14A, clearly indicate that at least two different components elute at the similar retention time. The spectrum at 1.56 min can be assigned as phenanthrene, because agreement between the filed and measured UV spectra was confirmed, as demonstrated in Fig. 15. The peak deconvolution can separate peak No. 15 into two components, as shown in Fig. 14B. However, identification of other peaks, Nos. 12, 14, 16, 17 and 18, is very difficult because of the complexity of the chromatogram and non-reliability of the obtained UV spectra.

To understand the complexity of the chromatogram in this region more clearly, the contour plot shown in Fig. 16 was called from the data filed in the microcomputer. From Fig. 16 about fifteen components can be found in this region, whose retention times in minutes are 0.63, 0.70, 0.93, 0.96, 1.04

174 K. Jinno

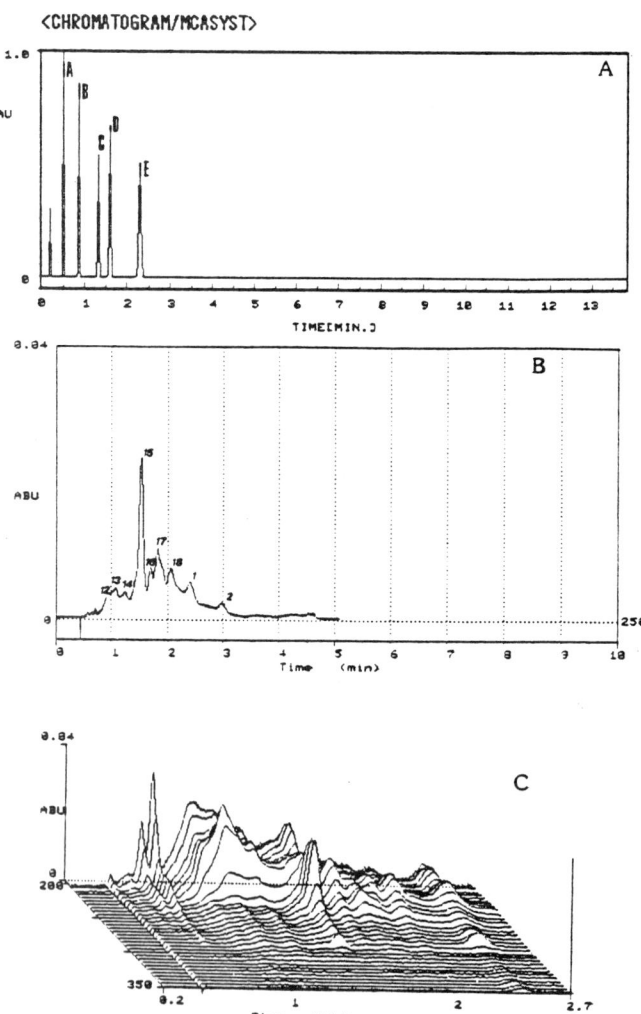

Figure 12. Predicted and measured chromatograms of PAHs with supercritical carbon dioxide as the mobile phase. A: Predicted chromatogram: mobile phase; carbon dioxide, pressure; 100 atm, temperature; 40°C; peak assignments are the same as in Fig. 5A. B: Measured chromatogram. C: Three-dimensional measured chromatogram. (Reproduced by permission from the American Chemical Society, 1986.)

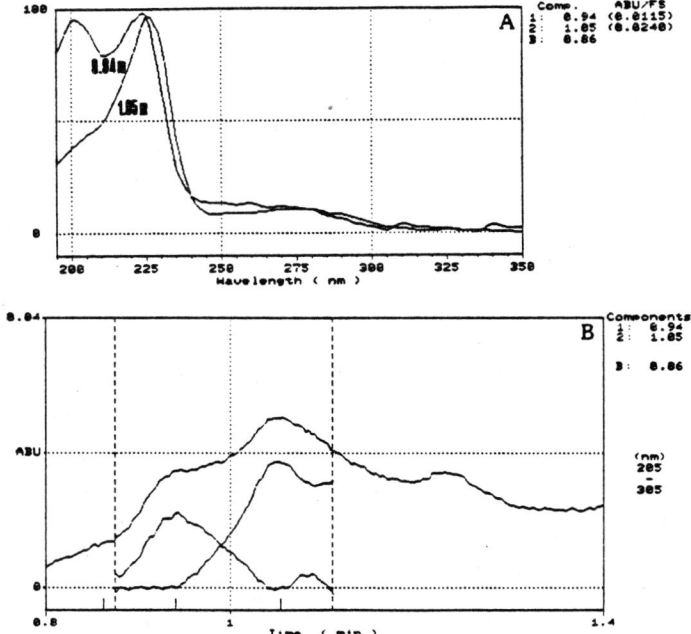

Figure 13. Peak deconvolution procedure for peak No. 13 in the chromatogram of Figure. 12B. A: UV spectra; B: superimposed chromatogram.

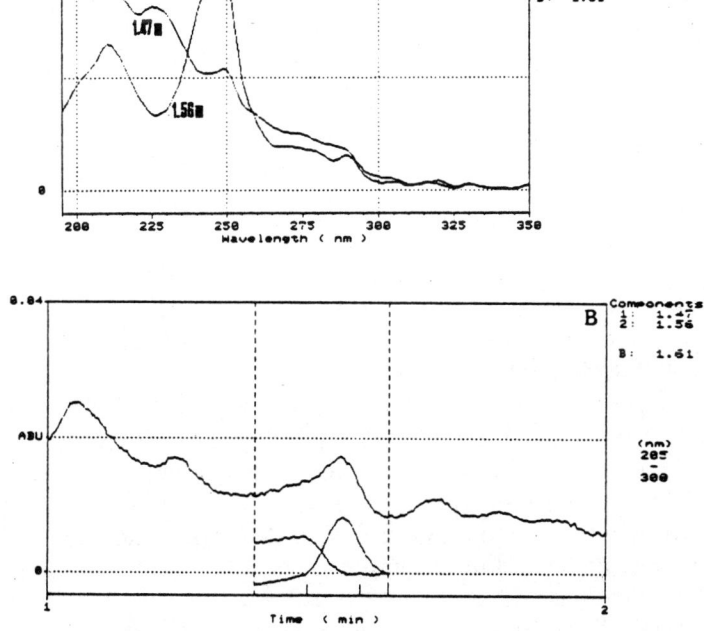

Figure 14. Peak deconvolution procedure for peak No. 15 in the chromatogram of Figure 12B. A: UV spectra; B: superimposed chromatogram.

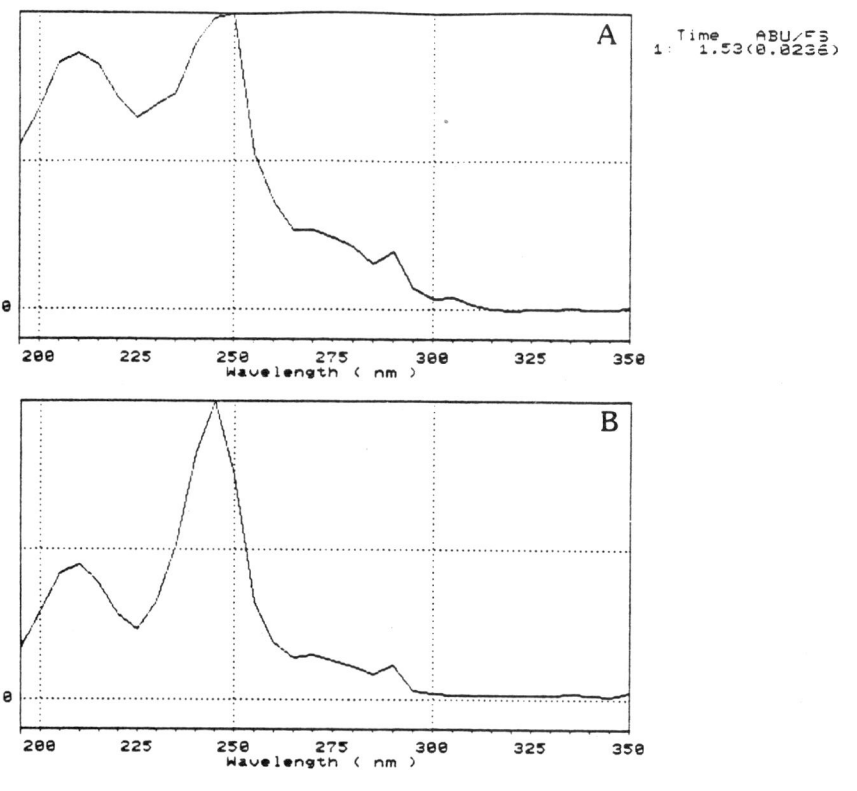

Figure 15. Comparison of UV spectra between measured and filed for a part of peak No. 15. A; measured, B; filed, C; computer output.

(fluorene), 1.07, 1.18, 1.22, 1.25, 1.34, 1.52 (phenanthrene), 1.69, 1.79, 1.82, and 1.92. The separation condition of higher column temperature and lower column pressure was then examined for this analysis, but no remarkable improvements in resolution in this small PAHs region were observed using this column. Therefore, to improve resolution in the analysis of small PAHs, one should use much longer and higher efficiency columns with more severe separation conditions, although the approach had not been attempted in this example.

As a conclusion, eleven priority pollutant PAHs and benzo[e]pyrene were

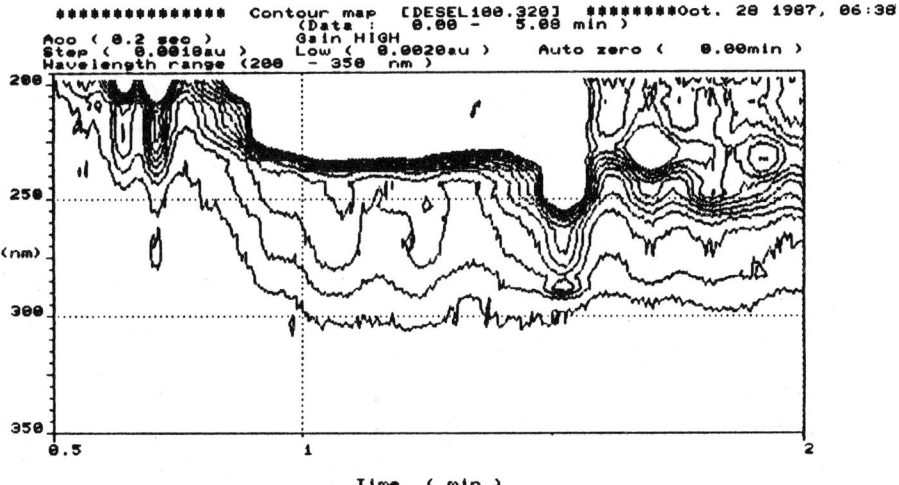

Figure 16. Coutour plot for the retention time range between 0.50 and 2.00 min.

identified in the extract sample using a very smart system in SFC. This example clearly indicates that the procedure using this system has very high potential for PAH analysis in practical samples.

CONCLUSION

The potential of a computer-assisted SFC separation system has been shown for the identification of PAHs in a practical environmental sample. This concept will be suitable for application to other compound groups by investigating basic QSRR studies.

REFERENCES

Berridge, J. C. (1986). *Techniques for the Automated Optimization of HPLC Separations*. John Wiley & Sons, Chichester, UK.

Bondi, A. (1964). Van der Waals volumes and radii. *J. Phys. Chem.* 68, 441–451.

Clar, E. (1964). *Polycyclic Hydrocarbons*. Academic Press, London, UK.

French, S. B. and Novotny, M. (1986). Xenon, a unique mobile phase for supercritical fluid chromatography. *Anal. Chem.* 58, 164–166.

Fujimoto, C., Hirata, Y., and Jinno, K. (1985). Supercritical fluid chromatography-infrared spectroscopy of oligomers: use of buffer-memory technique. *J. Chromatogr.* 332, 47–56.

Gere, D. R., Board, R., and McManigill, D. (1982). Supercritical fluid chromatography with small particle diameter packed columns. *Anal. Chem.* 54, 736–740.

Harvey, R. G. (Ed.) (1985). *Polycyclic Hydrocarbons and Carcinogenesis*. ACS Symposium Series No. 283. American Chemical Society, Washington, DC, USA.

Hirata, Y. and Nakata, F. (1984). Supercritical fluid chromatography with fused-silica packed columns. *J. Chromatogr.* 295, 315–322.

Hoshino, T., Senda, M., Hondo, T., Saito, M., and Tohei, S. (1984). Application of photo-diode array ultraviolet detector to unresolved peak analysis. *J. Chromatogr.* 316, 473–486.

Jinno, K. and Kawasaki, K. (1984a). Automated optimization of reversed-phase liquid chromatographic separations using a computer-assisted retention prediction. *J. Chromatogr.* 298, 326–335.

Jinno, K. and Kawasaki, K. (1984b). Computer-assisted retention prediction system for reversed-phase micro high-performance liquid chromatography. *J. Chromatogr. 316*, 1–23.

Jinno, K. and Kawasaki, K. (1986). Microcomputer-assisted retention prediction in reversed-phase liquid chromatography. In: *Chromatography and Separation Chemistry — Advances and Developments*, S. Ahuja (Ed.). ACS Symposium Series No. 297. American Chemical Society, Washington, DC, USA, pp. 167–187.

Jinno, K., Hoshino, T., Hondo, T., Saito, M., and Senda, M. (1986a). Computer enhanced spectrometric separation of coeluted components in supercritical fluid chromatography coupled with an ultraviolet multichannel detector. *Anal. Lett. 19*, 1001–1018.

Jinno, K., Saito, M., Hondo, T., and Senda, M. (1986b). Correlation between retention data of polycyclic aromatic hydrocarbons and several descriptors in supercritical fluid chromatography. *Chromatographia, 21*, 219–222.

Jinno, K., Hoshino, T., Hondo, T., Saito, M., and Senda, M. (1986c). Identification of polycyclic aromatic hydrocarbons in extracts of diesel particulate matter by supercritical fluid chromatography coupled with an ultraviolet multichannel detector. *Anal. Chem. 58*, 2696–2699.

Jinno, K., Nagoshi, T., Tanaka, N., Okamoto, M., Fetzer, J. C., and Biggs, W. R. (1987a). Elution behavior of peropyrene-type polycyclic aromatic hydrocarbons on various chemically bonded stationary phases in reversed-phase liquid chromatography. *J. Chromatogr. 386*, 123–132.

Jinno, K., Nagoshi, T., Tanaka, N., Okamoto, M., Fetzer, J. C., and Biggs, W. R. (1987b). Elution behavior of planar and nonplanar polycyclic aromatic hydrocarbons on various chemically bonded stationary phases in liquid chromatography. *J. Chromatogr. 393*, 75–81.

Kier, L. B. and Hall, L. H. (1976). *Molecular Connectivity in Chemistry and Drug Research*. Academic Press, New York, USA.

Klesper, E., Corwin, A. H., and Turner, D. A. (1962). High pressure gas chromatography above critical temperatures, *J. Org. Chem. 27*, 700–701.

Lamparczyk, H. and Radecki, A. (1983). Lack of evidence for dispersive interaction between polycyclic aromatic hydrocarbons and stationary phases in gas-liquid chromatography, *J. High Res. Chromatogr. Chromatogr. Commun. 6*, 390–391.

Miller, K. J. and Savchik, J. A. (1979). A new empirical method to calculate average molecular polarizabilities. *J. Am. Chem. Soc. 101*, 7206–7213.

Novotny, M., Springston, S. R., Peaden, P. A., Fjeldsted, J. C., and Lee, M. L. (1981). Capillary supercritical fluid chromatography. *Anal. Chem. 53*, 407A–414A.

Olesik, S. V., French, S. B., and Novotny, M. (1984). Development of capillary supercritical fluid chromatography/fourier transform infrared spectrometry. *Chromatographia 18*, 489–495.

Schabron, J. F., Hurtubise, R. J., and Silver, H. F. (1977). Separation of hydroaromatics and polycyclic aromatic hydrocarbons and determination of tetralin and naphthalene in coal-derived solvents. *Anal. Chem. 49*, 2253–2260.

Shafter, K. H. and Griffiths, P. R. (1983). On-line supercritical fluid chromatography/fourier transform infrared spectrometry. *Anal. Chem. 55*, 1939–1942.

Sie, S. T. and Rijnders, G. W. A. (1967a). High-pressure gas chromatography and chromatography with supercritical fluids. III: Fluid–liquid chromatography. *Sep. Sci. 2*, 729–753.

Sie, S. T. and Rijnders, G. W. A. (1967b). High-pressure gas chromatography and chromatography with supercritical fluids. IV: Fluid–solid chromatography. *Sep. Sci. 2*, 755–777.

Smith, R. D., Wright, B. W., and Udseth, H. R. (1986a). Advances in capillary supercritical fluid GC/MS. In: *Advances in Capillary Chromatography*, J. G. Nikelly (Ed.). Huethig, Heidelberg, FRG, pp. 56–94.

Smith, R. D., Wright, B. W., and Udseth, H. R. (1986b). Capillary supercritical fluid chromatography and supercritical fluid chromatography-mass spectrometry. In: *Chromatography and Separation Chemistry — Advances and Developments*. S. Ahuja (Ed.). ACS Symposium Series No. 297, American Chemical Society, Washington, DC, USA, pp. 260–293.

Wise, S. A., Bonnett, N. J., Guenther, F. R., and May, W. E. (1981). A relationship between reversed-phase C18 liquid chromatographic retention and the shape of polycyclic aromatic hydrocarbons. *J. Chromatogr. Sci. 19*, 457–465.

MICRO-HPLC

Progress in HPLC, Vol. 4, pp. 181—209.
Yoshioka *et al.* (Eds)
© 1989 VSP.

Fluorescent-HPLC for adenine nucleosides and nucleotides in life science

MASANORI YOSHIOKA, KAORU YAMADA, MEDHAT M.
ABU-ZEID, HIROYUKI FUJIMORI, AKIKO FUKE[1], KIKUO HIRAI[1],
ATSUO GOTO[2], MASAO ISHII[2], TSUNEAKI SUGIMOTO[2], and
HASAN PARVEZ[3]

Faculty of Pharmaceutical Sciences, Setsunan University, 45-1, Nagaotoge-
cho, Hirakata, Osaka 573-01;   [1] Faculty of Pharmaceutical Sciences,
University of Tokyo, 7-3-1, Hongo, Bunkyo-ku, Tokyo 113;   [2] Faculty of
Medicine, University of Tokyo, 7-3-1, Hongo, Bunkyo-ku, Tokyo 113,
Japan;   [3] Neuropharmacology Unit, University of Paris XI, CNRS, Bat
440, 91405 Orsay Cedex, France

INTRODUCTION

All organisms contain a variety of adenine-containing compounds. ATP and
ADP are utilized as high-energy compounds, precursors of RNA and others.
Cyclic AMP (cAMP), generated from ATP by adenylate cyclase, is a second
messenger of hormones and nerve transmitters. Adenosine is involved in
vasodilatation, platelet aggregation and immune responses. These adenine
nucleosides and nucleotides are enzymatically transformed and degraded.
Another form of adenine compounds are also found in DNA (as deoxy-
adenosine compounds), RNA, NAD, NADP, FAD, S-adenosylmethionine
and others. It is worth while to mention that NAD is not only a coenzyme in
the oxidation and reduction but also a donor of poly(ADP-ribose) which has
recently been characterized as a regulator of the nucleus. There are four
nucleic acid bases, adenine and guanine as purine, cytosine and uracil or
thymidine as pyrimidine. Most of the base derivatives are ionic, nonvolatile
or thermally labile and suitable for liquid chromatography (LC).

LC, and recently high-performance liquid chromatography (HPLC), have
been extensively used for the separation and detection of bases, nucleosides
and nucleotides (Brown, 1973; Scoble and Brown, 1983). The history of
these methods, not only for adenines but also the other base compounds,
will be described first.

The adenine compounds are much higher in concentration than the normal purine and pyrimidine ones in cells and body fluids. Our method, using

a specific fluorescent reaction for adenine compounds and separation by HPLC, will be described in detail.

HISTORY OF LC AND HPLC FOR NUCLEIC ACID COMPOUNDS

Analyses of nucleic acid bases, nucleosides and nucleotides are historically summarized in Table 1. Cohn (1949) developed a porous polystyrene—divinylbenzene ion exchange resin (nominal diameter of 100 μm), although particle sizes were distributed, for analysis of nucleic acids, nucleosides and nucleotides. The ion exchange resins gained popularity and were a great advance in nucleic acid research. Anderson et al. (1963) built up a flow-through cell which could successively monitor UV absorbance of an eluate. For HPLC as a new technology, many efforts have been made regarding preparation of resins and improvement of detectors and fittings. Anion- or cation-exchanger was coated on glass beads (50 μm), so called pellicular, made by Horvath and co-workers (1967, 1969). These resins were usually packed into long narrow tubes, e.g. 1 m × 1 mm or 3 m × 1 mm, described by Brown (1973). These kinds of anion exchange resins were prepared by many other researchers and shown to be suitable for the separation of nucleotides. Cation exchange resins were suitable for the separation of nucleic acid bases and nucleosides. The pellicular resins, however, had two disadvantages. The exchange capacity and mechanical and chemical resistance were relatively low. Kirkland (1969) investigated in detail the parameters for separation of materials in LC, and pointed out that surface porosity of resin, properties of supporters in resin, liquid film thickness, diameter of column and reproducibility of column preparation were critical. In 1970 he introduced a controlled-surface-porosity ion exchange resin for HPLC, and with DeStefano (1970) developed a chemically bonded porous one that was durable and possessed high exchange capacity.

Microparticulate ion exchange resins (5–10 μm), which were larger in capacity, and higher in stability and quality, were prepared by Hartwick and Brown (1975), and were packed in tubes of regular size. For ion exchange elution, a gradient concentration of salt or proton was generally used for the separation of a variety of polyanions in order to shorten analysis times.

Another principle of separation based on hydrophobic interaction was introduced by Anderson and Murphy (1976) with reversed-phase columns for the bases and nucleosides. For this method Hoffman and Liao (1977) added ion pairing of tetra-n-butylammonium hydrogen sulfate (TBHS), which was able to transfer nucleotides in hydrophobic phase. A mixture of 12 nucleotides and cAMP could be separated through a column of reversed phase by a gradient elution from a phosphate buffer of pH 3.9 to 3.4 with 0.025 M TBHS and 30% methanol for 40 min. A more promising principle is affinity. Thymidine-linked and phenylboronate columns were developed on the basis of the properties of base pairing with adenines and cis-diol boronate complex formation with ribonucleotides, respectively.

Detection of nucleic acid compounds is not so difficult because nucleic acid bases have large molar extinction coefficients. As shown in Table 2,

Table 1.
Analyses of nucleic acid compounds

Compound	Column (particle diameter: μm)	Separation	Mode	Reference
Bases or mononucleotides	Totally porous polystyrene–divinylbenzene copolymer, anion or cation (100 μm)	Ion exchange	Isocratic	Cohn (1949) Uziel et al. (1968)
Mono-, di-, tri-nucleotides	Pellicular anion exchange — glass bead (50 μm)	Ion exchange	Gradient	Horvath et al. (1967)
Bases or mononucleotides	Controlled surface porosity-glass bead, anion or cation (20–37 μm)	Ion exchange	Isocratic	Kirkland (1970)
Mono-, di-, tri-nucleotides	Chemically bonded porous ion, anion (30–40 μm)	Ion exchange	Gradient	Henry et al. (1973)
Nucleosides	Anion exchange (17.5 ± 2μm)	Exclusion	Isocratic	Singhal (1972) Singhal and Cohn (1973)
Mono-, di-, tri-nucleotides of purines and pyrimidines	Microparticulate ion, anion	Ion exchange	Gradient	Hartwick and Brown (1975) Breter et al. (1977) Eksteen et al. (1978) Cohen et al. (1980)
Bases or nucleosides	Thymidine-linked porous spherical resin (12–15 μm)	Affinity	Isocratic	Kato et al. (1977)
ATP, ADP, AMP, cAMP adenosine and adenine	ODS-reversed phase silica gel (40 μm)	Reversed phase	Gradient	Anderson and Murphy (1976) Rustum (1978) Brown and Grushka (1980) Ramos and Schoffstall (1983)
Mono-, di-, tri-nucleotides of purines and pyrimidines	ODS-reversed phase silica (10 μm)	Ion-pairing	Gradient	Hoffman and Liao (1977) Ingebretsen and Bakken (1982) Caronia et al. (1983)

Table 1. (*Cont'd*)

Compound	Column (particle diameter: μm)	Separation	Mode	Reference
o-Methyribose-nucleosides	Reversed-phase boronate	Boronate complex exchange	Batchwise	Singhal *et al.* (1980)
Deoxyribonucleotides	Anion and cation/reversed phase	Ion-pairing	Isocratic	Crowher and Hartwick (1982)

Table 2.
Detection of nucleic acid compounds

Compound	Detection	Detectable	Reference
Bases, nucleosides and nucleotides	UV at 260 and 280 nm	0.2 mM	Anderson *et al.* (1963)
Bases and nucleosides	UV at 254 nm	0.1 nmol	Horvath and Lipsky (1969)
Adenine and inosine compounds	UV at 254 nm	10 ng-level	Anderson and Murphy (1976)
Adenine, adenosine, cAMP and AMP	Fluorescence detector	1 pmol	Yoshioka and Tamura (1976)
Bases, nucleosides or mononucleotides	RI degradation and UV at 254 nm	25 nCi of ^{3}H and 2.5 nCi of ^{14}C	Bakay *et al.* (1978)
Twelve ribonucleotides	Inductively coupled plasma atomic emission spectrometry and UV at 260 nm	0.4 μg of Pi/ml	Yoshida *et al.* (1983)
Bases, nucleosides and mononucleotides	Conductivity detector and UV	10 nmol	Naikwadi *et al.* (1983)
Adenosine, adenine, deoxyadenosine, inosine hypoxanthin and guanosine	Electrochemical detector	Adenosine 0.05 pmol and deoxyadenosine 0.5 pmol	Henderson and Griffin (1984)

various kinds of flow-through cells in UV detectors were elaborated to enhance sensitivity. With a UV spectrophotometer, over 150 peaks from human urine were found by Scott *et al.* (1967). Biological samples such as brain (Shmukler, 1972), erythrocyte (Dean and Perrett, 1976; Schweinsberg and Loo, 1980), serum (Hartwick *et al.*, 1979; Seta *et al.*, 1981) and fibroblast (Bakay *et al.*, 1978) also contain a variety of UV-absorbing materials. Assignment of peaks should be carefully done by various means such as retention times, absorbance ratios at different wavelengths of eluted compounds, enzymic peak shifts and co-chromatography of a biological sample with known standard compounds. To avoid these inconveniences, a scanning UV detector drawing a three-dimensional chromatogram, and a multiwavelength UV detector monitoring absorbances at four different wavelengths simultaneously, were utilized by Denton *et al.* (1976) and Catterick (1983), respectively. To enhance selectivity, a heterodetection system of UV and conductivity which is insensitive to nucleosides and bases was employed for the determination of mixtures of nucleic acid bases, nucleosides and nucleotides. An inductively coupled plasma atomic emission spectrometer was introduced to detect phosphorus and was slightly less sensitive than the UV detectors (Yoshida *et al.*, 1983). Nucleic acid–bases are electrochemically oxidizable. Electrochemical detection was utilized to enhance sensitivity and selectivity. However, this detection will be interfered with by many oxidizable ingredients in biological samples.

Fluorimetry is relevant to improve the sensitivity. Kochetkov *et al.* (1971) found that chloroacetaldehyde reacted with adenine and cytosine bases to produce 1-N^6-etheno derivatives. Their fluorescence properties were characterized by Barrio *et al.* (1972) and Secrist *et al.* (1972). The fluorescent reaction was applied to determine 10^{-8} to 10^{-11} mole order of adenine nucleotides obtained after separation by LC (Avigad and Damle, 1972).

Yoshioka and Tamura (1976) established a method of fluorescent determination of adenine compounds by HPLC using chloroacetaldehyde. Kuttesch *et al.* (1978) reported a clinical application of HPLC and found an abnormal excretion of deoxyadenosine in patients with immunodeficiency diseases. The aldehyde was also used for HPLC determinations of adenosylmethionine, adenosylhomocysteine, methylthioadenosine, poly(ADP-ribose), adenine and cytosine in tRNA, and adenine compounds in marine phytoplankton by Shugart (1979), Sims *et al.* (1980), Juarez-Salinas *et al.* (1983), Krzyosiak *et al.* (1981) and Preston (1983), respectively.

Further, Yoshioka *et al.* (1984) improved the fluorescent reaction using bromoacetaldehyde which was higher in reactivity with adenine compounds, as shown in Fig. 1. The method led to substantial findings such as adenosine in synaptosomes from guinea pig cerebral cortex (Kobayashi *et al.*, 1981) and adenine nucleotide release with catecholamines from adrenal chromaffin cells (Kuroda *et al.*, 1980).

Fluorescent reaction-HPLC is highly specific and selective. As derivatizations of adenine compounds with bromoacetaldehyde are performed by heating before HPLC (pre-column reaction), there is a possibility of degra-

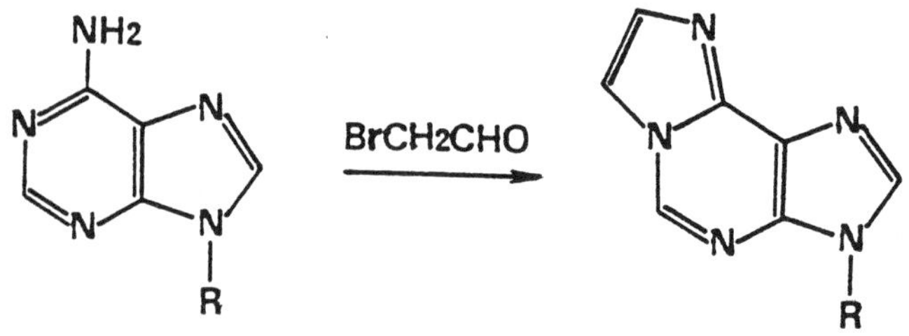

Figure 1. Fluorescent reaction of adenines.

dation of the compounds if conditions are not suitable. Taking this disadvantage and biochemical application into consideration, the analyzer for the adenine compounds was designed by way of post-column reaction with a single pump (Yoshioka *et al.*, 1985).

Judging from the overview of nucleic acid determination by HPLC, our method is the most specific and selective for adenine compounds. We describe the method for biological applications such as measurements of ATP in blood and culture cells, and enzymic activities of Na,K-ATPase, alkaline phosphatase and adenylate cyclase as shown in Fig. 2.

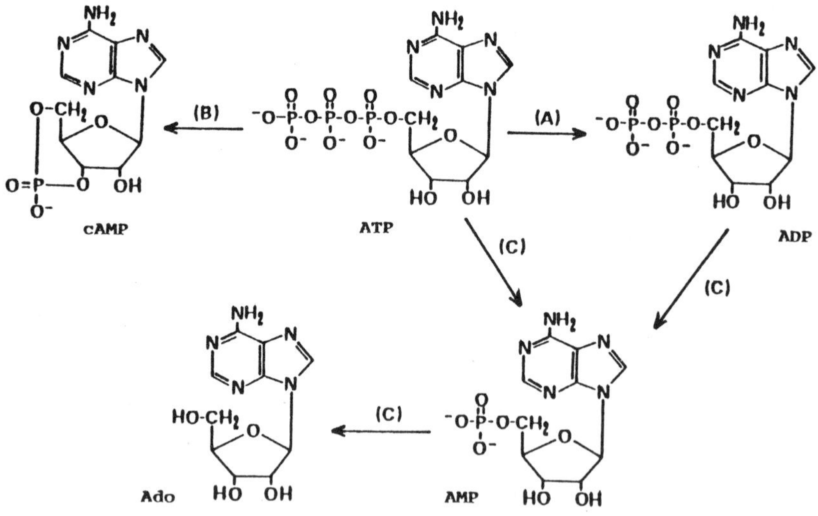

Figure 2. Enzymes involved in metabolism of ATP in this experiment. A: Na,K-ATPase, B: adenylate cyclase, C: phosphatase. Adenosine is abbreviated Ado in Figures and Table.

METHODS

Preparation of bromoacetaldehyde

Bromoacetaldehyde was prepared and crystallized according to the method of Schukovskaya *et al.* (1962). To make 1.9 M bromoacetaldehyde, 234 mg of the crystal ($3BrCH_2CHO \cdot H_2O$) was dissolved in 1 ml of water in a test tube by heating at 85°C for 15 min. The tube was glass-stoppered and storable at 4°C for more than 4 months to date.

Quantitative fluorescent reaction of adenine compounds with bromoacetaldehyde

Mixtures of adenosine, AMP, cAMP, ADP and ATP in 0.1 M phosphate buffer (pH 7.0) or 1 M acetate buffer (pH 5.0) at 0.5–50 μM each were routinely used as standard solutions and frozen at −80°C until use. To 100 μl of the solution thawed at 40°C was added 10 μl (for pH 7) or 5 μl (for pH 5) of 1.9 M bromoacetaldehyde in a Reacti-Vial (Pierce, Rockford, IL, USA). The vial was heated at 80°C for 15 min and kept at 4°C until HPLC analysis.

Packing of anion exchange resins

The packing is vital to enhance column efficiency. In our laboratory, ion exchange resins were routinely packed into columns as follows.

Fittings of a stainless tube (12.5 cm × 4.6 mm) for regular size or (25 cm × 1.5 mm) for semi-microbore, should be detached, sonicated in water, and rinsed with water and methanol.

The new resin of Hitachi gel No. 3013-N (diameter; 5 μm) or No. 3012-N (diameter; 7 μm) for HPLC are spherical porous copolymers of styrene-divinylbenzene chemically bonded $-\overset{\oplus}{N}R_3$ as functional group (Hitachi, Tokyo, Japan). Both resins were suspended in methanol. Fine powders from resins were removed by decantation. Each resin was further washed with 0.1 M HCl, water, 0.1 M NaOH and water successively, until the pH of the supernatant was around 7. The suspension was added to 50% methanol to make 30% slurry. To remove fine particles, again, the slurry was stirred, sonicated for 5 min, and centrifuged at $200g$ for 3 min. The turbid supernatant was decanted repeatedly. The resin was suspended in 50% methanol to make 50% slurry.

Four volumes of 50% slurry against the tube volume were poured into a packer (2 cm × 11 cm) connected with the tube. The packer was filled with 50% methanol. The resin sedimented by gravity for 30 min. A packing solvent was pumped into the tube through the packer at the constant pressure of 80–180 kg/cm^2 for 2 h.

HPLC

The packed column was maintained at 45°C. The mobile phase consisted of 0.025 M citric acid–0.4 M sodium chloride–0.05 M disodium hydrogen phosphate buffer (pH 5.0) and methanol. The mixing ratio was varied and finally selected as 1:1 (v/v). The flow rate was set at 0.3 ml/min with a Twinkle pump (Jasco, Tokyo, Japan) or Familic 300 pump (Jasco) connected to an Autosampler AS-L350 (Jasco). The eluate was monitored by a fluorescence spectrophotometer FP-110 (Jasco). In the spectrophotometer, light of 253.7 nm from a low-pressure mercury lamp was dispersed by an excitation concave grating and focused on a 3-μl flow-through cell. The fluorescence at 400 nm was collected by an emission concave grating.

Assay of Na,K-ATPase inhibition with ouabain

The reaction mixture contained 200 mM TES (N-Tris(hydroxymethyl)methyl-2-aminoethanesulfonic acid)-Tris (Tris(hydroxymethyl)aminomethane) (pH 7.4), 500 mM sodium chloride, 25 mM potassium chloride, 25 mM magnesium sulfate, 5 mM EGTA (ethylene glycol bis(β-aminoethylether)-N,N,N', N'-tetraacetic acid) and 200 mg/l of Na,K-ATPase from dog kidney (Sigma, USA). To 20 μl of the solution was added 40 μl of a sample and the mixture was preincubated at 37°C for 20 min. The enzyme reaction was initiated by adding 40 μl of 5 mM ATP at 37°C. After incubation for 20 min the reaction was stopped by boiling the incubate for 10 min. To 25 μl of the mixture was added 25 μl of 0.2 M phosphate buffer (pH 7.0) and 5 μl of 1.9 M bromoacetaldehyde. The fluorescence reaction was carried out as described above.

The solution was usually diluted thirty- to forty-fold with 0.1 M phosphate buffer (pH 7.0). Ten microliters of the diluted one was injected into the HPLC system.

Forty microliters of ouabain solutions at various concentrations was added to 20 μl of the reaction mixture. The percentage of inhibitory activity was calculated from the following equation. Inhibitory (%) = $(X_0 - X_i)/(X_0 - X_{100}) \times 100$, where X is the peak-height ratio of ADP/(ADP + ATP) in the HPLC chromatogram, X_{100} is X at 10^{-3} M ouabain, which inhibited the ATPase completely, X_0 is X without ouabain and X_i is X at given concentrations of ouabain or other effectors.

Assay of alkaline phosphatase activity against ATP

To 50 μl of 40 μM ATP in 0.1 M sodium hydrogen carbonate–sodium carbonate buffer (pH 10.4) was added 10 μl of alkaline phosphatase from bovine intestinal mucosa (Sigma) and the mixture was incubated at 30°C for 30 min. As the control, 10 μl of the buffer in place of the enzyme solution was added into 50 μl of the substrate solution. The reaction was stopped by adding 30 μl of 4 M perchloric acid chilled on ice. After removal of protein by centrifugation and perchloric acid with potassium hydroxide, to 25 μl of the supernatant was added 25 μl of 0.2 M phosphate buffer (pH 7.0) and 5 μl of 1.9 M bromoacetaldehyde. The fluorescence reaction was performed as described above.

Assay of adenylate cyclase activity

Fat cells were isolated from epididymal adipose tissue of Wistar rats (150 g) by the method of Rodbell (1964). Up to 1 g of the tissue was cut into small pieces and incubated in 3 ml of minimum essential Eagle medium (pH 7.4) containing 10 mg of collagenase and 4% of bovine serum albumin at 37°C for 1 h. The liberated fat cells were floated to the surface by centrifugation at 400*g* for 1 min. The cells were pipetted in a siliconized test tube. The remaining suspension was centrifuged again. The cells was pipetted in the tube. This procedure was repeated three times. Fat cell ghosts were prepared according to the method of Birnbaumer *et al.* (1969). The cells were pooled and washed with a lysing medium, which consisted of 2.5 mM ATP, 2.5 mM magnesium chloride, 0.1 mM calcium chloride and 1 mM potassium bicarbonate (pH 7.4). After lysis of the cells by 20 inversions for 1 min, the suspension was centrifuged at 200*g* for 1 min. The lysates in the turbid infranatant were collected. The remaining fat cells were suspended in the lysing medium again, and the procedure was repeated four times. The pooled lysates were centrifuged for 15 min at 900*g*. The pellet was resuspended in 1 mM potassium bicarbonate and the suspension was centrifuged at 900*g* for 15 min. The pellet, mainly consisting of fat cell ghosts was suspended in 1 mM potassium bicarbonate and was kept at 4°C until use.

The ghosts were suspended in the assay medium, consisting of 3.0 mM ATP, 3.6 mM magnesium chloride, 7.3 mM theophylline, 0.36 mM glycylglycine, 0.18 mM disodium sulfate, and 0.1% bovine serum albumin in 30 mM phosphate buffer (pH 7.6) at 37°C. The enzyme reaction was initiated by the addition of 10 µl of L-epinephrine or sodium fluoride solution. After incubation for 20 min the reaction was stopped by boiling the incubate for 30 s. The solution was applied to an alumina column (5 × 15 mm) equilibrated with water. The first 50 µl was discarded. The column was eluted with 150 µl of water. To 150 µl of the eluate was added 50 µl of 1 M acetate buffer (pH 5.0) and 20 µl of 1.9 M bromoacetaldehyde. The fluorescence reaction was carried out as described above. Five microliters of the solution was injected into HPLC system. The other HPLC was performed at 40°C with a column (50 cm × 2 mm) of Hitachi gel No. 3010 (diameter; 25 µm). The mobile phase consisted of 0.1 M phosphate buffer (pH 7.0) and methanol (7:3, v/v). The flow rate was 0.1 ml/min.

Analysis of adenine compounds in blood

One milliliter of the blood drawn from a human brachial vein was added into EDTA, which was tentatively used as an anticoagulant. The blood was centrifuged to remove cells at 1000*g* for 5 min. To 100 µl of the plasma was added 25 µl of 4 M perchloric acid. The mixture was centrifuged at 6000*g* for 30 min. After removal of the acid with potassium hydroxide, to 25 µl of the supernatant was added 25 µl of 0.2 M phosphate buffer (pH 7.0) and 5 µl of 1.9 M bromoacetaldehyde. The fluorescence reaction was carried out as described above.

Adenine compounds in human whole blood were determined as follows. One milliliter of the blood from a human brachial vein was added to 250 µl of 4 M perchloric acid chillded on ice and homogenized in a glass homogenizer. After removal of proteins by centrifugation at 6000*g* for 20 min, and the acid with potassium hydroxide, the adenine compounds in the supernatant were derivatized as described above.

Analysis of adenine compounds in neuroblastoma cells

Mouse neuroblastoma N1E 115 cells were kindly contributed by Dr Takeo Deguchi, Department of Medical Chemistry, Tokyo Metropolitan Institute for Neurosciences, and grown in Dulbecco's modified Eagle's minimal essential medium, supplemented with 8% fetal calf serum, in a humidified atmosphere of 10% CO_2 and 90% air at 37°C. Twelve milliliters of approximately 3×10^6 cells/ml was collected in phosphate-buffered saline (0.14 M sodium chloride, 3 mM potassium chloride and 0.01 M phosphate buffer, pH 7.4) and centrifuged at 200g for 4 min. Two procedures were performed for the determination of adenine compounds. In one procedure, to 400 μl of the suspension was added 100 μl of 4 M perchloric acid chilled on ice. The cells were homogenized in a glass homogenizer and centrifuged at 6000g for 20 min. After removal of potassium perchlorate by centrifugation, to 25 μl of the supernatant was added 25 μl of 0.2 M phosphate buffer (pH 7) and 5 μl of 1.9 M bromoacetaldehyde. The fluorescence reaction was carried out as described above.

In the second procedure, 20 μl of the aldehyde was added to 100 μl of the above-mentioned cell suspension since the aldehyde was hydrophobic and was expected to penetrate into the cells, and the mixture was heated at 85°C for 20 min. After removal of proteins by centrifugation at 300g for 20 min, 10 μl of the supernatant was injected into the HPLC system.

RESULTS

The adenine compounds were better separated by isocratic elution through No. 3013-N than No. 3012-N under the same conditions, previously reported by Yoshioka *et al.* (1984). The chromatogram is shown in Fig. 3.

Horvath and Lipsky (1969) and Brown (1973) recommended examining several factors for optimizing the separation of compounds in HPLC. The chromatogram of adenine compounds could be affected by properties of resin (size of particle, structure of polymer, etc.), chemical compositions, concentrations and pH values of mobile phases, column temperatures, sizes (length and diameter), flow rates, elution modes, packing pressure of resin and so on.

We examined the factors on No. 3013-N. For comparison of the columns of No. 3013-N and No. 3012-N, both resins, were packed into the same tubes of regular size in the same manner and the adenine compounds were eluted under the same condition. Two parameters of analysis time and resolution were adopted using a standard adenine mixture after reaction with bromoacetaldehyde. As summarized in Table 3, analysis time for No. 3013-N was about half of that for No. 3012-N and the resolutions afforded by No. 3013-N were higher than those afforded by No. 3012-N. The theoretical plate numbers of adenosine (3200), AMP (3700), cAMP (5300), ADP

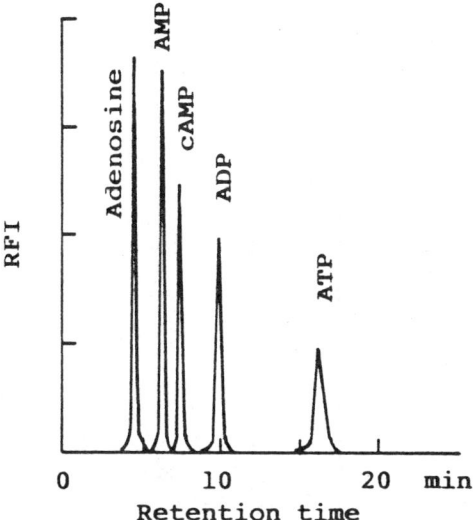

Figure 3. Chromatogram of authentic adenine compounds. Ten picomoles of each was injected. Relative fluorescence intensity (RFI) of the peak was drawn.

(3600) and ATP (3600) by No. 3013-N were larger than those of No. 3012-N. Thus the resin of No. 3013-N was recommended for the further experiments.

To examine effects of sodium chloride in the mobile phase on the capacity factors of AMP, cAMP, ADP and ATP, they were calculated using the retention time of adenosine eluted at void volume. The capacity factors of AMP and cAMP were gradually decreased as the final concentration of sodium chloride increased from 0.1 to 0.3 M. Those of polyanionic ADP and ATP were strongly affected by changes of the concentrations as shown in Fig. 4. The resolutions of adenosine and AMP, cAMP and ADP, and ADP and ATP became smaller, whereas that of AMP and cAMP was constant except in the case of 0.1 M sodium chloride, as shown in Fig. 5. In the experiment with 0.3 M sodium chloride, ATP and ADP were eluted at the same time. Judging from these results, 0.2 M sodium chloride was optimal.

As the ratio of the buffer and methanol in the mobile phase was changed from 7:3 to 4:6 (v/v), the resolutions of adenosine and AMP, cAMP and ADP, and ADP and ATP were increased, as shown in Fig. 6. That of AMP and cAMP was decreased by the low ratio but they were still completely separated at 60% methanol. The capacity factors of ADP and ATP were increased as a function of the methanol concentration, but those of AMP and cAMP were not affected. This suggested that the separation of adenine compounds on this resin could be affected not only by ion exchange but also by other mechanisms such as hydrophobic adsorption on the polystyrene-divinylbenzene.

The effect of flow rate on the determination of adenine compounds was examined. Their retention times became shorter as the flow rate was increased from 200 to 500 μl/min. The analysis times at the flow rate of 200,

Table 3.
Retention time and resolution of adenine compounds with the columns of No. 3013-N and No. 3012-N

Resin	Diameter (μm)	Retention time (min)					Resolution[a]			
		Ado	AMP	cAMP	ADP	ATP	Ado – AMP	AMP – cAMP	cAMP – ADP	ADP – ATP
No. 3013-N	5	5.3	6.9	7.9	10.2	16.1	2.33	1.20	2.75	6.10
No. 3012-N	7	5.3	8.4	10.1	15.1	31.4	1.68	1.03	1.32	2.53

[a] Resolution between the neighboring adenine compounds.

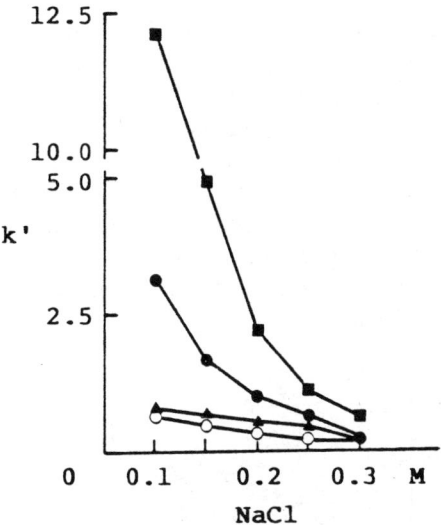

Figure 4. Effect of sodium chloride in ·mobile phase on capacity factor. Adenosine is considered to be eluted not retained. The capacity factor (k') is defined as $k' = t_R - t_0/t_0$, where t_0 is the retention time of adenosine and t_R is the retention time of the other compounds. Data points: $\bigcirc$ = ε-AMP; $\blacktriangle$ = ε-cAMP; $\bullet$ = ε-ADP; $\blacksquare$ = ε-ATP.

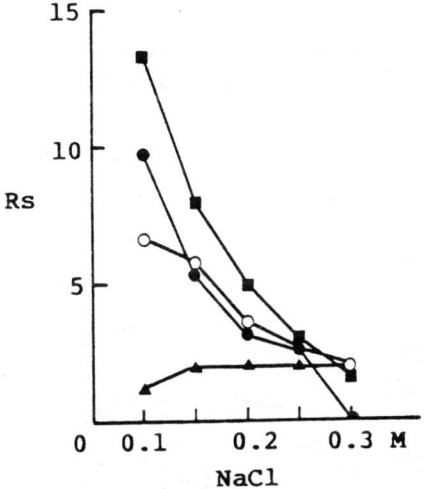

Figure 5. Effect of sodium chloride on resolution. Resolution between adenosine and AMP ($\bigcirc$), AMP and cAMP ($\blacktriangle$), cAMP and ADP ($\bullet$), and ADP and ATP ($\blacksquare$).

300, 400 and 500 µl/min were 25, 16, 13 and 8 min, respectively. No change was observed in terms of the capacity factors of each compound, and the resolutions between the neighbouring adenine compounds at the flow rates used.

The effects of temperature on the resolution of adenine compounds are shown in Fig. 7. The resolution of AMP and cAMP became smaller as a function of temperature. Resolutions of other combinations gradually increased. The capacity factors and retention times of adenine compounds were not changed except where those of ATP were increased.

The effect of the packing pressures of the resin on the determination of adenine compounds was examined. No significant change of retention times and capacity factors was observed with three different columns. The resolutions of adenine compounds by the column packed at 130 kg/cm^2 were evaluated to be the best among three, as shown in Fig. 8. The column of No. 3013-N at 130 kg/cm^2, as well as that of No. 3012-N at 80 kg/cm^2, was very stable after 1 year of use.

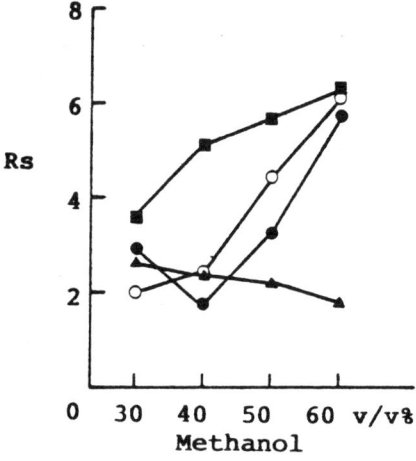

Figure 6. Effect of methanol in mobile phase on resolution. Resolution as in Fig. 5.

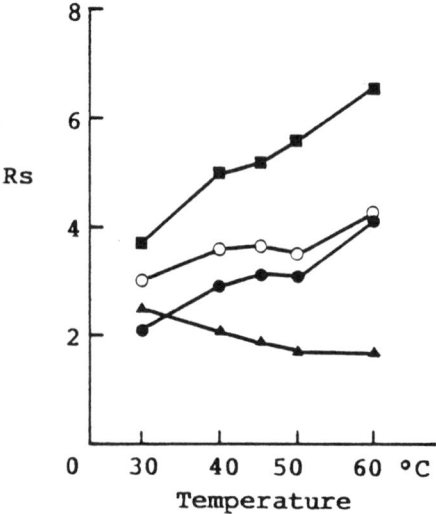

Figure 7. Effect of temperature of column on resolution. Resolution as in Fig. 5.

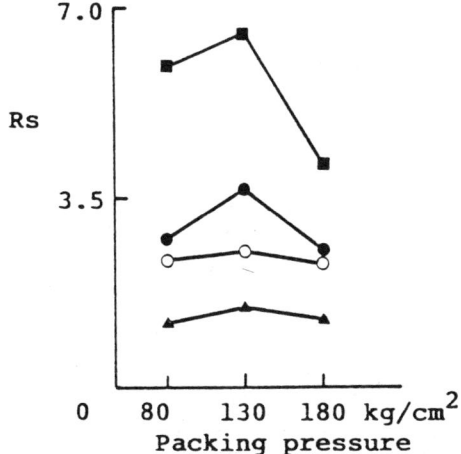

Figure 8. Effect of packing pressure on resolution. Resolution as in Fig. 5.

Recent investigations have demonstrated some advantages of the use of microbore columns which were pointed out in terms of higher column efficiency, higher speed of analysis, and lower volume of the mobile phase. We tried to use a semi-microbore column packed with No. 3013-N under the same conditions of the column temperature at 45°C, the chemical composition of the mobile phase as described in Methods and the flow rate of 70 μl/min. As shown in Fig. 9, the resolution of AMP and cAMP was very poor at any sodium chloride concentrations. The resolutions between the other compounds were improved as the concentrations of sodium chloride in the mobile phase were decreased from 0.2 to 0.1 M. The effect of methanol on the resolutions was examined at 0.15 M sodium chloride. As shown in Fig. 10, the resolution of AMP and cAMP was improved, but still not as good. The best condition for the semi-microbore HPLC in our system was 0.3 M sodium chloride−0.025 M citric acid−0.05 M disodium hydrogen phosphate (pH 5.0) and methanol (3:2, v/v) as the mobile phase, at a flow rate of 70 μl/min and at 45°C.

Under each optimal condition the results obtained from the semi-microbore were compared with those from the regular size, as shown in Table 4. Although the analysis time of adenine compounds with the semi-microbore was longer than that with the regular size, the resolution indices and theoretical plate numbers were worse. This fact indicated that the semi-microbore was not suitable for anion exchange.

Deoxyadenine compounds also exist in biological samples. The elution positions of deoxyadenine compounds should be checked for the application. The resolution of deoxy-AMP and deoxy-cAMP was very poor at the value of 0.51. For the systematic analysis of deoxyadenine compounds the best mobile phase consisted of 0.3 M sodium chloride−0.025 M citric acid−0.05 M disodium hydrogen phosphate buffer (pH 5.0) and methanol (3:2, v/v). The retention times of deoxy-adenosine, deoxy-AMP, deoxy-cAMP, deoxy-ADP and deoxy-ATP at a flow rate of 500 μl/min were 3.1, 4.1, 4.8, 7.0 and 14.2 min, respectively.

M. Yoshioka et al.

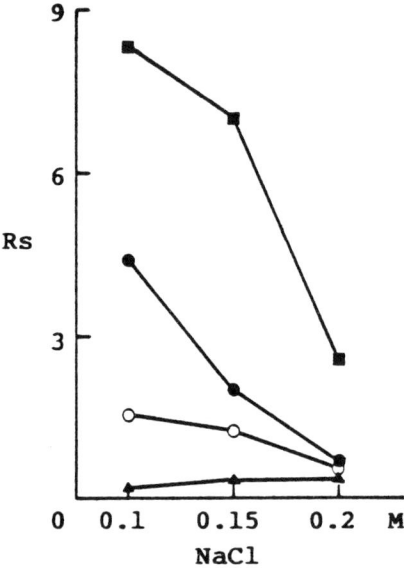

Figure 9. Effect of sodium chloride on resolution. Resolution as in Fig. 5.

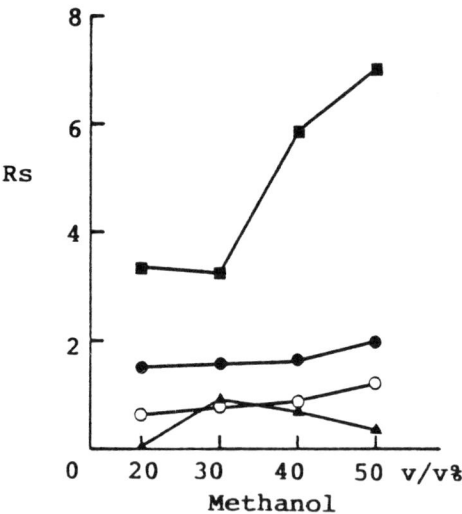

Figure 10. Effect of methanol on resolution. Resolution as in Fig. 5.

Table 4.
Comparison of resolution and retention time of adenine compounds by a semi-microbore and a regular HPLC column

Column	Flow rate (μl/min)	Resolution[a]				Analysis time (min)
		Ado − AMP −	cAMP −	ADP −	ATP	
Regular	300	2.5	1.5	2.6	6.5	17
Semi-microbore	70	0.9	0.7	1.6	5.9	30

[a] Resolution between the neighboring adenine compounds.

The retention times of adenosine and deoxy-adenosine, and deoxy-AMP, AMP and deoxy-cAMP, were nearly same, respectively, and the resolutions of deoxy-ADP and ADP, and deoxy-ATP and ATP, were very poor when using the mobile phase as described in Methods. Under the reduction of sodium chloride concentration from 0.2 to 0.15 M, the separations of deoxy-ADP from ADP, and deoxy-ATP from ATP were improved and the resolution indices became 1.0 and 1.0, respectively. However, it was impossible to separate deoxy-adenosine from adenosine, and deoxy-AMP and deoxy-cAMP from AMP simultaneously.

Brown *et al.* (1973) and Seta *et al.* (1981) showed with HPLC chromatograms by UV monitors that blood contained ATP, ADP and AMP but concentrations of adenine compounds in blood were obscure. The adenine compounds in human whole blood were determined as shown in Fig. 11. The concentrations of ATP and ADP were 700 and 100 µM, respectively. As shown in Fig. 12, ATP and ADP concentrations in plasma in the presence of 1 mg EDTA in 1 ml of blood were 3.2 and 2.4 µM, respectively. The time course of ATP levels in a human plasma in the presence of 25 mg EDTA in 1 ml of blood were examined by two procedures (Fig. 13). In one procedure, at 6-min intervals during 30 min the blood was centrifuged and then ATP in plasma was determined. The amount of ATP in the plasma decreased as a function of time and was not detectable after 30 min. The half-life of ATP in the plasma of the stored blood at 4°C was less than 4 min. In the second procedure, ATP in plasma, which was separated immediately after the blood was drawn, did not change significantly during 1 h. The latter procedure on the rapid separation of plasma from blood cells is better to obtain rather accurate values than the former, and determination as soon as possible after the separation will be recommended, to avoid spontaneous hydrolysis of ATP.

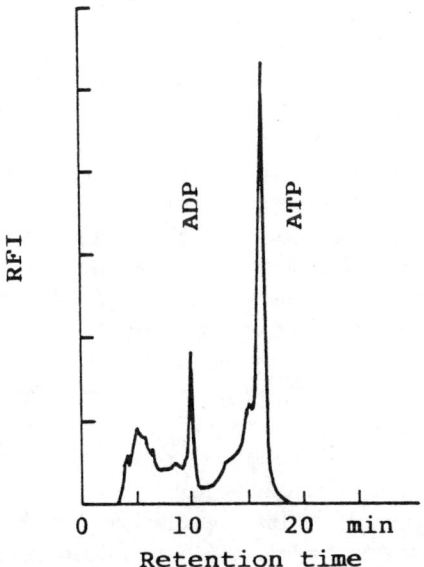

Figure 11. Chromatogram of human whole blood.

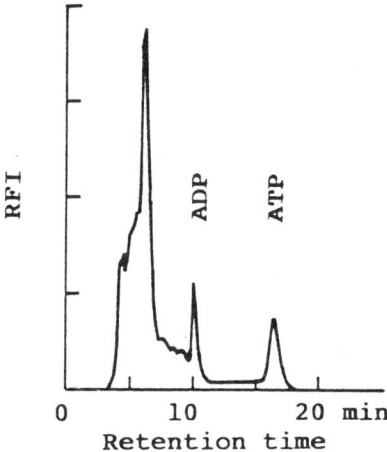

Figure 12. Chromatogram of human plasma.

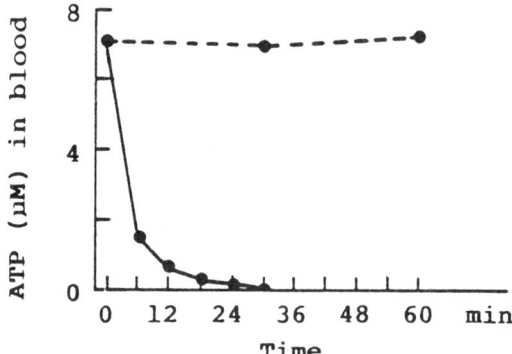

Figure 13. ATP concentrations in human plasma in the presence of EDTA. The blood cells were separated immediately (●– – –●) and at 6-min intervals (●——●).

The adenine compounds in neuroblastoma N1E 115 cells were measured by two different procedures. When the cells were allowed to react directly with bromoacetaldehyde, which penetrated into cells, two high peaks appeared and were followed by the relatively low peaks due to ADP and ATP, as shown in Fig. 14. The amounts of ADP and ATP were 41.0 ± 2.1 (SD) and 18.5 ± 0.74 nmol/10^7 cells ($n = 4$), respectively. In the second procedure the adenine compounds in the supernatant of the cells were determined as shown in Fig. 15. The amounts of ADP and ATP were 14.5 ± 0.62 and 67.4 ± 2.4 nmol/10^7 cells ($n = 4$), respectively. The recoveries of ADP and ATP added with 4 M perchloric acid were 93% ± 6.6 and 99% ± 3.3 ($n = 4$) from the neuroblastoma cells, respectively. The latter procedure can provide us with higher values of the compounds under the practical condition because free adenine compounds from cells are able to react quickly with bromoacetaldehyde before they become degraded. The latter is suitable for the determination of adenine compounds in cells.

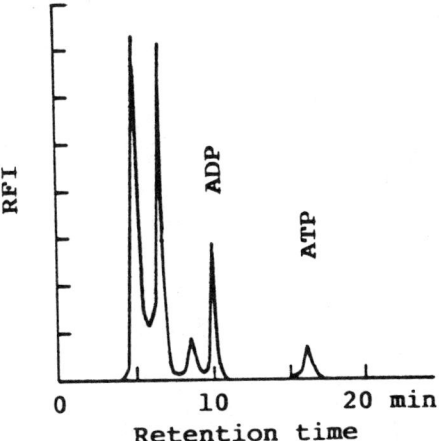

Figure 14. Chromatogram of neuroblastoma N1E 115 cells directly reacted with bromo-acetaldehyde.

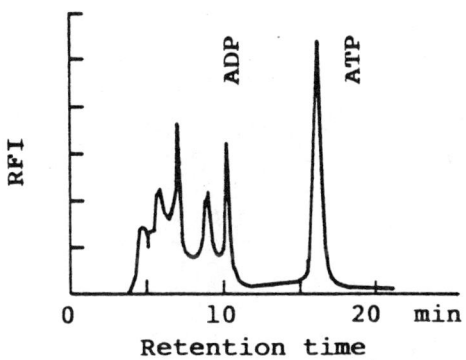

Figure 15. Chromatogram of neuroblastoma N1E 115 cells, homogenized and allowed to react with bromoacetaldehyde.

Na,K-ATPase was inhibited by ouabain, which blocks the interaction of potassium ions with the potassium sites of the enzyme. Josephson and Cantley (1977) reported that 50 μM of ouabain inhibited over 98% of the activity of the Na,K-ATPase from dog kidney. In the presence of the ATPase and the excess of ouabain, the highest peak of ATP appeared concomitantly with smaller amounts of ADP and AMP, spontaneously degraded or hydrolyzed by non-specific phosphatases, as shown in Fig. 16B. In the absence of ouabain the peak of ADP increased, whereas the peak of ATP decreased, as shown in Fig. 16A. The typical sigmoidal curve of the dose−response was obtained, ranging from 10^{-9} to 10^{-3} M ouabain as shown in Fig. 17.

Alkaline phosphatase activity was measured with using ATP as substrate. In the absence of the enzyme in the reaction mixture, the peak of ATP was contaminated with ADP after the incubation at pH 10.4 (Fig. 18B). Two high peaks of adenosine and AMP were referred to the enzymic products

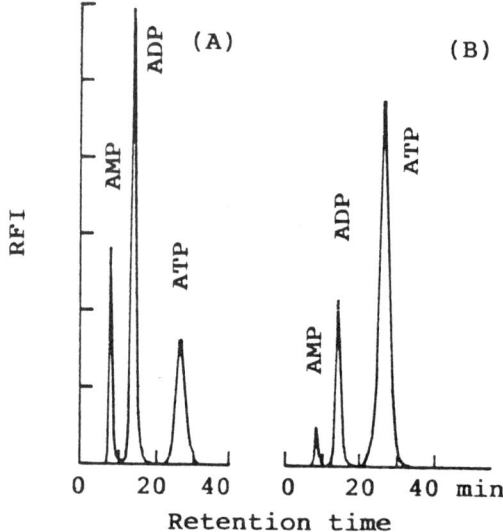

Figure 16. Peak shift of ATP to ADP by Na,K-ATPase. (A): Control, (B): 10^{-3} M of ouabain in the reaction medium.

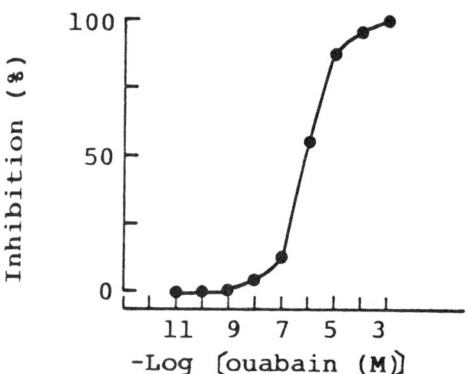

Figure 17. Effect of ouabain on Na,K-ATPase.

from ATP by the phosphatase (Fig. 18A). In Fig. 19 the total amount of adenosine, AMP and ADP was increased as a function of time. The total released amount as adenine molecule of adenosine and AMP was proportionally increased as the amount of alkaline phosphatase from 5 to 40 µg at 30 min (data not shown). The K_m value of the enzyme against ATP was determined as 185 µM from Lineweaver−Burk plots. The specific activity was 1.44 pmol/µg of protein/min. The activities of the enzyme against 5'-AMP and 2'(3')-AMP were over 4.44 pmol/µg of protein/min under the same assay condition. In addition, the activity of Mg^{2+}-ATPase in the alkaline phosphatase preparation was not detected under our assay condition, even if 3 mM ATP was used as substrate.

As shown in Fig. 20, cAMP was generated from ATP by adenylate cyclase from fat cell ghosts in the presence of L-epinephrine. In adsorption chroma-

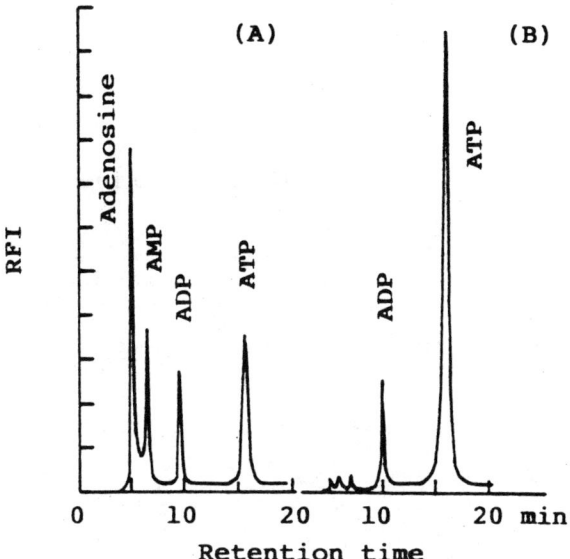

Figure 18. Peak shift of ATP by alkaline phosphatase. (A): Alkaline phosphatase (20 μg), (B): control.

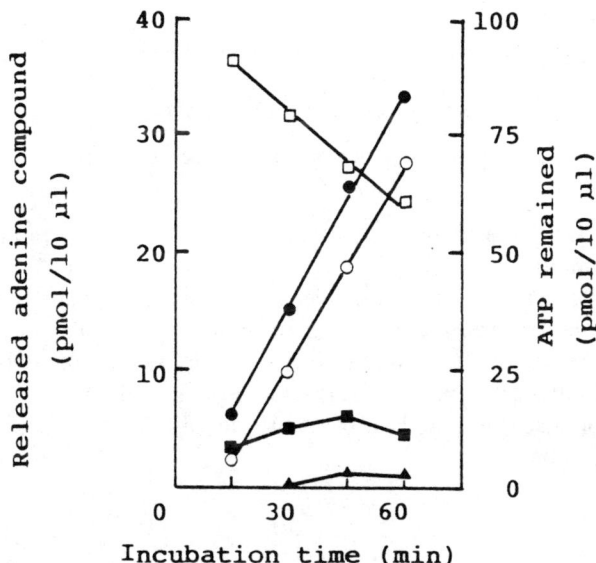

Figure 19. Time course of released adenine compounds and remaining ATP by alkaline phosphatase. Data points: ○ = Ado; ■ = AMP; ▲ = ADP; ● = total adenine compounds (Ado + AMP + ADP); □ = remaining ATP.

tography with Hitachi gel No. 3010, cAMP was eluted at 30 min, followed by adenosine and adenine, and AMP, ADP and ATP were eluted at void volume. Activation curves were drawn for 10^{-3} to 10^{-2} M of sodium fluoride or 10^{-7} to 10^{-3} M of L-epinephrine, respectively (Fig. 21). These results

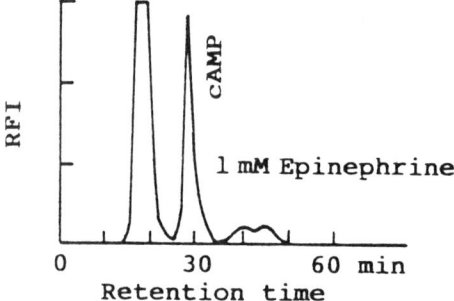

Figure 20. Chromatogram of cAMP generated by adenylate cyclase.

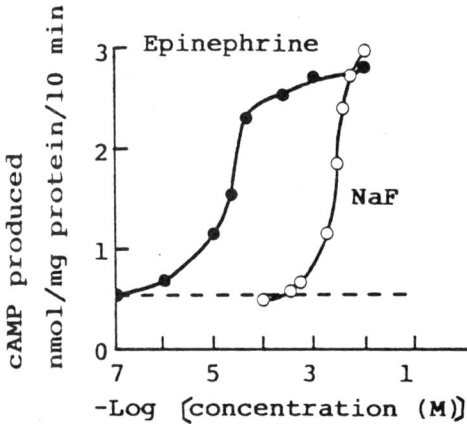

Figure 21. Effect of L-epinephrine and sodium fluoride on adenylate cyclase.

coincided with the data from a tracer experiment with a radioisotope, reported by Birnbaumer and Rodbell (1969).

Deoxy-ATP kept at 4°C was dissolved in 0.1 M phosphate buffer (pH 7.0), and the solution was reacted with bromoacetaldehyde. As shown in Fig. 22A, four peaks of deoxy-adenosine, an unknown peak, deoxy-AMP and deoxy-ADP appeared other than the main peak of deoxy-ATP. Deoxy-ADP preparation also contained deoxy-AMP, an unknown peak and deoxy-adenosine (Fig. 22B).

DISCUSSION

As described in the history of LC and HPLC for nucleic acid compounds, Yoshioka et al. (1976, 1984, 1985) have already established the systematic fluorescent-HPLC determination of adenine compounds without the aid of gradient elution using chloroacetaldehyde or bromoacetaldehyde, and a column of No. 3012-N or No. 3010. Judging from the reduction in analysis time, and enhancement of resolution, we have improved our previous determination of adenine compounds, which was performed on No. 3012-N, by introducing No. 3013-N.

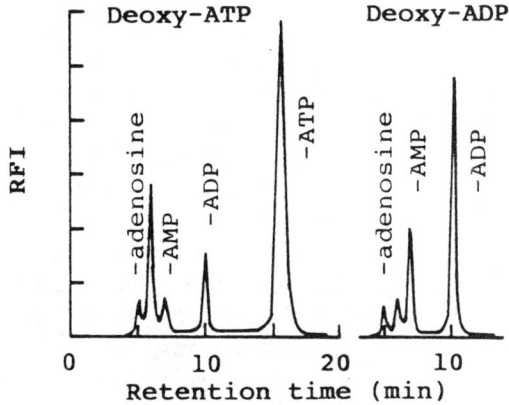

Figure 22. Chromatograms of deoxy-ATP and deoxy-ADP preparations.

We could not show any advantages in using a semi-microbore column (25 cm × 1.5 mm) packed with No. 3013-N for the determination of adenine compounds. As reviewed by Dezaro and Hartwick (1984), there will be the possibility of improving the column efficiency of a microbore if a narrower flow-through cell in the detector is constructed; a sample of 0.2−0.5 μl is injected and the flow rate is 10−80 μl/min. Yoshioka *et al.* (1984) demonstrated the usefulness of a microbore column (13 cm × 0.5 mm) to enhance the sensitivity, but not suitable for routine analyses. Up to now we agree with the view by Dezaro and Hartwick that the regular column is more efficient in analysis of nucleic acid compounds than the narrow-bore column in routine work.

There are already established methods for measuring the activities of Na,K-ATPase and adenylate cyclase. The former was measured by a radiochemical method (Noel and Godfraind, 1984), the Fiske−Subbarow method (Fiske and Subbarow, 1925) or luciferin−luciferase method (DeLuca and McElroy, 1978), which could determine ATP or ADP in enzymic solutions. The latter was determined by a radiochemical method or radioimmunoassy (Honma *et al.*, 1977) which needed either specific antibody or binding protein. Our method is more specific, and at least 1000 times more sensitive than that of Fiske and Subbarow. The enzymic method with luciferase has a high sensitivity, but is troubled with a high background and unable to measure ATP and ADP simultaneously. Our HPLC method has a sensitivity comparable to RIA, because the threshold of the detection via fluorescence is as low as 1 pmol. RIA was sometimes interfered with by salts, proteins and tissue culture medium (Albano *et al.*, 1974), whereas these factors do not interfere in the fluorescent reaction of adenine compounds with bromoacetaldehyde. Additionally, the appearance of unexpected peaks in the chromatogram can discriminate contamination(s) of the other enzyme(s) in an enzyme preparation.

Shimada *et al.* (1985) reported an assay procedure for Na,K-ATPase from guinea pig hearts by HPLC with a UV monitor. Our results in Fig. 17 were similar to theirs, but our method is more sensitive, specific, and suitable for small and unpurified samples. Recently, a circulating inhibitor of Na,K-ATPase (Hamlyn et al., 1982), which was assumed to be a digitalis-like factor (Gruber et al., 1980), was found to be increased in essential hypertension. We are now examining inhibitory effects of the factor from human urine on Na,K-ATPase with our method, and the details will be reported elsewhere.

Hartwick et al. (1979) determined 86 compounds in plasma including nucleosides, bases and other low-molecular weight UV-absorbing compounds on the basis of retention times, the peak−height ratios (280/254 nm) and their fluorescence. Peak identification was very critical. It is of significance for biochemists to obtain the definitive peaks of adenine nucleotides. The chromatograms of blood (Figs 11 and 12) and neuroblastoma N1E 115 (Figs 14 and 15) showed several peaks which were easily identifiable, because the reaction of bromoacetaldehyde is relatively specific for adenine bases. Although cytosine and guanine compounds also react with bromoacetaldehyde, the relative fluorescence intensity of the products of cytosines are negligible, and those of guanines are less than one-hundredth of those of adenine derivatives in the fluorescence detector. As concentrations of guanines and cytosines in biological materials, except RNA and DNA, are less than one-tenth to one-hundredth of those of adenines, they are virtually not found in the chromatograms.

Sample preparation prior to HPLC is essential to obtain the actual nucleotide content, as pointed out by Hartwick and Brown (1975). Recently, Born and Kratzer (1984) reported that the concentration of ATP in human blood was about 2 μM at 2−4 s and increased to 20 μM at 3−5 min after vascular injury. In our experiments the ATP concentration in human plasma varied from 1 to 10 μM, probably due to physiological conditions. The half-life of ATP in stored blood is a few minutes in the presence of EDTA (Fig. 13). To measure ATP levels accurately in plasma, blood cells should be collected and separated as soon as possible, even in the presence of EDTA, to prevent ATP in plasma from being taken up by blood cells and hydrolyzed by ecto-ATPases (Gordon, 1986). These results suggested that EDTA in plasma was useful as an anticoagulant and inhibitor of phosphatases which needed metal ions for their activations, but not suitable for the depression of blood cell functions. The shoulder of the ATP peak in Fig. 11 is assumed to be due to deoxy-ATP, the separation of which will be described elsewhere. The energy charge of neuroblastoma N1E 115 was calculated from the equation (ATP + 1/2 × ADP)/(ATP + ADP + AMP), proposed by Atkinson (1977). The value (0.86) from Fig. 15 was higher than that (0.35) from Fig. 14, indicating that the homogenization of cells is better in the procedure for the determination of adenine compounds than in the direct reaction of the cells with bromoacetaldehyde. As the peaks are easily identified on our chromatograms, energy charges of the cells calculated with our method are reliable.

Many adenine compounds are now commercially available. The purities of reagents vary from lot to lot and from company to company. Deoxy-ATP and deoxy-ADP preparations contained undesired deoxy-adenine compounds due to contamination or degradation (Fig. 22). Yoshioka *et al.* (1982) reported contamination of deoxy-ATP in commercial ATP samples. ATP can stimulate synthesis of high molecular weight DNA in isolated HeLa cells but deoxy-ATP cannot do so. Hanaoka *et al.* (1985) demonstrated that commercial deoxy-ATP stimulated the synthesis of DNA in HeLa cells due to contamination of ATP in the deoxy-ATP sample. Contamination of deoxy-ATP in ATP samples, and ATP in deoxy-ATP samples, was in the range of 0.01 to 0.9 mol%, and 0.1 to 0.43 mol%, respectively (Yoshioka *et al.*, 1982; Hanaoka *et al.*, 1985). As the most purified ATP sample was used as a standard in a series of our experiments, the errors of ATP values in various measurements were less than 1%. Taking variations due to sample preparations, mechanical errors, etc., into consideration, the precision of our method is higher than 95%.

We have described our method of pre-column derivatization and the results in this paper. Now that the fluorescent-HPLC method is extented to be useful and versatile to various samples, an analyzer with the post-column derivatization described by Yoshioka *et al.* (1985) should be used for routine analyses.

CONCLUSION

Earlier, we developed a method of fluorescent determination of adenine compounds by HPLC, using bromoacetaldehyde as a new fluorescent reagent and a column of Hitachi gel No. 3012-N.

The method was improved by introduction of a regular column packed with the finer beads, Hitachi gel No. 3013-N, which was found to be better than that of No. 3012-N. However, the semi-microbore column packed with No. 3013-N was not so effective in determining adenine compounds compared with a regular-sized column under our HPLC system.

For measuring enzymic activity, ADP, hydrolyzed from ATP by Na,K-ATPase, was determined quantitatively and cAMP from ATP by adenylate cyclase was also determined in the presence of various concentrations of L-epinephrine or sodium fluoride. ATP levels in human blood, and the cellular levels of ATP and ADP in neuroblastoma N1E 115, were determined by our fluorescent-HPLC.

Thus, the method proved to be useful and versatile for determination of adenine compounds in biological samples.

ACKNOWLEDGEMENTS

This work was supported by a Grant-in-Aid for Scientific Research No. 60480456 from the Ministry of Education, Science and Culture. A fellowship to M.Y. was afforded by INSERM-JSPS. We are deeply grateful to Shinichi Kikuchi, Isao Takishima, Masaaki Senda and Tadashi Miyazaki of Jasco, for their support. Hitachi Ltd, Tokyo, kindly supplied us with the Hitachi gels No. 3012-N and No. 3013-N used in this investigation.

REFERENCES

Anderson, F. S. and Murphy, R. C. (1976). Isocratic separation of some purine nucleotide, nucleoside and base metabolisms from biological extracts by HPLC. *J. Chromatogr. 121*, 251–262.

Anderson, N. G., Green, J. G., Barber, M. L., and Ladd, F. C. Sr (1963). Analytical techniques for cell fractions. III. Nucleotides and related compounds. *Anal. Biochem. 6*, 153–169.

Albano, J. D. M., Barnes, G. D., Maudsley, D. V., Brown, B. L., and Etkins, R. P. (1974). Factors affecting the saturation assay of cyclic AMP in biological systems. *Anal. Biochem. 60*, 130–141.

Atkinson, D. E. (Ed.) (1977). *Cellular Energy Metabolism and its Regulation*. Academic Press, New York.

Avigad, G. and Damle, S. (1972). Fluorometric assay of adenine and its derivatives. *Anal. Biochem. 50*, 321–323.

Bakay, B., Nissinen, E., and Sweetman, L. (1978). Analysis of radioactive and nonradioactive purine bases, nucleosides, and nucleotides by high-speed chromatography on a single column. *Anal. Biochem. 86*, 65–77.

Barrio, J. R., Secrist III, J. A., and Leonard, N. J. (1972). Fluorescent adenosine and cytidine derivatives. *Biochem. Biophys. Res. Commun. 46*, 597–605.

Birnbaumer, L., Pohl, S. L., and Rodbell, M. (1969). Adenyl cyclase in fat cells. I. Properties and the effects of adrenocorticotropin and fluoride. *J. Biol. Chem. 244*, 3468–3476.

Birnbaumer, L. and Rodbell, M. (1969). Adenyl cyclase in fat cells. II. Hormone receptors. *J. Biol. Chem. 244*, 3477–3482.

Born, G. V. R., and Kratzer, M. A. A. (1984). Source and concentration of extracellular adenosine triphosphate during haemostasis in rats, rabbits and man. *J. Physiol. (London). 354*, 419–429.

Breter, H., Seibert, G., and Zahn, R. K. (1977). Single-step separation of major and rare ribonucleosides and deoxyribonucleosides by high-performance liquid cation-exchange chromatography for the determination of the purity of nucleic acid preparations. *J. Chromatogr. 140*, 251–256.

Brown, P. R. (Ed.) (1973). *HPLC: Biochemical and Biomedical Application*. Academic Press, New York.

Brown, P. R. and Grushka, E. (1980). Structure–retention relations in the reversed-phase HPLC of purine and pyrimidine compounds. *Anal. Chem. 52*, 1210–1215.

Brown, P. R., Parks, R. E. Jr, and Herod, J. (1973). Use of high pressure liquid chromatography for monitoring nucleotide concentration in human blood: A preliminary study with stored blood cell suspensions. *Clin. Chem. 19*, 919–922.

Caronia, J. P., Crowther, J. B., and Hartwick, R. A. (1983). Reversed-phase separation of the major deoxyribonucleosides and their mononucleotides using tetra-butylammonium heterons. *J. Liq. Chromatogr. 6*, 1673–1691.

Catterick, T. (1983). The development of a multiwavelength detector for HPLC. *J. Chromatogr. 259*, 59–67.

Cohen, M. B., Maybaum, J., and Sadee, W. (1980). Analysis of purine ribonucleotides and deoxyribonucleotides in cell extracts by HPLC. *J. Chromatogr. 198*, 435–441.

Cohn, W. E. (1949). The separation of purine and pyrimidine bases and of nucleotides by ion exchange. *Science 109*, 377–378.

Crowther, J. B. and Hartwick, R. A. (1982). Chemically bonded multifunctional stationary phases for HPLC. *Chromatographia 16*, 349–353.

Dean, B. M. and Perrett, D. (1976). Studies on adenine and adenosine metabolism by intact human erythrocytes using HPLC. *Biochem. Biophys. Acta 437*, 1–15.

DeLuca, M. and McElroy, W. D. (1978). Purification and properties of firefly luciferase. *Methods Enzymol.* 57, 3−15.

Denton, M. S., DeAngelis, T. P., Yacynych, A. M., Heineman, W. R., and Gilbert, T. W. (1976). Oscillating mirror rapid scanning ultraviolet-visible spectrometer as a detector for liquid chromatography. *Anal. Chem.* 48, 20−24.

Dezaro, D. and Hartwick, R. A. (1984). Microbore columns. In: *HPLC in Nucleic Acid Research. Methods and Applications*, P. R. Brown (Ed.). Marcel Dekker, New York, pp. 113−137.

Eksteen, R., Kraak, J. C., and Linssen, P. (1978). Conditions for rapid separations of nucleobases and nucleosides by high-pressure anion-exchange chromatography. *J. Chromatogr. 148*, 413−427.

Fiske, C. H. and Subbarow, Y. (1925). The colorimetric determination of phosphorus. *J. Biol. Chem.* 66, 375−400.

Gordon, J. L. (1986). Extracellular ATP: effects, sources and fate. *Biochem. J.* 233, 309−319.

Gruber, K. A., Whitaker, J. M., and Buckalew, Jr. V. M. (1980). Endogenous digitalis-like substance in plasma of volume-expanded dogs. *Nature 287*, 743−745.

Hamlyn, J. M., Ringel, R., Schaeffer, J., Levinson, P. D., Hamilton, B. P., Kowarski, A. A., and Blaustein, M. P. (1982). A circulating inhibitor of $(Na^+ + K^+)$ATPase associated with essential hypertension. *Nature 300*, 650−652.

Hanaoka, F., Takahashi, M., Enomoto, T., Yamada, M., Tanaka, S., Wataya, Y., and Hayatsu, H. (1985). Contamination by ATP of commercial dATP samples causing erroneous results in studies of DNA replication in isolated HeLa cell nuclei. *J. Biochem.* 97, 383−386.

Hartwick, R. A. and Brown, P. R. (1975). The performance of microparticle chemically-bonded anion-exchange resins in the analysis of nucleotides. *J. Chromatogr. 112*, 651−662.

Hartwick, R. A., Assenza, S. P., and Brown, P. R. (1979). Identification and quantitation of nucleosides, bases and other UV-absorbing compounds in serum, using reversed phase HPLC. *J. Chromatogr. 186*, 647−658.

Henderson, R. J. Jr. and Griffin, C. A. (1984). Electrochemical detection of adenosine and other purine metabolites during high-performance liquid chromatographic analysis. *J. Chromatogr. 298*, 231−242.

Henry, R. A., Schmit, J. A., and Williams, R. C. (1973). Application of a new chemically bonded, superficially porous anion-exchange packing to nucleotide analysis. *J. Chromatogr. Sci. 11*, 358−365.

Hoffman, N. E. and Liao, J. C. (1977). Reversed phase high performance liquid chromatographic separation of nucleotides in the presence of solvophobic ions. *Anal. Chem. 49*, 2231−2234.

Honma, M., Satoh, T., Takezawa, J., and Ui, M. (1977). An ultrasensitive method for the simultaneous determination of cyclic AMP and cyclic GMP in small-volume samples from blood and tissue. *Biochem. Med. 18*, 257−273.

Horvath, C. G. and Lipsky, S. R. (1969). Rapid analysis of ribonucleosides and bases at the picomole level using pellicular cation exchange resin in narrow bore columns. *Anal. Chem. 41*, 1227−1234.

Horvath, C. G., Preiss, B. A., and Lipsky, S. R. (1967). Fast liquid chromatography: An investigation of operating parameters and the separation of nucleotides on pellicular ion exchangers. *Anal. Chem. 39*, 1422−1428.

Ingebretsen, O. C. and Bakken, A. M. (1982). Determination of adenine nucleotides and inosine in human myocard by ion-pair reversed-phase HPLC. *J. Chromatogr. 242*, 119−126.

Josephson, L. and Cantley, Jr. L. C. (1977). Isolation of a potent (Na-K)ATPase inhibitor from striated muscle. *Biochemistry 16*, 4572−4578.

Juarez-Salinas, H., Mendoza-Alvarez, H., Levi, V., Jacobson, M. K., and Jacobson, E. L. (1983). Simultaneous determination of linear and branched residues in poly (ADP-ribose). *Anal. Biochem. 131*, 410−418.

Kato, Y., Seita, T., Hashimoto, T., and Shimizu, A. (1977). Separation of nucleic acid bases and nucleosides by high-performance affinity chromatography. *J. Chromatogr. 134*, 204−206.

Kirkland, J. J. (1969). High-speed liquid chromatography with controlled-surface-porosity supports, *J. Chromatogr. Sci. 7*, 7−12.

Kirkland, J. J. (1970). High speed separations of nucleotides and nucleic acid bases by column chromatography using controlled surface porosity ion exchangers. *J. Chromatogr. Sci. 8*, 72−75.

Kirkland, J. J. and DeStefano, J. J. (1970). Controlled surface porosity supports with chemically-bonded organic stationary phases for gas and liquid chromatography. *J. Chromatogr. Sci. 8*, 309−314.

Kobayashi, K., Kuroda, Y., and Yoshioka, M. (1981). Change of cyclic AMP level in synaptosomes from cerebral cortex; Increase by adenosine derivatives. *J. Neurochem. 36*, 86−91.

Kochetkov, N. K., Shibaev, V. N., and Kost, A. A. (1971). New reaction of adenine and cytosine derivatives, potentially useful for nucleic acids modification. *Tetrahedron Lett. 22*, 1993−1996.

Krzyosiak, W. J., Biernat, J., Ciesiolka, J., Gulewicz, K., and Wiewiorowski, M. (1981). The reactions of adenine and cytosine residues in tRNA with chloroacetaldehyde. *Nucl. Acids Res. 9*, 2841−2851.

Kuroda, Y., Yoshioka, M., Kumakura, K., Kobayashi, K., and Nakajima, T. (1980). Effects of peptides on the release of catecholamines and adenine nucleotides from cultured adrenal chromaffin cells. Mastoparan-induced release. *Proc. Japan Acad. 56*, 660−664.

Kuttesch, J. F., Schmalstieg, F. C., and Nelson, J. A. (1978). Analysis of adenosine and other adenine compounds in patients with immunodeficiency diseases. *J. Liq. Chromatogr. 1*, 97−109.

Leigh, C. P. H. and Cashion, P. J. (1980). Rapid separation of nucleoside mono-, di- and triphosphates on ion-exclusion/exchange columns. *J. Chromatogr. 192*, 490−493.

McKeag, M. and Brown, P. R. (1978). Modification of high-pressure liquid chromatographic nucleotide analysis. *J. Chromatogr. 152*, 253−254.

Naikwadi, K. P., Rokushika, S., and Hatano, H. (1983). Ion chromatography of nucleobases, nucleosides and nucleotides using a dual-detection system. *J. Chromatogr. 280*, 261−269.

Noel, F. and Godfraind, T. (1984). Heterogeneity of ouabain specific binding sites and sodium-potassium ATPase inhibition in microsomes from rat heart. *Biochem. Pharmacol. 33*, 47−53.

Pennington, S. M. (1971). 3′, 5′-cyclic adenosine monophosphate phosphodiesterase assay using high speed liquid chromatography. *Anal. Chem. 43*, 1701−1703.

Peters, G. J., Kraal, I., Laurensse, E., Leyva, A., and Pinedo, H. M. (1984). Separation of 5-fluorouracil and uracil by ion-pair reversed-phase HPLC on a column with porous polymeric packing. *J. Chromatogr. 307*, 464−468.

Preston, M. R. (1983). Determination of adenine, adenosine and related nucleotides at the low picomole level by reversed-phase HPLC with fluorescence detection. *J. Chromatogr. 275*, 178−182.

Ramos, D. L. and Schoffstall, A. M. (1983). Reversed-phase high-performance liquid chromatographic separation of nucleosides and nucleotides. *J. Chromatogr. 261*, 83−93.

Ratech, H. and Thorbecke, G. J. (1980). Identification of adenosine and eight modified adenine nucleosides using reversed-phase HPLC and enzymatic peak shift with adenosine deaminase. *J. Chromatogr. 183*, 499−504.

Ritter, E. J. and Bruce, L. M. (1978). The quantitative determination of deoxy-ribonucleoside triphosphates using HPLC. *Biochem. Med. 21*, 16−21.

Rodbell, M. (1964). Metabolism of isolated fat cells. I. Effects of hormones on glucose metabolism and lipolysis. *J. Biol. Chem. 239*, 375−380.

Rustum, Y. M. (1978). High-pressure liquid chromatography. I. Quantitative separation of purine and pyrimidine nucleosides and bases. *Anal. Biochem. 90*, 289−299.

Schukovskaya, L. L., Ushakov, S. N., and Galania, N. K. (1962). Synthesis of haloacetaldehyde hydrates. *Izv. Akad. Nauk SSSR. Otd. Khim. Nauk.* 1692−1693.

Schweinsberg, P. D. and Loo, T. L. (1980). Simultaneous analysis of ATP, ADP, AMP, and other purines in human erythrocytes by HPLC. *J. Chromatogr. 181*, 103−107.

Scoble, H. A. and Brown, P. R. (1983). Reversed-phase chromatography of nucleic acid fragments. In: *HPLC: Advances and Perspectives*, Vol. 3, C. Horvath (Ed.). Academic Press, New York, pp. 1–47.

Scott, C. D., Attrill, J. E., and Anderson, N. G. (1967). Automatic high-resolution analysis of urine for its ultraviolet-absorbing constituents. *Proc. Soc. Expt. Biol. Med. 125*, 181–184.

Secrist III, J. A., Barrio, J. R., Leonard, N. J., and Weber, G. (1972). Fluorescent modification of adenosine-containing coenzymes. Biological activities and spectroscopic properties. *Biochemistry 11*, 3499–3506.

Seta, K., Washitake, M., Anmo, T., Takai, N., and Okuyama, T. (1981). High-resolution anion-exchange chromatography of ultraviolet-absorbing constituents of human erythrocytes. *J. Liq. Chromatogr. 4*, 129–143.

Shimada, K., Ohishi, K., and Nambara, T. (1985). A high performance liquid chromatographic method for the assay of Na^+,K^+-adenosine triphosphatase inhibition. *J. Pharmacobio-Dyn. 8*, 64–68.

Shmukler, H. W. (1972). The rapid chromatographic analysis of the free nucleotides from rat brain. *J. Chromatogr. Sci. 10*, 38–40.

Shugart, L. (1979). Identification of fluorescent derivatives of adenosylmethionine and related analogues with high-pressure liquid chromatography. *J. Chromatogr. 174*, 250–253.

Sims, J. L., Juarez-Salinas, H., and Jacobson, M. K. (1980). A new highly sensitive and selective chemical assay for poly (ADP-ribose). *Anal. Biochem. 106*, 296–306.

Singhal, R. P. (1972). Ion-exclusion chromatography: Analysis and isolation of nucleic acid components, and influence of separation parameters. *Archiv. Biochem. Biophys. 152*, 800–810.

Singhal, R. P. and Cohn, W. E. (1973). Cation-exclusion chromatography on anion exchangers: Application to nucleic acid components and comparison with anion-exchange chromatography. *Biochemistry 12*, 1532–1537.

Singhal, R. P., Bajaj, R. K., Buess, C. M., Smoll, D. B., and Vakharia, V. N. (1980). Reversed-phase boronate chromatography for the separation of *o*-methylribose nucleosides and aminoacyl-tRNAs. *Anal. Biochem. 109*, 1–11.

Uberti, J., Lightbody, J. J., and Johnson, R. M. (1977). Determination of adenosine deaminase activity using high-pressure liquid chromatography. *Anal. Biochem. 80*, 1–8.

Uziel, M., Koh, C. K., and Cohn, W. E. (1968). Rapid ion-exchange chromatographic microanalysis of ultraviolet-absorbing materials and its application to nucleosides. *Anal. Biochem. 25*, 77–98.

Yoshida, K., Haraguchi, H., and Fuwa, K. (1983). Determination of ribonucleoside 5'-mono-, 5'-di-, and 5'-triphosphates by liquid chromatography/inductively coupled plasma atomic emission spectrometry. *Anal. Chem. 55*, 1009–1012.

Yoshioka, A., Tanaka, K., Wataya, Y., and Hayatsu, H. (1982). dATP content in commercial ATP samples. *Chem. Pharm. Bull. 30*, 2651–2654.

Yoshioka, M. and Tamura, Z. (1976). Fluorimetric determination of adenine and adenosine and its nucleotides by HPLC. *J. Chromatogr. 123*, 220–224.

Yoshioka, M., Nishidate, K., Iizuka, H., Nakamura, A., El-Merzabani, M. M., Tamura, Z., and Miyazaki, T. (1984). Sensitive fluorimetry of adenine-containing compounds with HPLC. *J. Chromatogr. 309*, 63–71.

Yoshioka, M., Tamura, Z., Senda, M., and Miyazaki, T. (1985). Analyser of adenine nucleotides. *J. Chromatogr. 344*, 345–350.

Progress in HPLC, Vol. 4, pp. 211—227.
Yoshioka *et al.* (Eds)
© 1989 VSP.

Microbore HPLC for biological samples: catecholamines, peptides and proteins

KOHICHI KOJIMA, HASAN PARVEZ,[1] SIMONE PARVEZ[1] and TOSHIHARU NAGATSU[2]

Hatano Research Institute, Food and Drug Safety Center, Hadano, Kanagawa 257, Japan; [1]University of Paris XI, Orsay, France; [2]Department of Biochemistry, Nagoya University School of Medicine, Nagoya 466, Japan

INTRODUCTION

Conventional HPLC has made great advances and wide application in the studies of biological fields over the past 10 years. In contrast, although microbore HPLC has received a great deal of attention in life sciences in the past few years, only a few practical applications have been made. There are several intrinsic advantages of microbore columns over the standard conventional columns that make them more attractive for use in analytical liquid chromatography. Three basic advantages of microbore HPLC are high linear mobile phase velocities, low solvent consumption and high mass sensitivity. Owing to relatively low solvent volumetric flow rates, microbore HPLC is more economical in use and higher in sensitivity than conventional HPLC. These advantages of microbore HPLC with high efficiency, high speed and high sensitivity can become very apparent when dealing with the separation and analysis of samples of biological origin. Biologically active compounds such as catecholamines, peptides and proteins are present in tissues or biological fluids at comparatively low levels. Moreover, the mass of biological samples is often limited. Thus microbore HPLC is ideally suited to their analysis. As examples of the application of microbore HPLC in life sciences, the microanalyses of catecholamines, biologically active peptides and proteins are described in this chapter.

ANALYSIS OF CATECHOLAMINES

The coupling of HPLC with electrochemical detection has become the bes method of choice for the determination of biogenic amines. Other techni ques, such as gas chromatography–mass spectrometry and radioenzymatic assays, have similar limits of detection in the low picogram range. However, these other techniques lack the versatility of HPLC with electrochemical detection. In spite of the numerous technical improvements in chromatographic supports and electrochemical detectors, the limits of detection for HPLC with electrochemical detection have remained essentially unchanged. However, recent advances in the development of microbore HPLC columns have given a tremendous impetus to the miniaturization of HPLC, and much of the interest in microbore HPLC has been the result of increased mass sensitivity. Although more limited in application, amperometric detection offers several distinct advantages over spectroscopic technique when coupled with microbore HPLC. Most spectrophotometric detectors fail to achieve the predicted signal enhancement. Lowering the detector volume of UV or fluorescence detectors in order to eliminate extra-column dispersion has generally increased the noise as a result of diffraction and decreased the sensitivity due to a reduction in path length. In contrast, the cell volume for most commercially available amperometric cells is already quite small, and may be decreased to less than one microliter without difficulty and with no loss in signal by decreasing the spacer thickness.

Table 1 shows examples of catecholamine analysis using microbore HPLC. Hirata et al. (1980) designed a low-volume electrochemical cell (approximately 0.1 µl). The detector is less sensitive than an electrochemical cell (approximately 2 µl volume) designed for use with conventional HPLC columns, if the reduced area of the electrode and the lower flow rates were employed. Nevertheless, the electrode is sufficiently sensitive to detect directly a number of compounds in human urinary samples without preconcentration. In addition, they showed that the electrochemical method is far more sensitive than fluorescence or UV detection.

An electrochemical detector having two working electrodes (a dual electrochemical detector) was designed for use in microbore HPLC by Goto et al. (1982) and was successfully utilized for the selective detection of catecholamines. Their system consists of a mixing junction for adjusting the sample to pH 8.5, a micro-alumina precolumn for enriching catecholamines in the sample, and a dual electrochemical detector for selectively detecting catecholamines based on their electrochemical reversibility. These three processes in this system are able to operate with directly injected body fluids and to determine simultaneously all of the four catecholamines (norepinephrine, epinephrine, dopamine and dopa) with high precision.

Goto et al. (1982) reported the cyclic semi-differential voltammetric determination of applied potentials for the dual electrochemical detector in microbore HPLC, using catecholamines in human urine as the test samples. Cyclic semi-differential voltammetry provides higher sensitivity and better resolution than ordinary cyclic voltammetry. Hydrodynamic voltammetry measurements can determine precisely the optimum applied potentials for electrochemical detectors in HPLC, but require several hours or more for completion, owing to the time required for the background current to stabilize after each

Table 1.
Examples of microbore HPLC for analysis of catecholamines

Substance	Sample source	Column, particle size (i.d. × length, mm)	Mobile phase	Flow rate (μl/min)	Detector (condition)	Detection limit	Cell size	Reference
Methoxyhydroxyphenyl glycol, p-hydroxyphenylacetic acid, 5-hydroxyindoleacetic acid, homovanillic acid, and vanillic acid		LiChrosorb SI-100, 30 μm	0.2 M acetate buffer (pH 4.0)	1	Electrochemical (+1.0 V)		0.15 μl	Hirata *et al.* (1980)
Norepinephrine, epinephrine, dopamine, and L-dopa		Yanapak ODS, 5 μm (0.5 × 150)	Britton−Robinson buffer (pH 1.8) containing 0.5 mM l-heptanesulfonic acid sodium salt	8.3	Electrochemical (V vs Ag/AgCl, +0.80 and +0.20)			Goto *et al.* (1981)
Norepinephrine, epinephrine, dopamine, and L-dopa		(0.5 × 150)	Britton−Robinson buffer (pH 1.8) containing 0−0.5 mM sodium l-heptane sulfonate and 0−1 mM EDTA (disodium salt)	8.3	Electrochemical (V vs Ag/AgCl, −0.1 to +0.8)			Goto *et al.* (1982)

Table 1. (Cont'd)

Substance	Sample source	Column, particle size (i.d. × length, mm)	Mobile phase	Flow rate (µl/min)	Detector (condition)	Detection limit	Cell size	Reference
Serotonin, 5-hydroxyindoleacetic acid, 5-hydroxytryptophan, norepinephrine, epinephrine, dopamine, and 3,4-dihydroxyphenyl-acetic acid	rat brain	For hydroxyindoles Whatman microbore C_{18}, 10 µm spherical (1 × 250)	0.1 M Ammonium acetate, 0.05 M acetic acid, 50 mg/EDTA and 10% (v/v) acetonitrile	100–125	Electrochemical, carbon paste electrode (+0.600 V, vs Ag/AgCl)	50–200 fg		Caliguri and Mefford (1984)
		for catecholamines Alltech microbore C_{18} 5 µm spherical (1 × 250)	0.1 M Sodium acetate, 0.02 M citric acid, 50 mg/l EDTA, 100 mg/l sodium acetyl sulfate and 4.5% (v/v) acetonitrile	40–50				
Dopamine, dihydroxyphenylacetic acid, 5-hydroxyindoleacetic acid, and homovanillic acid	A perfusate of rat brain dialysis	Spherisorb S5 ODS-2 (1 × 250)	0.15 M NaH_2PO_4, 0.5 M sodium octanesulfonate, 0.1 mM EDTA and 20% methanol (pH 4.0)	50	Electrochemical, saturated calomel reference electrode (+0.70 V, vs SCE)	0.15 pg (Dopamine)	30 nl	Carlsson and Lundström (1985)
3,4-Dihydroxyphenyl-glycol, norepinephrine, 3,4-dihydroxyphenyla-lanine, epinephrine, 3,4-dihydroxypheny-lacetic acid, and dopamine		Hypersil C_{18}, 3 µm (1.2 × 5–10)	0.2 M NaH_2PO_4, 30 mg/l sodium octyl sulfate, 50 mg/l EDTA and 0.6% acetone (pH 3.2)	200–250	Electrochemical (+6.0 V, vs Ag/AgCl)	1 pg or less	210 nl	Caliguri et al. (1985)

Table 1. (*Cont'd*)

Substance	Sample source	Column, particle size (i.d. × length, mm)	Mobile phase	Flow rate (μl/min)	Detector (condition)	Detection limit	Cell size	Reference
Norepinephrine, epinephrine, and dopamine	Human plasma, human CSF and rat hypothalamus	Shandon Hypersphere, 3 μm (1.2 × 7–10)	0.10 M Sodium acetate, 0.02 M citric acid, 250 mg/l sodium octyl sulfate, 50 mg/l EDTA and 2.0% acetone (v/v)	200–250	Electrochemical (+0.60 V, vs Ag/AgCl)	280 fg for norepinephrine 400 fg for epinephrine 800 fg for dopamine		Durkin *et al.* (1985)
Norepinephrine, epinephrine, dopamine, 3,4-dihydroxypheny-lethyleneglycol, 3,4-dihydroxyphenyla-lanine and 3,4-dihydroxypheny-lacetic acid	Mouse brain		0.2 M NaH$_2$PO$_4$ 2% methanol, 32 mg/l sodium octyl sulfate (pH 3.2)					
Norepinephrine and epinephrine	Human plasma	Hypersil ODS, 5 μm (1.2 × 200)	Water–methanol (3:1, v/v) containing 0.15 M acetic acid buffer (pH 4.7), 0.04–0.06% (w/v) sodium dodecyl sulfate, 0.01% (w/v) EDTA and 0.01% (w/v) sodium chloride	1 mm/s	Fluorimetric (ex:278 nm, em:317 nm)	5 pg for both catecholamine (3 times the SD)	3 μl	Kamperman and Kraak (1985)

change of electrode potential. On the other hand, cyclic voltammetry measurements can be made much more rapidly, requiring less than 1 min for one cycle, and provide both the potentials for oxidation and re-reduction of the samples at the same time. Cyclic semi-differential voltammetry is a more suitable technique than cyclic voltammetry, because the peaks observed in cyclic semi-differential voltammetry are sharp and symmetrical, whereas those in cyclic voltammetry are broad and asymmetric.

In studies mentioned above microbore or capillary HPLC has been coupled with an originally designed electrochemical detector. These works did not mention any marked signal enhancement. Caliguri and Mefford (1984) achieved the lowering of these detection limits to the $50-100 \times 10^{-15}$ g $(250-500 \times 10^{-18}$ mol) range of indoles and catecholamines. This is accomplished using microbore HPLC with a commercially available amperometric detector aparatus. Reproducibility of $1.6-8.1\%$ CV was reported at the $1-5$ pg level. The utility of this approach is demonstrated for analysis of brain tissues of 1 μg or less.

The use of short $(7-10$ cm) microbore columns packed with 3 μm material allows one to take advantage of both signal enhancement and high sample resolution. The work described by Caliguri et al. (1985) indicated the performance of high-speed microbore columns using electrochemical detection, as well as packing procedures and detector modifications. The separation achieved in this work is accomplished in approximately 5 min. This will allow routine analysis of six catechol compounds extracted with aluminum oxide simultaneously. When coupled with amperometric detection, high-speed microbore HPLC offers limits of detection for biogenic amines of 1 pg or less, the low cost of these columns, and fast analysis of biogenic amines. Using the same size of columns Durkin et al. (1985) showed the analysis of catecholamines in small tissue samples and human blood plasma and cerebrospinal fluid. After extraction for preconcentration and clean-up of these samples, marked signal enhancement was observed due to the smaller column volume as well as the increased coulometric yield which results from the lower flow rates used with this technique. Detection limits of 0.2 to 0.5 pg were obtained. This will allow the analysis of catecholamines in extremely small tissue samples or small volumes of cerebrospinal fluid or plasma.

Carlsson and Lundström (1985) described a miniaturized electrochemical flow cell for amperometric detection in liquid chromatography which has been designed to meet the requirement of low volumetric dispersion when using microbore columns. The volume of the thin-layer cell is in the range $10-50$ nl, depending on the spacer dimensions, with the electrode area being about 1 mm². The magnitude of the current response of an electrochemical detector is a function of certain operating conditions, such as solution velocity, solute concentration and cell dimensions. The limit of detection for dopamine, defined as twice the peak-to-peak noise level, is estimated to be less than 0.15 pg. The applicability of the detector was demonstrated with the detection of dopamine, dihydroxyphenylacetic acid and homovanilic acid, and of Met- and Leu-enkephalin (summarized in Table 2) in samples of biological origin.

Table 2.
Examples of microbore HPLC for separation or purification of peptides and proteins

Substance	Sample source	Column, particle size (i.d. × length, mm)	Mobile phase	Flow rate (µl/min)	Detector	Detection limit	Cell size	Reference
Unidentified	Serum	RP-18, 10 µm (1 × 1000)	A: Methanol-water (75:25) B: 100% Methanol or A: Methanol-water (50:50) B: 100% Methanol	50 / 40	UV			Scott and Kucera (1979)
Epidermal growth factor, cytochrome C, myoglobin, lysozyme, ribonuclease A, and insulin	Commercial or purified proteins	Hypersil ODS, 5 µm (2.1 × 75, 1 × 75)	0.9% NaCl (pH 2.1) and acetonitrile-0.9% sodium chloride (pH 2.1) (60:40)	100	Fluorometer (ex: 215 nm, em: 340 nm) / UV	5 ng (Signal to noise ratio of 5) / 1 ng (signal to noise ratio of 5; at 210 nm)	25 µl / 4.6 µl	Nice et al. (1984)
Met-enkephalin and Leu-enkephalin	Rat striatum extract	Partisil 10 ODS-3, Micro B (1 × 250)	Acetonitrile-methanol-25 mM phosphate buffer (pH 6.9) (25:25:150, v/v)	50	Electrochemical (+1.10 V)			Carlsson and Lundström (1985)
Tryptic peptides	Murine transferrin receptor	Hypersil C_8, 5 µm (2.1 × 100) Vydac C_4, 5 µm (2.1 × 75)	A: Water containing 0.1% (v/v) trifluoroacetic acid B: Acetonitlile/40% water containing 0.1% (v/v) trifluoroacetic acid	50–100	UV (210, 254, 280, 200–240 nm at 2 nm band width)			Grego et al. (1985)

Table 2. (Cont'd)

Substance	Sample source	Column, particle size (i.d. × length, mm)	Mobile phase	Flow rate (µl/min)	Detector	Detection limit	Cell size	Reference
Cytochrome C, α-lactoalbumin, myoglobin, lysozyme, ribonuclease A, bovine serum albumin, and SCM-plastocyanin	Commercial proteins	Brownlee RP-300 (2.1 × 30)	A: Water containing either 0.9% (v/v) NaCl (titrated to pH 2.1 with HCl), 0.1 M Na_2HPO_4, pH 6.9 or 0.15% (v/v) TFA; B: Acetonitrile or 60% acetonitrile/40% water containing 0.1% (v/v) TFA	100	UV	1–10 ng (at 210 nm)		Nice et al. (1985)
Epidermal growth factor	Murine	Brownlee AX-300	10 mM and 500 mM ammonium bicarbonate gradient	100				
Ribonuclease, cytochrome C, carbonic anhydrase and ovalbumin	Commercial proteins	TMS-250 (1.0 × 50)	A: 0.05% TFA in water; B: Acetonitrile with about 0.03% TFA	50–400	UV (at 214 nm)	1 ng	14 µl or 1.9 µl	van der Zee and Welling (1985)
Trp-Tyr, bombesin insulin, xenopsin and glucagon	Commercial proteins	Microbore-1, C_{18}–5 protein column (1 × 300)	A: 0.1% TFA; B: 60% Acetonitrile/40% water and 0.1% TFA		UV (at 210 nm)		0.5 µl	Schachterle and Alfredson (1986)

Table 2. *(Cont'd)*

Substance	Sample source	Column, particle size (i.d. × length, mm)	Mobile phase	Flow rate (μl/min)	Detector	Detection limit	Cell size	Reference
Glutamate dehydrogenase, lactate dehydrogenase, enolase, adenylate kinase, hexokinase, cytochromec, catalase, and aldehyde dehydrogenase	Commercial proteins	AF-102 (0.35 × 940, 0.36 × 1060) TSK gel 3000 SW XL (0.35 × 825, 0.35 × 810, 0.35 × 2030) TSK gel 3000 SW (0.35 × 940)	0.1 M Phosphate buffer−0.2 M sodium chloride (pH 7)	1.0 0.96 0.94 0.76	UV (at 220 nm)			Takeuchi *et al.* (1986)
Commercial or purified peptides and proteins		Hypersil C_8, 5 μm (2.0 × 100) Vydac C_4, 5 μm (2.0 × 75) Brownlee C, 10 μm (2.0 × 30)	A: 0.1% TFA in water B: 0.1% TFA in acetonitrile−water (60:40) or A: Water containing 25 mM NH_4HCO_3 (pH 7.8) B: acetonitrile−water (50:50) containing 35 mM NH_4HCO_3 (pH 7.8)	100	UV	about 10 pmol		Grego *et al.* (1986a)
Leu-enkephalin, Met-enkephalin, Tyr-Gly-Gly-Phe, Tyr-Cys-Gly-Phe-Cys, Gly-Gly-Phe-Leu, and Phe	Commercial peptides	RSil C_3 or C_{18}, 10 μm (0.5 × 250)	2-Propanol−water containing (60:40) 1 mM TFA-TEA buffer (pH 2.0) or 2-propanol-water containing (70:30) 1 mM TFA-TEA buffer (pH 2.0)		Mass	25 ng		Beavis *et al.* (1986)

Table 2. (*Cont'd*)

Substance	Sample source	Column, particle size (i.d. × length, mm)	Mobile phase	Flow rate (µl/min)	Detector	Detection limit	Cell size	Reference
Phospholipase A2, cytochrome C, ribonuclease, myoglobin, chymotrypsinogen A, ovalbumin, and bovine serum albumin	Commercial proteins	Sephadex G-75 (superfine) or Bio-gel P-60 (>400 mesh) (1 × 250)		4–8	UV and VIS		1 µl	Zimina et al. (1986)
Triptic peptides	Murine plasma cell antigen PC-1	Hypersil C8, 5 µm (2.1 × 100) Vydac C4, 5 µm (2.1 × 75)	A: Water containinng 0.1% (v/v) trifluoroacetic acid B: 60% acetonitrile/40% water containing 0.1% (v/v) trifluoroacetic acid	100	diode array (210, 254, 280, 200–240 nm)			Grego et al. (1986b)

There is a need for a fast and sensitive method to analyze the concentration of catecholamines in body fluids, preferably in plasma, associated with certain metabolic disorders. With plasma samples, in which the common catecholamine concentrations are very low and moreover the available sample quantity is limited, a large scattering in the results is found. Kamperman and Kraak (1985) reported the results of an investigation into the applicability of microbore columns for the analysis of catecholamines in 0.5−1 ml of plasma, using on-column concentration and fluorometric detection. The reason why they chose fluorometric detection instead of electrochemical detection is its greater selectivity, which simplifies the sample pretreatment and is quite suitable to combine with microbore columns. The developed method was applied to plasma samples of essentially healthy persons and patients suffering from hypoglycemia. The detection limit, defined as three times the standard deviation of the noise, was determined to be 5 pg for adrenaline and noradrenaline. In the development of analytical methods using HPLC for biological samples, one of the most time-consuming steps, which introduces considerable sources of error, is sample pretreatment and enrichment prior to injection into the chromatography. Approaches towards on-column sample enrichment for direct injection of biological samples in HPLC will be the best way to resolve this problem.

ANALYSIS AND PURIFICATION OF PEPTIDES AND PROTEINS

HPLC has become a generally applied tool for the separation and purification of peptides and/or proteins from complex mixtures. Such methods have been shown to offer unrivalled advantages in terms of speed, resolution, sensitivity and recovery when used for the separation or concentration from relatively large volumes of a dilute protein or peptide solution. The advantages of microbore HPLC for application of peptide and protein research are the same as described in the section on catecholamines. The majority of publications were concerned with the technique itself rather than its application. Even so, the desire to illustrate the value of the technique on practical problems led to the generation of valuable application data (Table 2).

For the satisfactory separation of higher molecular weight materials such as peptides and proteins by HPLC, columns of intrinsic high efficiency are required to deal with the complex nature of the sample, and an appropriate gradient-elution technique must be employed to cope with the diverse chemical nature of the components. Microbore columns have been shown to have high intrinsic efficiency and, if used with an appropriately designed gradient-elution apparatus, should effectively separate mixtures containing substances covering a wide range of chemical types. However, due to the fact that very low flow rates are normally employed with microbore columns, the gradient-elution system has to be specially constructed.

Scott and Kucera (1979) made a mixing T using swagelok T and a 0.1 in. o.d., 0.010 in. i.d. stainless-steel tube. They also made an appropriate on-line concentration system. Using the sample concentration system in conjunction with microbore columns, chromatograms of low molecular weight, nonionic materials in blood can be separated and measured. The sampling device has been shown to separate blood serum with a high reproducibility, and significant differences between normal and abnormal blood serum have been clearly demonstrated.

In the study of Van der Zee and Welling (1985), it was shown that, by using microbore columns with otherwise conventional gradient HPLC equipment, much smaller peak volumes (about 5 times) and thus higher sensitivity of detection (5−6 times better) can be obtained, without loss of resolution or speed when compared with a conventional analytical column. The sensitivity of detection was further increased by a reduction in solvent peaks, resulting in a 20-fold overall increase (about 1 ng).

Schachterle and Alfredson (1986) evaluated the application of a preformed gradient concept to gradient-elution microbore HPLC using a single piston reciprocating pump and standard hydraulics without major modifications. To the system hydraulics were added an autopurge valve and a check valve for use in the preformed gradient techniques. The chromatographic performance was tested to separate peptides (2−30 residues) and PTH amino acids.

One of the major applications of HPLC is the separation of complex mixtures into fractions that can then be assayed for biological or chemical activity, or used as a stage in a purification process. This latter application has yet to exploit the advantages of microbore HPLC. The difficulty in using microbore columns for this purpose is the extremely small volumes of liquid that must be collected from a microbore column to take advantages of the column's resolution: volumes in the order of 1−10 μl.

This feature of microbore columns has also attracted researchers' interest in coupling HPLC to mass spectrometry. Various methods for coupling liquid chromatography and mass spectrometry have been investigated, involving direct introduction accompanied by chemical ionization, a thermospray method, a moving-belt method, etc.

Beavis et al. (1986) developed an off-line HPLC−mass spectrometry interface, for their time-of-flight secondary ion mass spectrometer, leading to the construction of a system for the automated collection of dry fractions from a microbore HPLC column. A criterion that they have found useful is that the solvents used should have the lowest amount of dissolved solids to get lower backgrounds and higher resolution. The chromatogram of a mixture of peptides (up to five residues) was shown to demonstrate the capabilities of their off-line microbore HPLC−mass spectrometry system.

The introduction of rapid-scanning UV−VIS detectors based on linear diode arrays provides the biochemist with a powerful analytical tool when used in conjunction with HPLC. A major avantage of multichannel detectors is that spectra of eluting compounds can be stored in digital form and then manipulated by a variety of algorithms to ascertain the purity of an eluting peak. Peptides and proteins exhibit UV-absorption spectra characteristic of their component amino acids, particularly the aromatic residues, phenylalanine, tyrosine and tryptophan. Grego *et al.* (1986a) described the characteristic zero- and second-order derivative spectra of aromatic residues and used to identify aromatic residues contained in sub-microgram amounts of polypeptides and proteins during their elution from reversed-phase short microbore columns under gradient conditions. The sub-microgram level of sensitivity was due to the small peak volumes and consequent elevated solute concentrations obtained on short (<10 cm), microbore (2 mm i.d.) reversed-phase columns.

The chromatographic efficiency associated with analytical HPLC columns means that proteins are typically recovered in eluent volumes of about 1 ml. The resultant protein concentration (μg/ml) of these eluents is not ideal for subsequent microsequencing analysis, radioiodination and so on. Unfortunately, concentration procedures such as lyophilization or evaporation frequently result in severe loss of sample. This problem is accentuated when working with sub-microgram amounts of materials. An important feature of microbore column technology is that it enables low microgram amounts of materials, in large volumes (>1 ml), to be concentrated by trace enrichment onto column supports and subsequently be recovered in high yield (>90%) in small volumes (40–60 μl) suitable for the following steps. The sample loading capacity of these columns is in excess of 1 mg, and the chromatographic resolution is similar to that achieved for conventional columns. Four examples were shown in Table 2. Nice *et al.* (1984) demonstrated the potential of short microbore columns (7.5 cm × 2.1 or 1.0 mm i.d.) for the chromatography and chromatographic concentration of proteins available in ng–μg quantities. Samples loaded in a volume of 2 ml or greater can be recorded after chromatography on the 2.1 mm column in a volume of less than 100 μl, representing an about 20-fold increase in concentration demonstrating the trace enrichment potential of these columns. There was also a significant increase in the level of detectability (about 5-fold for the 2.1 mm i.d. column) compared with the use of 4.6 mm i.d. columns of the same length in the same equipment. Grego *et al.* (1985) described a very similar short microbore HPLC system to fractionate and purify a number of tryptic peptides generated from approximately 200 pmol purified murine transferrin receptor. The use of reversed-phase microbore columns permitted the recovery of sub-microgram amounts of purified polypeptides in high yield (greater than 90%) in small eluent volumes (20–60 μl). In this manner, purified polypeptides can be loaded directly onto the gas-phase sequencer without further manipulation. Nice *et al.* (1985) also studied microbore HPLC guard columns (3 cm × 2 mm i.d.) and obtained almost similar results to those mentioned above (Nice *et al.*, 1984). Grego *et al.* (1986b) also applied their system (Grego *et al.*, 1985) to fractionate and purify a number of tryptic peptides generated from approximately 600 pmol of purified murine plasma cell antigen PC-1.

Proteins have been separated by reversed-phase chromatography, ion-exchange chromatography, size-exclusion chromatography (SEC), etc. The method of SEC gives us information regarding the molecular size of peaks in the chromatogram, which is an added advantage of SEC. It is for this reason that SEC has been employed for the separation of proteins, even though it has lower peak capacity compared with that of other chromatographic separation modes. Takeuchi et al. (1986) described the preparation procedures of SEC columns with about 0.35 mm i.d. for the separation and resolution of proteins. They found that resolution is improved by increasing the column length and using fine particles (5 μm). Although it took longer, their column resolved more constituents compared with commercially available columns.

The other application of SEC was described by Zimina et al. (1986). Problems related to the determination of physicochemical constants of samples available in micro-amounts often arise in molecular biology. Microcolumn chromatography, which makes it possible to reduce the amount of the substances investigated by about two orders of magnitude, is an invaluable method for the investigation of deficient proteins. SEC is traditionally used for the separation of substances, but it can also be applied to the study of physicochemical properties of reacting systems, for example, of protein association or micelle formation in surfactants. Zimina et al. (1986) investigated the relationship between association and enzymatic activity, taking phospholipase A_2 as an example. A procedure for determination of the critical concentration of micelle formation in surfactants was developed.

AN EXAMPLE OF THE EXPERIMENT USING MICROBORE HPLC

Goto et al. (1981) developed a new microbore HPLC system with a dual electrochemical detector for the simultaneous determination of the catecholamines in body fluids.

The direct injection analytical system of catecholamines in human urine is schematically shown in Fig. 1. The system is composed of: a microprecolumn of alumina for the sample enrichment; a microbore column for chromatographic separation of catecholamines; and a dual electrochemical detector. The first sample of human urine is taken with a 100-μl microsyringe and injected into the sample loop of the sample injector. By switching each valve as in Fig. 1, the procedures for conditioning the microprecolumn followed by sample enrichment, and for conditioning the microbore separation column, are performed at the same time. The microprecolumn is conditioned with the buffer solution of pH 8.8 (Tris buffer containing 0.25% Na_2 EDTA and 0.05% $NaHSO_3$) delivered at a flow rate of 33 μl/ml. The sample is delivered by water at a flow rate of 33 ul/min for enrichment by the microprecolumn, solid particles in the sample being removed through the solution filter, and the sample is mixed with a flow of the buffer solution of pH 8.8 in the mixing joint to adjust the sample to pH 8.5. The sample is injected into the microprecolumn for 15 min with a mixed flow of the water and the buffer solution, and then the microprecolumn is washed for a further 15 min with only a flow of water by stopping the flow of the buffer solution. The flow line from the mixing joint to the microprecolumn is made of a PTFE tube (12 cm × 0.5 mm i.d.) to mix the sample with the buffer

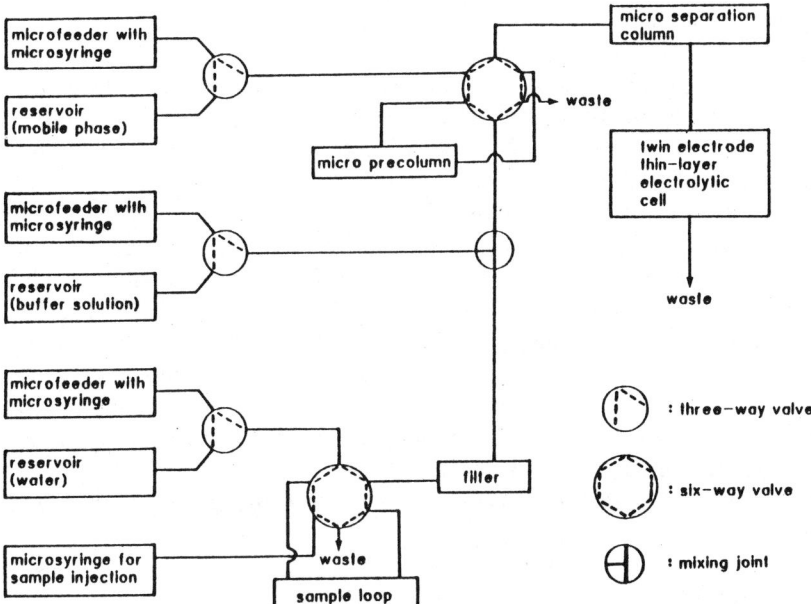

Figure 1. The schematic diagram of the microbore HPLC system with a microprecolumn of alumina for enrichment and a dual electrochemical detector for determination (Goto *et al.*, 1981).

solution completely. Parallel to the enrichment procedure the microbore separation column is conditioned with the mobile phase (Britton-Robinson buffer, pH 1.8, containing 0.5 mM 1-heptanesulfonic acid, sodium salt, as the ion-pair reagent) delivered at a flow rate of 8.3 µl/min. Next, by switching the six-way valve, the mobile phase is introduced into the microbore separation column through the microprecolumn. In this procedure the adsorbed compounds are eluted from the microprecolumn and simultaneously separated by the microbore separation column. Parallel to the chromatographic procedure the next sample is taken by switching the valve of the sample injector again. After the adsorbed compounds are completely eluted from the microprecolumn the next sample enrichment is performed during the separation process of catecholamines by again switching each valve. There is no need to change the alumina precolumn in a series of at least 100 analyses with direct injection of the urine samples. The catecholamines separated by the microbore separation column are introduced into the twin electrode thin-layer electrolytic cell, in which the anode and cathode are set at the potentials (V vs. Ag/AgCl) of +0.80 and +0.20, respectively. The catecholamines are selectively detected by monitoring the reduction current at the cathode. The dual electrochemical detector can selectively detect catecholamines from many electroactive species coexistent in urine on the basis of their electrochemical reversibility. The relative standard deviations for repetitive determination of catecholamines were 0.6, 0.9, 1.8 and 5.1% for noradrenaline, adrenaline, dopamine and dopa, respectively.

CONCLUSION

While the greater part of applications of microbore HPLC have been performed as part of the exploration of a new technique, microbore HPLC was used effectively for the analysis, separation or purification of catecholamines, peptides and proteins.

In many biochemical or biomedical applications the amount of sample is limited and high mass sensitivity and selectivity are needed. Further enhancement may be available in the future through pretreatment techniques of samples, detector design and additional improvements in chromatographic technology. The routine application of microbore HPLC of biological samples described here awaits the availability of commercial columns and instruments. These advancements with microbore HPLC should pave the way for new possibilities in biomedical and/or biochemical research or clinical routine use for low-cost and fast analysis.

For all these attractive reasons it is likely that the use of microbore HPLC will develop over the coming years.

REFERENCES

Beavis, R. C., Bolbach, G., Ens, W., Main, D. E., Schueler, B., and Standing, K. G. (1986). Automated dry fraction collection for microbore high-performance liquid chromatography-mass spectrometry. *J. Chromatogr.* *359*, 489–497.

Caliguri, E. J. and Mefford, I. N. (1984). Femtogram detection limits for biogenic amines using microbore HPLC with electrochemical detection. *Brain Res.* *296*, 156–159.

Caliguri, E. J., Capella, P., Bottari, L., and Mefford, I. N. (1985). High-speed microbore liquid chromatography with electrochemical detection using 3-μm C_{18} packing material. *Anal. Chem.* *57*, 2423–2425.

Carlsson, A. and Lundström, K. (1985). An amperometric detector for use with small-bore liquid chromatographic columns. *J. Chromatogr.* *350*, 169–178.

Durkin, T. A., Caliguri, E. J., Mefford, I. N., Lake, D. M., Macdonald, I. A., Sundstrom, E., and Jonsson, G. (1985). Determination of catecholamines in tissue and body fluids using microbore HPLC with amperometric detection. *Life Sci.* *37*, 1803–1810.

Goto, M., Nakamura, T., and Ishii, D. (1981). Micro high-performance liquid chromatographic system with micro pre column and dual electrochemical detector for direct injection analysis of catechoramines in body fluids. *J. Chromatogr.* *226*, 33–42.

Goto, M., Sakurai, E., and Ishii, D. (1982). Dual electrochemical detector for micro high-performance liquid chromatography and its application to the selective detection of catecholamines. *J. Chromatogr.* *238*, 357–366.

Grego, B., van Driel, I. R., Stearne, P. A., Goding, J. W., Nice, E. C., and Simpson, R. J. (1985). A microbore high-performance liquid chromatography strategy for the purification of polypeptides for gas-phase sequence analysis. Structural studies on the murine transferrin receptor. *Eur. J. Biochem.* *148*, 485–491.

Grego, B., Nice, E. C., and Simpson, R. J. (1986a). Use of scanning diode array detector with reversed-phase microbore columns for the real-time spectral analysis of aromatic amino acids in peptides and proteins at the sub-microgram level. Applications to peptide and protein microsequencing. *J. Chromatogr.* *352*, 359–368.

Grego, B., van Driel, I. R., Goding J. W., Nice, E. C., and Simpson, R. J. (1986b). Use of microbore high-performance liquid chromatography for purifying subnanomole levels of polypeptides for microsequencing. Structural studies on the murine plasma cell antigen PC-1. *Int. J. Peptide Protein Res.* *27*, 201–207.

Hirata, Y., Lin, P. T., Novotny, M., and Wightman, R. M. (1980). Small volume electro-chemical detector for microcolumn liquid chromatography. *J. Chromatogr. 181*, 287–294.

Kamperman, G. and Kraak, J. C. (1985). Simple and fast analysis of adrenaline and noradren-aline in plasma on microbore high-performance liquid chromatography columns using fluori-metric detection. *J. Chromatogr. 337*, 384–390.

Nice, E. C., Lloyd, C. J., and Burgess, A. W. (1984). The role of short microbore high-performance liquid chromatography columns for protein separation and trace enrichment. *J. Chromatogr. 296*, 153–170.

Nice, E. C., Grego, B., and Simpson, R. J. (1985). Application of short microbore HPLC 'guard' columns for the preparation of samples for protein microsequencing. *Biochem. Int. 11*, 187–195.

Schachterle, S. and Alfredson, T. (1986). Preformed gradient technique for microbore high-performance liquid chromatography. *Anal. Chem. 58*, 1368–1372.

Scott, R. P. W. and Kucera, P. (1979). Use of microbore columns for the separation of substances of biological origin. *J. Chromatogr. 185*, 27–41.

Takeuchi, T., Saito, T., and Ishii, D. (1986). Microcolumn high-performance size-exclusion chromatographic separation of proteins. *J. Chromatogr. 351*, 295–301.

Van der Zee, R. and Welling, G. W. (1985). Microbore reversed-phase chromatography of proteins with conventional gradient equipment for high-performance liquid chromatography. *J. Chromatogr. 325*, 187–194.

Zimina, T. M., Maltsev, V. G., and Belenkii, B. G. (1986). Study of protein association and micelle formation in surfactants by microcolumn exclusion chromatography. *J. High Res. Chromatogr. Chromatogr. Commun. 9*, 111–115.

Progress in HPLC, Vol. 4, pp. 229–271.
Yoshioka *et al.* (Eds)
© 1989 VSP.

Review: High-performance liquid chromatography of metabolites of catecholamines and serotonin in urine, plasma, cerebrospinal fluid and brain tissue. I. Analytical methodology

AKIRA YOSHIDA[1], MASANORI YOSHIOKA[2]* and HASAN
PARVEZ[3]

[1] Chemicals Laboratory, Mitsubishi Petrochemical Co., Toho-cho,
Yokkaichi-shi, Mie 510, Japan;   [2] Faculty of Pharmaceutical Sciences,
Setsunan University, Nagaotogecho, Hirakata-shi, Osaka 573-01, Japan;
[3] Unité de Neuropharmacologie, Université de Paris XI, Centre d'Orsay —
Bat. 440, 91405 Orsay Cedex, France

ABBREVIATIONS

CA, catecholamines; CNS, central nervous system; CSF, cerebrospinal fluid; COMT, catechol-
o-methyltransferase; DA, dopamine; DOPMA, 3,4-dihydroxymandelic acid; DOPA, 3,4-dihy-
droxyphenylalanine; DOPAC, 3,4-dihydroxyphenylacetic acid; DHPE, 3,4-dihydroxypheny-
lethanol; DHPG, 3,4-dihydroxyphenylethylene glycol; E, epinephrine; EC, electrochemical;
EDTA, ethylenediaminetetraacetic acid; EHPG, 3-ethoxy-4-hydroxyphenylethylene glycol; FL,
fluorescence; 5-HIAA, 5-hydroxyindole-3-acetic acid; 5-HIPA, 5-hydroxyindole-3-propionic
acid; HPLC, high-performance liquid chromatography; 5-HT, 5-hydroxytryptamine (serotonin);
5-HTP, 5-hydroxytryptophan; HVA, homovanillic acid (3-methoxy-4-hydroxyphenylacetic
acid); IAA, indole-3-acetic acid; ILA, indolelactic acid; IPA, indolepyruvic acid; IS, internal
standard; iso-HVA, iso-homovanillic acid (3-hydroxy-4-methoxyphenylacetic acid); iso-VMA,
iso-vanillylmandelic acid (3-hydroxy-4-methoxymandelic acid); MAO, monoamineoxidase;
MHBA, 3-methoxy-4-hydroxybenzoic acid; MHPE, 3-methoxy-4-hydroxyphenylethanol;
MHPG, 3-methoxy-4-hydroxyphenylethylene glycol; MN, metanephrine; 3-MT, 3-methoxy-
tyramine; NE, norepinephrine; NMN, normetanephrine; TCA, trichloroacetic acid; TRP,
tryptophan; UV, ultraviolet; VLA,vanillactic acid (3-methoxy-4-hydroxyphenyllactic acid);
VMA, vanillylmandelic acid (3-methoxy-4-hydroxymandelic acid); VPA, vanilpyruvic acid (3-
methoxy-4-hydroxyphenylpyruvic acid).

INTRODUCTION

CA such as DA, NE and E, and 5-HT perform a number of important
neural and hormonal functions in mammalian systems.

CA in the body are degraded via two reactions, as shown in Fig. 1: the *o*-
methylation at the 3-position of the catechol group and oxidative deamina-
tion of the alkylamine side chain. The two enzymes for the above reactions,

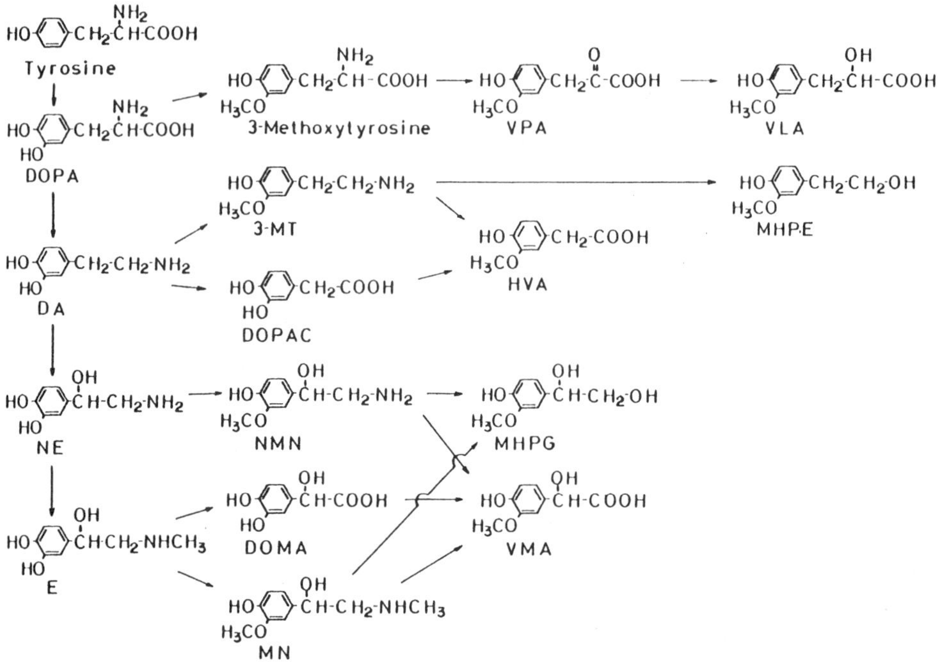

A. Yoshida et al.

Tyrosine

3-Methoxytyrosine   VPA   VLA

DOPA

3-MT   HVA   MHP.E

DA   DOPAC

NE   NMN   MHPG

E   DOMA   VMA

MN

Figure 1. Synthetic and metabolic pathways of CA.

COMT and MAO, are widely distributed throughout the body. The main metabolites from the DA are 3-MT, DOPAC and HVA, and those from the NE and/or E are NMN, MN, VMA and MHPG. On the other hand, 5-HT is metabolized to 5-HIAA by MAO in the CNS and the peripheral system, and to melatonin through *N*-acetylserotonin in the pineal body and the retina (Fig. 2).

Measurement of the levels of the metabolites is of great importance in the diagnosis of neuroblastoma (Greenberg and Gardner, 1960; von Studnitz, 1960, 1963; Williams and Greer, 1963; Bell, 1968; Voorhess, 1974; LaBrosse *et al.*, 1980; Sawada, 1981), phaeochromocytoma (von Studnitz, 1960; Voorhess, 1974) or carcinoid tumor (Sjoerdsma *et al.*, 1955; Udenfriend *et al.*, 1955; Dreux *et al.*, 1973) and in the course of treatment of these diseases and Parkinson's disease (Calne *et al.*, 1969; Sandler *et al.*, 1974). Furthermore, the levels of the metabolites in various physiological fluids or tissues are good indices in the fields of neurology and psychiatry (Joseph *et al.*, 1976) and in obtaining information about the turnover of CA and 5-HT (Co *et al.*, 1982).

Determination of the metabolites in body fluids and tissue samples requires the use of an efficient separation technique and sensitive detection devices, because of the complexity of biological materials and the low levels of endogenous compounds. Various techniques have been used for their determination such as spectrometry (Sunderman *et al.*, 1960; Pisano *et al.*,

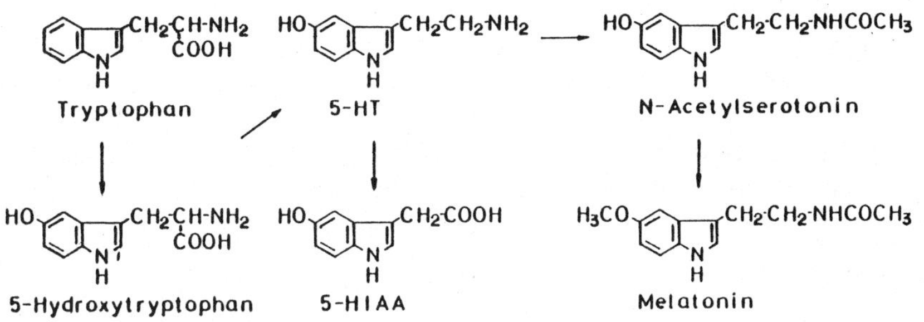

Figure 2. Synthetic and metabolic pathways of 5-HT.

1962; Udenfriend *et al.*, 1955; Anton and Sayre, 1967), fluorometry (Genefke, 1972), paper chromatography (Stott *et al.*, 1975) and thin-layer chromatography (Sankoff and Sourkes, 1963; Annino *et al.*, 1965). These procedures are, however, time-consuming and unreliable. Gas chromatography with packed columns (Karoum *et al.*, 1960; Dziedzic *et al.*, 1973; Muskiet *et al.*, 1977) and glass capillary columns (Chauhan and Darbre, 1980), and mass fragmentography (Gordon *et al.*, 1974; Takahashi *et al.*, 1978; Bacopoulos *et al.*, 1979; Muskiet *et al.*, 1980) have also been used. These methods are specific and sensitive, but complicated for routine analysis. An immunoassay method based on monoclonal antibodies will be used in the future. HPLC with UV, fluorometric and/or EC detection is a very promising method at present in the analysis of these substances, because of the simplicity of procedure and the relatively high reliability.

In this review we describe the liquid chromatographic methodology of the metabolites of CA and 5-HT in urine, plasma, CSF and brain tissue; in the following paper (Yoshida *et al.*, in submission), the clinical applications of HPLC methods for analysis of the metabolites will be reviewed.

TREATMENT OF THE PHYSIOLOGICAL FLUIDS AND TISSUE PRIOR TO CHROMATOGRAPHIC SEPARATION

Collection and storage of specimens

Because of the low levels of CA and 5-HT metabolites in various physiological fluids, extreme care must be taken during sample collection and storage.

Urine, usually collected in a bottle containing HCl, can be stored for several months at $-70°C$ to $-80°C$ (Jouve *et al.*, 1983). If samples are to be analyzed within a few weeks after collection, storage at $-20°C$ is sufficient (Yoshida *et al.*, 1976b; Hjemdahl *et al.*, 1979).

Plasma, CSF and brain tissue samples can be stored as raw samples, protein-free specimens or extracts at $-70°C$ to $-80°C$ in the presence of HCl, EDTA or sodium bisulfite (Hallman *et al.*, 1978; Davis *et al.*, 1978). These also can be stored at $-20°C$, as long as a few weeks (Hjemdahl, 1979; Lasley *et al.*, 1984). It is desirable that indole compounds such as 5-HIAA are stored at pH 3 to pH 5.6. At $-18°C$ under pH 2, a small loss of 5-HIAA (15%) is observed if stored longer than 6 days (Sato *et al.*, 1963; Geeraerts *et al.*, 1980).

A detailed discussion of the storage of biological samples is beyond this review, and reference should be made to the papers cited.

Isolation of the metabolites from physiological fluids and tissue

Three· principal methods for the isolation of the acidic or alcoholic meta-bolites are currently in use. The first is a liquid−liquid extraction method (Fig. 3), usually with ethyl acetate and often with diethyl ether and/or ethyl acetate−acetone. The second method is based on solid-phase extraction (Fig. 3) using Sep-Pak C_{18} (Waters Assoc., Milford, MA, USA), Bond-Elut C_{18} (Analytichem International, Harbor City, CA, USA) or Baker C_{18} (J. T. Baker, Phillipsburg, NJ, USA) columns pre-packed with C_{18} bonded silica. A third method for the isolation of the acidic or alcoholic constituents is based on anion exchange. Dowex AG, QAE- or DEAE-Sephadex appears to be the most widely used ion-exchange material.

For the isolation of the catechol compounds such as DOPAC, DOMA, DHPE and DHPG adsorption on alumina or boric acid gel is also useful. Schematic representations of the adsorption mechanism are shown in Figs 4 and 5.

Initial isolation of the basic CA metabolites, i.e. 3-*o*-methylated CA such as 3-MT, NMN and MN, proved to be more selective with cation exchange than with liquid−liquid or solid-phase extraction.

One more type of metabolites is the sulfate and glucuronide conjugates. These compounds contain either *o*-sulfate or *o*-glucuronide esters (e.g. $MHPG-SO_4$). Upon either acid or enzymatic hydrolysis the esters are cleaved to the parent metabolites.

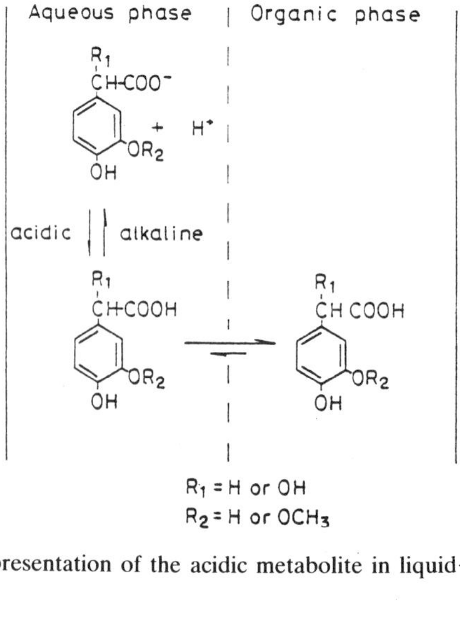

Figure 3. Schematic representation of the acidic metabolite in liquid−liquid equilibrium.

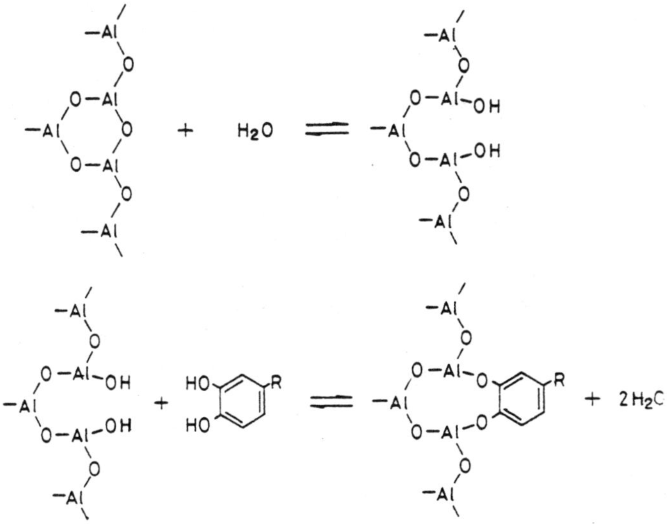

Figure 4. Schematic representation of the adsorption of catechol compounds on alumina.

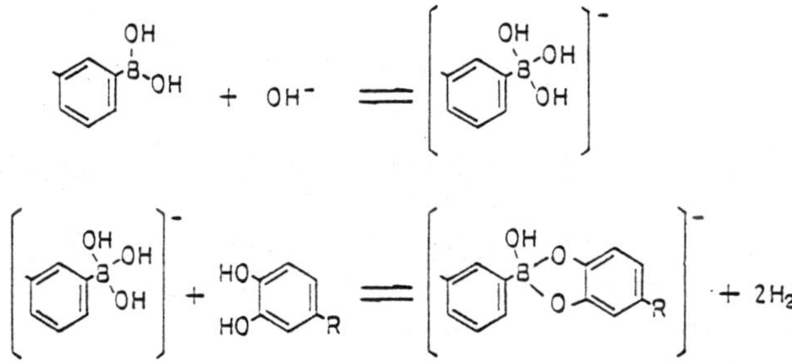

Figure 5. Scheme (Higa *et al.*, 1977) for the adsorption of catechol compounds on boric acid gel.

DETECTION METHODS

UV detection

The detection method most utilized for HPLC is UV detection. The wavelength utilized for a typical UV detection system is 280 nm. The UV detector is superior in stability, though lacking in selectivity and sensitivity. This detection system is therefore specially convenient for use in analysis of the acidic metabolites, e.g. HVA, VMA and 5-HIAA in urine, contained with concentrations of mg order per liter or higher.

Fluorometric detection

Fluorometic detection was the only method capable of monitoring picomole amounts of CA, 5-HT and their metabolites prior to the advent of the EC detector. Because of the relatively small number of compounds having natural FL, the use of this detection affords selectivity to the analytical system; the FL detector is stable compared with the EC detector. This detection method would therefore be useful in routine analysis for screening of CA secreting tumors such as neuroblastoma or phaeochromocytoma.

EC detection

EC detection is becoming increasingly popular for the determination of CA, 5-HT and their metabolites. This detection system is the most sensitive of the detectors currently in use. EC detection, however, is inconvenient compared with UV detection for routine use in a clinical laboratory because the sensitivity decreases with fouling of the surface of the glassy carbon electrode: the maintenance of an EC detector is more complicated than that of a UV or FL detector.

Although it was presumed for a long time that CA and vanil compounds were converted into the corresponding orthoquinones by EC oxidation (Kissinger et al., 1979; Kissinger, 1983), vanillin and 2,2'-dihydroxy-3,3'-dimethoxybiphenyl-5,5'-diacetic acid were actually identified as EC oxidation products from VMA and HVA, respectively, as shown in Fig. 6 (Yoshida et al., 1985a).

HPLC METHODS OF ANALYSIS FOR THE METABOLITIES OF CA AND 5-HT

The first use of liquid chromatography of the metabolites of CA and 5-HT in body fluids was demonstrated by Weise et al. in 1961, and concerned the separation of HVA and VMA from other organic acids using an anion exchange resin. The Body Fluids Analyses Group at Oak Ridge National Laboratory developed high-resolution anion-exchange liquid chromatographic systems that could separate many constituents in physiologic body fluids, including HVA, VMA and 5-HIAA (Scott et al., 1967; Scott, 1968; Mrochek et al., 1971; Chilcote and Mrochek, 1972; Chilcote, 1972; Mrochek et al., 1973). Forty hours were required for the separation of 50 constituents

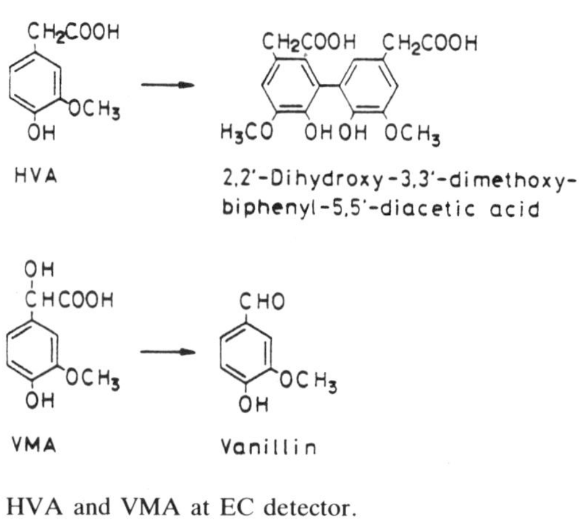

Figure 6. Oxidation of HVA and VMA at EC detector.

in urine: the method was good, but of little practical use because it was so time-consuming.

Änggård *et al.* (1970) separated methyl esters of phenolic acids and alcohols derived from CA and 5-HT using a column of Sephadex LH-20; Lange *et al.* (1970), a standard mixture of CA metabolites by cation-exchange chromatography; Persson and Karger (1974), biogenic amines and their metabolites by ion-pair HPLC; Molnár and Horváth (1976), CA and their metabolites by reversed-phase HPLC. The methods, however, were not applied to biological materials.

A practical method for the simultaneous determination of HVA and VMA in urine by HPLC was first proposed by the authors of this paper (Yoshida *et al.*, 1976a, b). A column of porous polystyrene was used in a gradient elution mode with increasing pH of the eluent, as shown in Fig. 7; absorbance at 280 nm as monitored for quantitative evaluation. Subsequently, Graffeo and Karger (1976) developed a HPLC with fluorometric detection for urinary 5-HIAA; Felice and Kissinger (1976), a HPLC with EC detection for urinary HVA; Stout *et al.* (1976), a HPLC for separating radiolabeled CA metabolites in cellular extracts derived from mammalian cells grown in tissue culture. The methods used chemically bonded reversed-phase packings. It is proved from these studies that HPLC is very useful for the determination of the metabolites of CA and 5-HT in body fluids.

Until recently, a number of HPLCs for metabolites have been developed and applied also to plasma, CSF and brain tissue samples.

Urine

Abnormal excretion of the metabolites is believed to be associated with many psychological and/or physiological diseases such as intrinsic hypertension, neuroblastoma, phaeochromocytoma and Parkinson's disease. Urinary levels of the metabolite are of great importance, especially in the diagnosis of CA or 5-HT secreting tumors.

HPLC methods for the determination of the metabolites of CA and 5-HT in human urine are summarized in Table 1. Pretreatment procedures for the acidic and alcoholic metabolites are easier to carry out than those for CA and 5-HT. Efficient and conventional purification procedures for these meta-

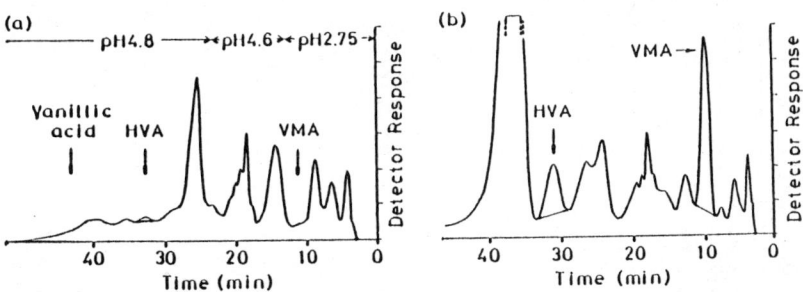

Figure 7. Liquid chromatograms (Yoshida *et al.*, 1976a) of urine specimens from (a) a normal subject and (b) a patient with neuroblastoma.

Table 1.
Literature of HPLC of CA and 5-HT metabolites in human urine

Compound	Sample preparation	Column	Mobile phase	Detection	Reference
HVA, VMA	Ethyl acetate extracction	Hitachi Gel No.3010 (porous polystyrene)	0.05 M Tartrate buffer/methanol (4:1); two-step gradient elution using the buffer of pH 2.75, 4.6 and 4.8	UV, 280 nm	Yoshida et al. (1976a)
HVA, VMA	Ethyl acetate extraction	Hitachi Gel No.3010	Primary, 0.05 M tartrate buffer, pH 3.15/methanol (17:3); secondary, 0.05 M tartrate buffer, pH 5.25/methanol (17:3); gradient, see article	UV, 280 nm	Yoshida et al. (1976b)
HVA	Ethyl acetate extraction	Vydac bonded phase pelicular anion exchanger	0.025 M Acetate buffer, pH 4.7/ 0.0025 M citrate buffer, pH 5.3 (4:1)	EC, +0.75 V	Felice and Kissinger (1976)
5-HIAA	Deproteinization with $ZnSO_4$–NaOH; centrifugation; filtration	μBondapak C_{18}	0.01 M Acetate buffer, pH 4.6/ methanol (9:1)	FL of 310 nm excited at 280 nm	Graffeo and Karger (1976)
VMA	Ethyl acetate extraction; K_2CO_3 back-extraction; conversion into vanillin; toluene extraction	μBondapak C_{18}	0.5 M Acetate buffer, pH 4.7	EC, +0.80 V	Felice and Kissinger (1977)
5-HIAA	Diethyl ether extraction	Hitachi Gel No.3010	Primary, 0.05 M tartrate buffer, pH 3.15/methanol (17:3); secondary, 0.05 M tartrate buffer, pH 5.25/methanol (17:3); gradient, see article	UV, 280 nm	Yoshida et al. (1977)
5-HIAA, IAA, 5-HIPA (IS)	Ethyl acetate extraction	Nucleosil C_{18}	0.01 M Acetate buffer, pH 5.5/ methanol (90.5:9.5)	FL of 310 nm excited at 272 nm	Beck et al. (1977)

Table 1. (*Cont'd*)

Compound	Sample preparation	Column	Mobile phase	Detection	Reference
DOPAC, HVA, VMA, 5-HIAA	Ethyl acetate extraction	Lichrosorb RP-18	Primary, 0.1 M phosphate buffer pH 2.1; secondary, acetonitrile; gradient, linear from 0 to 40% of secondary solvent in 30 min	UV, 280 nm	Molnár and Horváth (1977)
DOPAC	Ethyl acetate extraction	Zipax anion exchanger	0.1 M Acetate buffer, pH 4.7	EC, +0.60 V	Felice et al. (1977)
3-MT, NMN, MN	Hydrolysis with HCl; Bio-Rex 70 (weak cation exchanger) chromatography; ethyl acetate–acetone extraction	μBondapak C_{18}	0.1 M Citric acid/0.1 M Na_2HPO_4/methanol (10:10:1)	EC, +0.85 V	Shoup and Kissinger (1977)
DOPAC, HVA, VLA, vanillic acid, DA, DOPA, 3-methoxytyrosine	Hydrolysis with Glusulase; alumina adsorption	μBondapak C_{18}	Water/acetic acid (99:1) +5 mM 1-heptanesulfonic	UV, 280 nm	Mitchell and Coscia (1978)
5-HIAA	Dinitrolphenyl-coupled thiolated Sephadex G-25 chromatography	Partisil ODS	110 ml Acetonitrile +890 ml water +0.4 ml conc. sulfuric acid +0.1 g sodium lauryl sulfate	UV, 280 nm	Fornstedt (1978)
VPA, HVA, VMA	Ethyl acetate extraction	Hitachi Gel No.3010	Primary, 0.05 M tartrate buffer, pH 3.15/methanol (17:3); secondary, 0.05 M tartrate buffer, pH 5.25/methanol (17:3); gradient, see article	UV, 280 nm	Yoshida et al. (1978)
VMA	Du Pont Prep I anion-exchange column extraction	Du Pont Octyl silica	0.05 M Phosphate buffer, pH 3.0	Post-column periodate oxidation; UV, 360 nm	Rosano and Brown (1979)

Table 1. (*Cont'd*)

Compound	Sample preparation	Column	Mobile phase	Detection	Reference
VMA	Filtration	μBondapak C_{18}	0.01 M Phosphate buffer, pH 2.7/acetonitrile (97:3)	Post-column periodate oxidation; UV, 360 nm	Flood et al. (1979)
VMA	Ethyl acetate extraction	Lichrosorb RP-18	0.05 M Acetate buffer, pH 3.0	EC, +1.3 V	Morrisey and Shihabi (1979a)
HVA	Dilution	Lichrosorb RP-18	0.06 M Phosphate buffer, pH 3.0/methanol (77:23)	EC, +1.2 V	Morrisey and Shihabi (1979b)
MHPG, MHPG-SO$_4$, MHPG-Glu	Hydrolysis with aryl sulfatase or β-glucuronidase; conversion into vanillyl alcohol; ethyl ether extraction	μBondapak C_{18}	0.5 M Acetate buffer, pH 4.7/methanol (17:3)	EC, 0.85 V	Buchanan et al. (1979)
VMA	Ethyl acetate extraction	μBondapak C_{18}	Primary, 0.1 M phosphate buffer, pH 2.50; secondary, methanol/water (3:2); gradient, linear from 0 to 100% of secondary solvent in 20 min	UV, 280 nm	Bertani-Dziedzic et al. (1979)
5-HIAA[1], 5-HT[1], TRP[2]	Amberlite CG-50, Dowex AG-50, Sephadex G-10 chromatography	μBondapak C_{18}	[1] 0.5 M Acetate buffer, pH 5.1/methanol (17:3) [2] Mcllvaine buffer, pH 4.0/methanol (4:1)	EC, [1]+0.5 V [2]+1.00 V	Koch and Kissinger (1979)
VMA, MHPG	Ethyl acetate extraction	μBondapak C_{18}	Primary, 0.1 M KH_2PO_4; secondary, acetonitrile/water (3:2); gradient, linear from 0 to 60% of secondary solvent in 45 min	EC, +1.00 V	Krstulovic et al. (1979)

Table 1. (*Cont'd*)

Compound	Sample preparation	Column	Mobile phase	Detection	Reference
DOPAC, DOMA, HVA, VMA, DHPE, DHPG, MHPE, MHPG, 3-MT, NMN, MN, DA, NE, E	Hydrolysis with Glusulase; Dowex 50W-X4 chromatography; alumina chromatography	µBondapak C$_{18}$	See article	UV, 254 nm	Hoeldtke and Stetson (1980)
HVA, VMA	Dowex, AG1-X4 chromatography	µBondapak C$_{18}$	0.1 M Phosphate buffer, pH 6.7 + 0.5 mM tetrabutylammonium phosphate	EC, +0.76 V	Soldin and Hill (1980)
5-HIAA	Dichloromethane-t-amyl alcohol extraction	Ultrasphere C$_{18}$	0.01 M Phosphate buffer, pH 2.2/methanol (83:17)	EC, +1.0 V	Shihabi and Scaro (1980)
5-HIAA, p-nitrobenzoic acid (IS)	Ether extraction; pH 9.2 buffer back-extraction	Lichrosorb RP-18	Acetate buffer, pH 3.5/methanol (19:1)	UV, 254 nm	Draganac et al. (1980)
VMA	Anion-exchange resin column extraction	High-capacity cation-exchange resin (Dionex)	10 mM HCl + 140 mM acetonitrile	EC, +1.0 V	Rich et al. (1980)
MHPG, conjugated MHPG	Hydrolysis with Glusulase; ethyl acetate extraction	µBondapak C$_{18}$	Primary, 0.1 M phosphate buffer, pH 2.5; secondary, acetonitrile/water (3:2); gradient, linear from 0 to 60% of secondary solvent in 45 min	EC, +1.00 V	Krstulovic et al. (1980)
5-HIAA, N-acetyltryptophan (IS)	Deproteinization with sulfasalicyclic acid	Partisil SAX	5 mM Phosphate buffer, pH 2.30	FL of 338 nm excited at 301 nm	Garnier et al. (1981)

Table 1. (*Cont'd*)

Compound	Sample preparation	Column	Mobile phase	Detection	Reference
DOPAC, HVA, VMA, MHPG, 5-HIAA	Hydrolysis with aryl sulfatase and β-glucuronidase; ethyl acetate extraction	Hypersil ODS	Primary, 0.1 M phosphate buffer, pH 3.0/methanol (19:1); secondary, 0.1 M phosphate buffer, pH 3.0/methanol (17:3); single-step gradient from primary solvent to secondary solvent	EC, +0.72 V	Joseph et al. (1981)
NMN, MN	Acid hydrolysis; Dowex CG-50 chromatography	Ultrasphere ODS	Primary, 0.1 M KH_2PO_4; secondary, methanol/water (3:2); gradient, linear from 0 to 60% of secondary solvent in 35 min	EC, +1.00 V	Bertani-Dziedzic et al. (1981)
5-HIAA	Filtration or centrifugation	Lichrosorb RP-18 coated with tri-n-butyl phosphate	0.1 M Phosphate buffer, pH 6.0/ methanol (9:1)	FL of 370 nm excited at 280 nm	Wahlund and Edlén (1981a)
5-HIAA, IAA	Filtration or centrifugation	Lichrosorb RP-18 coated with tri-n-butyl phosphate	0.1 M Phosphate buffer, pH 6.0/ methanol (9:1)	UV, 280 nm; FL of 370 nm excited at 280 nm	Wahlund and Edlén (1981b)
HVA, VMA, 5-HIAA	Sep-Pak C_{18} column extraction	Lichrosorb SI-10 coated with triethanolammonium perchlorate phosphoric acid aqueous solution	n-Heptane/isopropyl alcohol/ isobutyl alcohol (6:3:1) saturated with stationary phase	UV, 280 nm	Ghebregzabher et al. (1981)
VMA, p-hydroxybenzoic acid (IS)	Ethyl acetate extraction; K_2CO_3 back-extraction; pre-column dansylation; chloroform extraction	SS-10 silica gel	Ligroin (b.p. range 80–100°C)/ chloroform/ethyl acetate (1:4:8)	FL of 505 nm excited at 360 nm	Yamada et al. (1981)

Table 1. (*Cont'd*)

Compound	Sample preparation	Column	Mobile phase	Detection	Reference
MHPG	Hydrolysis with Glusulase; ethyl acetate extraction; K_2CO_3 washing	Du Pont CLC-2	0.01 M Acetate buffer, pH 5.1/ acetonitrile (92.5:7.5)	FL of 310 nm excited at 265 nm	Taylor et al. (1981)
HVA	Dilution with formic acid; adjustment to pH 2.9	Du Pont CLC-1	Acetonitrile/formic acid/water (35:1:465)	Post-column potassium ferricyanide oxidation; FL of 420 nm excited at 320 nm	Rosano et al. (1981)
MHPG	Hydrolysis with Glusulase; ethyl acetate extraction; $KHCO_3$ washing	μBondapak C_{18}	0.05 M Phosphate buffer, pH 5	EC, +0.70 V	Alonso et al. (1981)
HVA, VMA	Ethyl acetate extraction	Hitachi Gel No.3010-O (hydroxymethylated porous polystyrene)	0.05 M Tartrate buffer, pH 3.20/acetonitrile (4:1)	UV, 280 nm	Yoshida et al. (1982)
MHPG	Precipitation of anionic compounds with $BaCl_2$; hydrolysis with Glusulase; ethyl acetate extraction	Hypersil ODS	0.05 M Na_2HPO_4 + 1.34 mM EDTA–2Na; → pH 3	EC, +0.8 V	Moleman and Borstrok (1982)
5-HIAA, ILA, IPA, IAA, 5-methoxytryptamine, tryptamine	Hydrolysis with HCl; chloroform or ether extraction; pH 7.0 buffer back-extraction	Hitachi Gel No.3053	1 M Sodium acetate/1 M acetic acid/1 M Na_2SO_4/methanol/ water (1:4:10:25:60)	UV, 280 nm	Yamaguchi et al. (1982)
HVA, 5-HIAA	Sephadex G-10 chromatography	Nucleosil C_{18}	0.1 M Trichloacetic acid-sodium acetate, pH 4.0/methanol (17:3)	EC, +0.7 V	Westerink et al. (1982)
5-HIAA, 5-hydroxyindole-2-carboxylic acid (IS)	Ethyl acetate extraction	μBondapak C_{18}	0.08 M Sodium acetate-0.297 mM EDTA-2Na (→ pH 4.5) + 1–3% methanol	EC, +0.5 V	Petruccelli et al. (1982)

Table 1. (Cont'd)

Compound	Sample preparation	Column	Mobile phase	Detection	Reference
HVA, VMA	Ethyl acetate extraction; 0.2 M tris(hydroxymethyl)aminomethane back-extraction; DEAE-cellulose chromatography	μBondapak C_{18}	200 ml 0.05 M Citric acid + 800 ml 0.05 M Na_2HPO_4 + 20–60 ml acetonitrile	EC, +0.65 V	Bauersfeld et al. (1982)
HVA	Filtration	μBondapak C_{18}	0.1 M Phosphate buffer, pH 4.0/ methanol (9:1)	EC, +0.80 V	Seegal et al. (1983)
3-MT, NMN, MN	Hydrolysis with HCl; Dowex 50W-X2, Bio-Rex 70 chromatography	μBondapak C_{18}	0.02 M Citric acid/0.02 M Na_2HPO_4 (2:1) + 0.025 M sodium octylsulfonate + 0.5 mM EDTA-2Na	EC, +0.90 V	Jouve et al. (1983)
VMA	Graphitized carbon black chromatography	Perkin-Elmer C18/10	0.01 M Phosphate buffer, pH 3.0/methanol (99:1)	UV, 280 nm	Laganà and Rotatori (1983)
5-HIAA[1], TRP[2]	Filtration	Polygosil C_{18} coated with tri-n-butyl phosphate	0.05 M Phosphate buffer, pH 1.95/methanol (4:1) + 0.14 M ClO_4^-	EC, +0.70 V;[1] FL of 333 nm excited at 296 nm [2] FL of 354 nm excited at 282 nm	De Jong et al. (1983)
HVA, VMA	Dowex AG1-X4 chromatography	Nucleosil C_{18}	Primary, 5 mM H_3PO_4 + 25 μM EDTA; secondary, methanol; gradient, linear from 0 to 30% of secondary solvent in 50 min	UV, 280 nm	Betto et al. (1983)

Table 1. (*Cont'd*)

Compound	Sample preparation	Column	Mobile phase	Detection	Reference
HVA	Direct injection	Hitachi Gel No. 3011-N	0.5 M Borate buffer, pH 10.0–0.25 M KCl/acetonitrile (4:1)	Post-column electrochemical oxidation; FL of 420 nm excited at 325 nm	Momose et al. (1983)
NMN, MN	Hydrolysis with HCl; Bio-Rad catecholamine isolation column chromatography; ethyl acetate-acetone extraction	Radial-Pak C_8	Triethylamine-phosphoric acid buffer, pH 3.0–5 mM sodium heptane sulfonate/acetonitrile (47:3)	EC, +0.85 V	Orsulak et al. (1983)
HVA[1], VMA[1], 5-HIAA[2]	Ethyl acetate extraction	Yanapak ODS-T	[1] Primary, 0.2 M phosphate buffer, pH 3.0; secondary, 0.2 M phosphate buffer, pH 3.0/acetonitrile (9:1); single-step gradient [2] 0.2 M phosphate buffer, pH 3.0/acetonitrile (9:1)	EC, +0.6 V	Fujita et al. (1983)
VMA	Ethyl acetate extraction; pH 8.5 buffer back-extraction	Hypersil ODS	0.1 M Na_2HPO_4 + 1.49 mM EDTA-2Na; →pH 2.7	EC, +0.8 V	Moleman and Borstrok (1983)
HVA, VMA, 5-HIAA, iso-VMA (IS)	Ethyl acetate extraction, $NaHCO_3$ back-extraction	μBondapak C_{18}	9 mM Citric acid-89 mM sodium acetate buffer, pH 4.7/methanol (47:3)	EC, +0.80 V	Frattini et al. (1983)
HVA, VMA	Deproteinization with acetonitrile	Yanako NB-5801	3% Formic acid + 50 μM EDTA-2Na	EC, +0.80 V	Kodama et al. (1984)
HVA, VMA, 5-HIAA	QAE-Sephadex A-25 chromatography; ethyl acetate extraction	Nucleosil C_{18}	0.05 M Citric acid/acetonitrile (7:1)	UV, 280 nm; EC, +0.75 V	Yoshida et al. (1984)

Table 1. (*Cont'd*)

Compound	Sample preparation	Column	Mobile phase	Detection	Reference
5-HIAA	Sephadex G-10 chromatography	Licrosorb RP-18	0.05 M Acetate buffer, pH 4.4 + 0.1 M Na_2SO_4 + 0.5 μM $CoCl_2$	Post-column derivatization with Na_2CO_3-NaOH-$CoCl_2$; FL of 460 nm excited at 360 nm	Iinuma *et al.* (1984)
HVA, VMA	Direct injection	Shodex Axpak N-422	5 M Acetate buffer, pH 4.6	EC, +0.65 V	Tokuda *et al.* (1984)
MHPG	Hydrolysis with Glusulase; mixed-bed anion−cation (Bio-Rad AG-1-X4 + Bio-Rad AG-50W-X4) exchange chromatography; ethyl acetate extraction	Spherisorb ODS-2	0.01 M Acetate buffer, pH 4.0 + 1 mM EDTA	EC, +0.9 V	Shea and Howell (1984)
5-HIAA, 5-HT, TRP	Centrifugation; filtration	Unicil C_{18}	Primary, stock solution/water (1:1); secondary, stock solution/water/methanol (4:3:1); stock solution, 4.8 g acetic acid, 1.64 g sodium acetate, 28 g Na_2SO_4 in 1 liter water; single-step gradient	FL of 340 nm excited at 285 nm	Iwatani and Nakamura (1984)
HVA, VMA, 5-HIAA	Sep-Pak C_{18} column extraction	Radial-Pak C_{18}	See article	FL of 315 nm excited at 285 nm	Wielders and Mink (1984)
5-HIAA, 5-hydroxyindole-2-carboxylic acid (IS)	Direct injection	μBondapak C_{18}	0.1 M Acetate buffer, pH 4.5/ methanol (43:7)	FL of 345 nm excited at 295 nm	Skrinska and Hahn (1984)

Table 1. (*Cont'd*)

Compound	Sample preparation	Column	Mobile phase	Detection	Reference
HVA[1], VMA[2], iso-vanillic acid[1] (IS), iso-VMA[2] (IS)	Anion-exchange chromatography	Bio-Rad ODS	[1] 0.05 M Phosphate buffer, pH 2.30/ethanol/isopropyl alcohol (500:50:1) [2] 0.05 M phosphate buffer, pH 3.00/ethanol (125:2)	EC, [1] +0.46 V [2] +0.30 V	Binder and Sivorinovsky (1984)
MHPG, EHPG (IS)	Hydrolysis with Glusulase; anion-exchange chromatography; ethyl acetate extraction	µBondapak C$_{18}$	0.1 M H$_3$PO$_4$ + 0.27 mM EDTA-2Na + 2% acetonitrile; →pH 3.1	EC, +0.85 V	Brown *et al.* (1984)
MHPG	Hydrolysis with sulfatase; PH Bond-Elut column extraction	µBondapak C$_{18}$	0.05 M Citrate-phosphate buffer, pH 3.5/methanol (10:1)	EC, +0.85 V	Karege (1984)
HVA, VMA, 5-HIAA, MHBA (IS)	Ethyl acetate extraction	Zorbax SP-150 C$_8$	0.9% KH$_2$PO$_4$/methanol (24:1); →pH 3.1	EC, +0.65 V	Dutrieu and Delmotte (1984)
HVA, 5-HIAA	Hydrolysis with HCl; 10-fold dilution; µBondapak C$_{18}$ HPLC	µBondapak C$_{18}$	0.1 M Phosphate buffer, pH 6.0/methanol (17:3) + 86 µM EDTA	FL of 360 nm excited at 254 nm	Anderson *et al.* (1985b)
5-HIAA, 5-HIPA (IS)	Baker C$_{18}$ column extraction	Biophase ODS	50.7 ml conc. NH$_4$OH–64.7 ml acetic acid–0.2 g EDTA–2Na–1760 ml water (→pH 5.1) + 325 ml methanol	EC, +0.55 V	Chou and Jaynes (1985)
VMA, iso-VMA (IS)	µBondapak C$_{18}$ preparatory column extraction	µBondapak C$_{18}$	0.1 M Na$_2$HPO$_4$–86 µM EDTA–2 mM octyl-triethylammonium phosphate (→pH 5.0) + 41.7 ml methanol	FL of 325 nm excited at 285 nm	Anderson *et al.* (1985a)

bolites include extraction into an organic solvent (or a mixture of organic solvents) (Yoshida *et al.*, 1976a, b, 1977, 1978, 1982; Felice and Kissinger, 1976; Beck *et al.*, 1977; Molnár and Horváth, 1977; Felice *et al.*, 1977; Morrisey and Shihabi, 1979a; Bertani-Dziedzic *et al.*, 1979; Krstulovic *et al.*, 1979; Petruccelli *et al.*, 1982; Fujita *et al.*, 1983; Dutrieu and Delmotte, 1984) and successive back-extraction into an alkaline aqueous solution frequently followed by re-extraction into an organic solvent (or a mixture of organic solvents) (Fellice and Kissinger, 1977; Draganac *et al.*, 1980; Yamada *et al.*, 1981; Yamaguchi *et al.*, 1982; Bauersfeld *et al.*, 1982; Moleman and Borstrok, 1983; Frattini *et al.*, 1983). A C_{18} column extraction method may be simple as compared with organic solvent extraction (Ghebregzabher *et al.*, 1981; Wielders and Mink, 1984; Chou and Jaynes, 1985; Anderson *et al.*, 1985a). Karege (1984) reports that column extraction gives very clean sample extraction as well as good recovery, compared to liquid–liquid extraction. Use of an anion-exchange column extraction techniques is also effective (Rosano and Brown, 1979; Koch and Kissinger, 1979; Soldin and Hill, 1980; Rich *et al.*, 1980; Betto *et al.*, 1983; Yoshida *et al.*, 1984; Shea and Howell, 1984; Binder and Sivorinovsky, 1984; Brown *et al.*, 1984). Alumina adsorption is one of the better methods for DOPAC (Mitchell and Coscia, 1978; Hoeldtke and Stetson, 1980). For 3-MT, NMN and MN cation-exchange column chromatography is used (Shoup and Kissinger, 1977; Hoeldtke and Stetson, 1980; Bertani-Dziedzic *et al.*, 1981; Jouve *et al.*, 1983; Orsulak *et al.*, 1983).

Absorbance at 280 nm, natural fluorescence of 310 nm to 370 nm excited at 265 nm to 300 nm or EC response is monitored for quantitative evaluation of urinary metabolites.

Felice and Kissinger (1977) oxidize VMA to vanillin with periodate in a pre-column and detect the vanillin by EC detector. Rosano and Brown (1979) and Flood *et al.* (1979) convert VMA in the chromatographic effluent to vanillin with periodate and continuously monitor the periodate oxidation products at 360 nm. For HVA, Rosano *et al.* (1981) and Momose *et al.* (1983) report a method that detects the ferricyanide oxidation product and the electrochemical oxidation product by fluorescence, respectively. These post-column reaction chromatographies do not require complicated pretreatment procedures.

We applied HPLC to the determination of HVA, VMA and 5-HIAA in urine of dog, rat, mouse and muskrat (Yoshioka *et al.*, 1985); Seegal *et al.* (1986) also separated HVA and 5-HIAA from rat urine. The urinary excretion of the acids in the experimental animals had not been determined by a reliable method such as HPLC.

Plasma

Plasma concentrations of DOPAC and HVA have been suggested to reflect changes in the brain dopaminergic system (Bacopoulos *et al.*, 1979; Kendler *et al.*, 1981). Plasma levels of VMA and MHPG may closely reflect central and peripheral noradrenergic activity (Schildkraut *et al.*, 1978; De Met *et al.*, 1982; Jimerson *et al.*, 1983; Ko *et al.*, 1983); that of 5-HIAA, serotonergic activity. Therefore, determination of the metabolites of CA and 5-HT in plasma is an important diagnostic and investigatory aid for pharmacologists and clinicians.

HPLC methods for the determination of the metabolites in human plasma are summarized in Table 2. Conventional purification procedures for plasma samples are deproteinization with $HClO_4$ or the other acids (Petruccelli *et al.*, 1982; Martinez *et al.*, 1983; Chang *et al.*, 1983; Javaid *et al.*, 1983; Ishimitsu and Hirose, 1985) and successive extraction into an organic solvent or on an ion-exchange column (Koch and Kissinger, 1979; Harris *et al.*, 1984; Schinelli *et al.*, 1985; Shea and Howell, 1984). A direct adsorption on alumina may be effective for DOPAC and DHPG (Mefford *et al.*, 1981; Jackman *et al.*, 1982; Rossetti *et al.*, 1983; Howes *et al.*, 1985). A C_{18} column extraction method is also very effective because of the simplicity (Minegishi and Ishizaki, 1984; Yoshida *et al.*, 1985b; Huber-Smith *et al.*, 1986).

Because of low concentrations of the metabolites in plasma, EC detection is used for monitoring.

Mais *et al.* (1981) determined the 5-HT metabolites in mouse plasma by HPLC with fluorescence detection; Morita *et al.* (1981) developed HPLC with both fluorometric and EC detection for the 5-HT metabolites in rat or rabbit plasma.

CSF

The assessment of monoamine metabolism in the CNS is usually deduced from the metabolite levels of the parent brain amines in CSF, which is the source that probably best reflects neural activity in the CNS as opposed to urine and plasma (Garelis *et al.*, 1974). The study of the CA and 5-HT metabolites in CSF is therefore important for clinical investigation of diseases of the central nervous system.

HPLC methods for the determination of the metabolites in human CSF are summarized in Table 3. Pretreatment procedures for CSF samples are centrifugation (Anderson *et al.*, 1979; Langlais *et al.*, 1980; Anderson *et al.*, 1981a), filtration (Scheinin *et al.*, 1983; Botttiglieri *et al.*, 1984), deproteinization with $HClO_4$ or acetonitrile (Petruccelli *et al.*, 1982; Bockstaele *et al.*, 1983; Javors *et al.*, 1984) and ethyl acetate extraction (Beck *et al.*, 1977; Krstulovic *et al.*, 1981, 1982). These are simple to carry out. Direct injection into an analytical column is tried by Anderson and Purdy (1979), Semerdjian-Rouquier *et al.* (1981), Narasimhachari *et al.* (1982) and Gagnieu *et al.* (1984). In the major reports, EC detection is used for monitoring.

HPLC methods with EC detection are also effective for the determination of the metabolites in CSF of experimental animals: monkey, (Scheinin *et al.*, 1983; Seegal *et al.*, 1986); rabbit (Wightman *et al.*, 1977; Bockstaele *et al.*, 1983); rat (Anderson and Purdy, 1979; Wagner *et al.*, 1982; Le Quan-Bui *et al.* (1982).

Table 2.
Literature of HPLC of CA and 5-HT metabolites in human plasma

Compound	Sample preparation	Column	Mobile phase	Detection	Reference
5-HIAA[1], 5-HT[1], TRP[2]	Deproteinization with $HClO_4$; Amberlite CG-50, Dowex AG-50, Sephadex G-10 chromatography	μBondapak C_{18}	[1] 0.5 M Acetate buffer, pH 5.1/ methanol (17:3) [2] McIlvaine buffer, pH 4.0/ methanol (4:1)	EC, [1] +0.50 V [2] +1.00 V	Koch and Kissinger (1979)
DOPAC, DA, NE, E, 3,4-dihydroxybenzylamine (IS)	Alumina adsorption	Ultrasphere ODS	0.1 M Sodium acetate–0.2 M citric acid–0.431 mM sodium octyl sulfate–0.17 mM EDTA/methanol (9:1)	EC, +0.60 V	Mefford et al. (1981)
DHPG	Alumina adsorption	Spherisorb ODS	10 mM $HClO_4$	EC, +0.82 V	Jackman et al. (1982)
5-HIAA, 5-HT, 6-fluoro-5-HT (IS)	Deproteinization with $HClO_4$; centrifugation	μBondapak C_{18}	0.08 M Sodium acetate–0.297 mM EDTA–2Na (→pH 4.5) + 1–3% methanol	EC, +0.5 V	Petruccelli et al. (1982)
VMA, MHPG, hydroquinone (IS)	Bio-gel P-10 chromatography; ethyl acetate extraction	Spherisorb ODS	0.07 M Phosphate buffer, pH 2.5/methanol (97:3) + 0.297 mM EDTA-2Na	EC, +0.8 V	Ong et al. (1982)
5-HIAA[1], IAA[2]	Deproteinization with $HClO_4$ or TCA	μBondapak C_{18}	[1] 30 mM Citric acid–60 mM Na_2HPO_4 buffer, pH 4.8/ methanol (47:3) [2] 5 mM 1-pentanesulfonic acid/methanol/water (10:7:3)	[1] EC, +0.55 V [2] FL of 340 nm excited at 280 nm	Martinez et al. (1983)
HVA, 5-fluoro-HVA (IS)	Deproteinization with $HClO_4$	Supelco LC8	0.08 M Sodium acetate/ methanol/acetonitrile (45:4:1) + 0.25 mM EDTA–2Na; →pH 4.20	EC, +0.75 V	Chang et al. (1983)

Table 2. *(Cont'd)*

Compound	Sample preparation	Column	Mobile phase	Detection	Reference
HVA, 5-HIAA, vanillic acid (IS)	Deproteinization with $HClO_4$; centrifugation	Analytical C_8	0.1 M Acetate buffer, pH 4.5–1 mM EDTA/methanol (9:1)	EC, +0.95 V	Javaid et al. (1983)
DOPAC, DA, NE, E, 3,4-dihydroxybenzylamine (IS)	Alumina adsorption	Biosil ODS	0.1 M Citrate–acetate buffer, pH 4.35–0.513 mM EDTA-0.647 mM sodium octyl sulfate/methanol (96:5:3.5); flow-gradient, linear from 0.5 to 2.0 ml/min in 15 min	EC, +0.700 V	Rossetti et al. (1983)
MHPG	Deproteinization with formic acid; heptane-chloroform washing; mixed-bed anion–cation (Bio-Rad AG-1-X4 + Bio-Rad AG-50W-X4) exchange chromatography; ethyl acetate extraction	Spherisorb ODS-2	0.01 M Acetate buffer, pH 4.0 + 1 mM EDTA	EC, +0.9 V	Shea and Howell (1984)
DOPAC[1], HVA[2], 5-HIAA[1], 5-hydroxyindole-2-carboxylic acid (IS)	Bond-Elut C_{18} column extraction	Yanapak ODS-A	0.1 M Phosphate buffer, pH 3.2–10 μM EDTA–2Na/methanol (41:9)	EC, [1]+0.6 V [2]+0.75 V	Minegishi and Ishizaki (1984)
HVA	Hexane washing; deproteinization with $HClO_4$; ethyl acetate extraction	Biophase ODS	0.05 M Sodium acetate–0.342 mM EDTA/methanol (22:3)	EC, +0.72 V	Harris and Bacopoulos (1984)
MHPG	Hydrolysis with sulfatase; PH Bond-Elut column extraction	μBondapak C_{18}	0.05 M Citrate-phosphate buffer, pH 3.5/methanol (10:1)	EC, +0.85 V	Karege (1984)

Table 2. (*Cont'd*)

Compound	Sample preparation	Column	Mobile phase	Detection	Reference
3-MT, NMN, MN, DA, NE, E, DOPA, 3-methoxytyrosine, o-tyrosine (IS)	Deproteinization with TCA; centrifugation	Yanapak ODS	Primary, 0.05 M phosphate buffer, pH 3.1; secondary, methanol; 0–6 min, primary solvent; 6–36 min, gradient, linear from 0 to 15% of secondary solvent in 30 ml	EC, 0–17 min: +0.6 V 17 min–: +0.9 V	Ishimitsu and Hirose (1985)
MHPG, iso-MHPG (IS)	Deproteinization with $HClO_4$; ethyl acetate extraction	Ultrasphere ODS	0.09 M Sodium acetate–0.09 M citric acid buffr, pH 5.0/ methanol (9:1)	EC, +0.75 V	Schinelli et al. (1985)
DHPG, NE, α-methyl-NE (IS)	Alumina adsorption	Spherisorb silica	0.07 M Phosphate buffer, pH 3.0 + 1.85 mM octanesulfonic acid + 13.4 mM EDTA	EC, +0.66 V	Howes et al. (1985)
MHPG, iso-MHPG (IS)	Ethyl acetate extraction; K_2CO_3 washing	Rainin ODS	0.1 M Sodium acetate/methanol (9:1); →pH 5.0	EC, detector 1: +0.1 V detector 2: +0.4 V	Molyneux and Franklin (1985)
HVA	Sep-Pak C_{18} column extraction QAE-Sephadex A-25 chromatography; ethyl acetate extraction	Lichrosorb C_{18}	0.05 M Phosphate buffer, pH 2.2/methanol (4:1)	EC, +0.70 V	Yoshida et al. (1985b)
MHPG	Ethyl acetate extraction; DuPont Type OD column extraction	μBondapak C_{18}	0.089 M Acetate buffer, pH 4.15/methanol (24:1) + 0.1 mM EDTA	EC, +0.73 V	Huber-Smith et al. (1986)
MHPG	Hydrolysis with sulfatase; deproteinization with $ZnSO_4$; ethyl acetate extraction	μBondapak C_{18}	0.05 M Na_2HPO_4–1.34 mM EDTA–2Na (→pH 3.0)/ methanol (49:1)	EC, +0.75 V	Sharpless et al. (1986)

Table 2. (*Cont'd*)

Compound	Sample preparation	Column	Mobile phase	Detection	Reference
HVA, 5-HIAA, 5-HT	Deproteinization with TCA; centrifugation	Supelcosil LC-18-DB	0.1 M NaH$_2$PO$_4$/methanol (21:4) + 2.6 mM sodium octyl sulfate + 0.1 mM EDTA−2Na + 0.25 mM TEA; →pH 3.85	EC, +0.8 V	Seegal *et al.* (1986)

Table 3.
Literature of HPLC of CA and 5-HT metabolites in human CSF

Compound	Sample preparation	Column	Mobile phase	Detection	Reference
5-HIAA, IAA, 5-HIPA (IS)	Ethyl acetate extraction	Nucleosil C_{18}	0.01 M Acetate buffer, pH 5.5/ methanol (90.5:9.5)	FL of 370 nm excited at 272 nm	Beck et al. (1977)
HVA, 5-HIAA, TRP, tyrosine	Centrifugation	μBondapak C_{18}	0.01 M Acetate buffer, pH 4.0/ methanol (17:3)	EC, +0.8 V; FL, see article	Anderson et al. (1979)
5-HIAA[1], TRP[1], IAA[2], IPA[2]	Direct injection	μBondapak C_{18}	[1] 0.01 M Acetate buffer, pH 4.3/acetonitrile (9:1) [2] 0.01 M Acetate buffer, pH 4.0/acetonitrile (7:3)	FL, see article	Anderson and Purdy (1979)
MHPG[1], DOPAC[2], HVA[2], VMA[2], MHPE[2], 5-HIAA[2]	Centrifugation	μBondapak C_{18}	[1] Citrate—phosphate buffer, pH 2.7 [2] tetrabutyl ammonium phosphate, pH 7.5	EC, +0.80 V	Langlais et al. (1980)
HVA, 5-HIAA	Direct injection	Spherisorb ODS	0.1 M KH_2PO_4—citric acid buffer, pH 4.7/methanol (19:1) + 0.1 mM EDTA	EC, +0.6 V	Semerdkjian-Rouquier et al. (1981)
MHPG	Centrifugation	μBondapak C_{18}	Citrate—acetate buffer, pH 5.15	EC, +0.75 V	Anderson et al. (1981a)
MHPG, MHPG-SO_4	Hydrolysis with sulfatase; ethyl acetate extraction	μBondapak C_{18}	Primary, 0.1 M phosphate buffer, pH 2.50; secondary, methanol/water (3:2); gradient, linear from 0 to 60% of secondary solvent in 45 min	EC, +1.0 V or +0.7 V	Krstulovic et al. (1981)

Table 3. (*Cont'd*)

Compound	Sample preparation	Column	Mobile phase	Detection	Reference
DOPAC, HVA, VMA, MHPG	Ethyl acetate extraction	Ultrasphere ODS	Primary, 0.1 M phosphate buffer, pH 2.50; secondary, methanol/water (3:2); gradient, linear from 0 to 100% of secondary solvent in 35 min	EC, +0.700 V	Krstulovic et al. (1982)
HVA, 5-HIAA, 5-hydroxyindole (IS)	Direct injection	Bioanlytical Systems ODS	0.34 M Acetate buffer, pH 4.75/ methanol/acetonitrile (85:7.5:7.5)	EC, +0.75 V	Narasimhachari et al. (1982)
5-HIAA, 5-HT, 5-HTP, 6-fluoro-5-HT (IS)	Deproteinization with HClO$_4$	µBondapak C$_{18}$	0.08 M Sodium acetate–0.297 mM EDTA–2Na (→pH 4.5) + 1–3% methanol	EC, +0.5 V	Petruccelli et al. (1982)
HVA, MHPG, 5-HIAA, 5-hydroxyindole-2-carboxylic acid (IS)	Deproteinization with HClO$_4$	RSil C18 HL	0.07 M NaH$_2$PO$_4$–0.1 mM EDTA–2Na/methanol (9:1); →pH 5.4	EC, +0.8 V	Bockstaele et al. (1983)
HVA, MHPG, 5-HIAA, 5-fluoro-HVA (IS)	Filtration	Ultrasphere ODS	0.1 M Sodium acetate–0.01 M citric acid–0.25 mM EDTA–2Na/methanol (41:4)	EC, +0.75 V	Scheinin et al. (1983)
HVA, 5-HIAA	Direct injection	µBondapak C$_{18}$	1.5 M Acetate buffer, pH 4.0–0.05 mM EDTA/ethanol (197:3)	EC, +0.75 V	Gagnieu et al. (1984)
HVA, MHPG, 5-HIAA	Filtration	Hypersil ODS	0.1 M Acetate buffer, pH 5.15 + 0.27 mM EDTA	EC, +0.85 V	Bottiglieri et al. (1984)
HVA[1], MHPG[1], 5-HIAA[2], VLA[1] (IS)	Deproteinization with acetonitrile	Ultrasphere-IP	75 mM Sodium citrate–H$_3$PO$_4$ buffer, pH 5.25/methanol (24:1)	EC, [1]+0.40 V [2]+0.23 V	Javors et al. (1984)

Brain tissue

Changes in the turnover and functional activity of the monoamine transmitters in the brain are known to be associated with a number of psychiatric
and neurological disorders (depression, schizophrenia, Parkinson's disease,
Huntington's chorea). The brain concentrations of monoamine metabolites
have been shown to reflect the activity of their parent neurons, and have
been used as an index of monoamine turnover (Cross and Joseph, 1981).
Thus, to the neuropharmacologist requiring the actual rate of CA or 5-HT,
monitoring the metabolites is far more informative than monitoring the
parent amines themselves (Kissinger, 1983). Most reports are determination
methods for rats, in which sampling is easy to perform as opposed to
human. HPLC methods for the determination of the metabolites of CA and
5-HT in rat brain tissue are summarized in Table 4. Semerdjian-Rouquier
et al. (1981), Cross and Joseph (1981) and Anderson *et al.* (1982) developed
methods for metabolites in human brain tissue; Falkowski and Wei (1981),
Towell and Erwin (1981), Ishikawa and McGaugh (1982) and Nabeshima
et al. (1982), in mouse brain tissue.

For the determination of the metabolites in brain tissue, the tissue samples,
punched from slices of brain tissue and weighing $1-2$ mg, are homogenized
ordinarily in $HClO_4$ or TCA solution, and the proteins are precipitated by
centrifugation. The aqueous layer is directly injected on the LC column;
occasionally, for the clean-up, the acid and neutral metabolites are extracted
into ethyl acetate (Hefti, 1979; Kempf and Mandel, 1981; Saraswat *et al.*,
1981; Cross and Joseph, 1981; Towell and Erwin, 1981; Shibuya *et al.*, 1982;
Ishikawa and McGaugh, 1982) or on a C_{18} column (Falkowski and Wei,
1981). In most of these methods, detection is performed by using an EC
detector as shown in Table 4.

SUMMARY

HPLC, coupled with UV, fluorometric or EC detection is a very useful
technique for the isolation and determination of pg$-$µg amounts of substances of neurochemical interest, such as CA, 5-HT and their metabolites.
During the past decade, great efforts have been made to improve and
simplify the analytical methodology for routine use in clinical laboratories or
neuropharmacological studies, where this technique is necessary. Perhaps no
other technique can rival it in sensitivity, simplicity, precision, accuracy and
economy in analyses of CA and 5-HT metabolites from biological samples.
The purpose of this review was to describe the current analytical methodology of HPLC in the field of the metabolites of CA and 5-HT: the review
suggests applications of HPLC technique to clinical analysis or clinical
pharmacology.

REFERENCES

Alonso, R., Gibson, C. J., and McGill, J. (1981). Determination of 3-methoxy-4-hydroxy-
   phenylglycol in urine by high-performance liquid chromatography with amperometric detection. *Life Sci.* 29, 1689–1696.

Table 4.
Literature of HPLC of CA and 5-HT metabolites in rat brain tissue

Compound	Sample preparation	Column	Mobile phase	Detection	Reference
5-HIAA[1], 5-HT[1], TRP[2]	Homogenization in $HClO_4$; centrifugation; Amberlite CG-50, Dowex AG-50, Sephadex G-10 chromatography	µBondapak C_{18}	[1] 0.5 M Acetate buffer, pH 5.1/ methanol (17:3) [2] McIlvaine buffer, pH 4.0/ methanol (4:1)	EC, [1]+0.50 V [2]+1.00 V	Koch and Kissinger (1979)
DOPAC, HVA, vanillic acid (IS)	Homogenization in $HClO_4$; centrifugation; ether extraction	µBondapak C_{18}	0.05 M Acetate buffer, pH 5.0	EC, +0.7 V	Hefti (1979)
5-HIAA, TRP, 5-HTP, 5-HT, melatonin	Homogenization in $HClO_4$; centrifugation	Vydac 201 TP	0.1 M Sodium acetate−0.1 M citric acid, pH 4.1	EC, +0.58 V	Mefford and Barchas (1980)
DOPAC, DA, NE, E, 3,4-dihydroxybenzylamine (IS)	Homogenization in $HClO_4$; alumina adsorption	Vydac 201 TP	0.1 M Sodium acetate−0.02 M citric acid, pH 5.2 + 0.862 mM sodium octyl sulfate	EC, +0.50 V	Mefford et al. (1980)
DOPAC, HVA, DA, epinine (IS)	Homogenization in $HClO_4$; centrifugation	Nucleosil C_{18}	Citrate buffer, pH 4.25/ methanol (23.2) 1.7 mM sodium hexyl sulfate	EC, +0.60 V	Magnusson et al. (1980)
5-HIAA, 5-HT	Homogenization in $HClO_4$; centrifugation; Amberlite CG-50, Sephadex G-10 chromatography	µBondapak C_{18}	0.5 M Acetate buffer, pH 5.1/ methanol (17:3)	EC, +0.50 V	Koch and Kissinger (1980)
5-HIAA, 5-HT	Homogenization in 75% ethanol; centrifugation; Bio-Rex 70 chromatography	µBondapak C_{18}	0.5 M Ammonium acetate/ methanol (17:3); →pH 5.1	EC, +0.5 V	Lyness et al. (1980)
5-HIAA, 5-HT	Homogenization in $HClO_4$; centrifugation	µBondapak C_{18}	0.1 M Acetate buffer, pH 4.7/ methanol (47:3)	EC, +0.7 V	Reinhard et al. (1980)
5-HIAA, 5-HT, 5-HTP	Homogenization in 10% $ZnSO_4$; centrifugation	Bio-Sil ODS	0.1 M Citric acid−0.2 M Na_2HPO_4 buffer, pH 4.8/ water/methanol (23:23:4)	EC, +0.5 V	Lackovic et al. (1981)

Table 4. (*Cont'd*)

Compound	Sample preparation	Column	Mobile phase	Detection	Reference
DOPAC, HVA, 5-HIAA, 3,4-dihydroxyhydrocinnamic acid (IS)	Homogenization in $HClO_4$; centrifugation	µBondapak C_{18}	0.1 M NaH_2PO_4–0.1 mM EDTA–2NA (→pH 2.9)/methanol (19:1)	EC, +0.78 V	Kempf and Mandel (1981)
DOPAC, HVA, 5-HIAA, DA, NE, DOPA	Homogenization in $HClO_4$; centrifugation; Sephadex G-10 chromatography	Nucleosil C_{18}	McIlvaine buffer, pH 3.5/methanol (22:3)	EC, see article	Westerink and Mulder (1981)
5-HIAA, 5-HT, 5-HTP	Homogenization in $HClO_4$; centrifugation	Spherisorb ODS	0.1 M KH_2PO_4–citric acid buffer, pH 4.7/methanol (19:1) + 0.1 mM EDTA–2Na	EC, +0.6 V	Semerdjian-Rouquier et al. (1981)
DOPAC, HVA, 3,4-dihydroxycinnamic acid (IS)	Homogenization in TCA; centrifugation; ethyl acetate extraction	Zorbax ODS	0.1 M Phosphate–0.05 M citrate buffer, pH 4.3/methanol (9:1)	EC, +0.61 V	Sarawat et al. (1981)
HVA[1], 5-HIAA[1], 5-HT[1], TRP[2], 5-hydroxytryptophol[2], melatonin[2]	Homogenization in $HClO_4$; centrifugation	µBondapak C_{18}	1 0.01 M Acetate buffer, pH 4.25/methanol (22:3) 2 0.01 M acetate buffer, pH 4.25/methanol (17:3)	EC, +0.70 V; FL of 360 nm excited at 254 nm	Anderson et al. (1981b)
DOPAC, HVA, 5-HIAA, DA, 5-HTP, 5-HT, N-methyl-5-hydroxytryptamine (IS)	Homogenization in mobile phase; centrifugation	Spherisorb ODS	0.1 M Citrate–0.075 M Na_2HPO_4–75 mM sodium heptane sulfonate/methanol (43:7); →pH 3.9	EC, +0.75 V	Kilts et al. (1981)
NMN, DA, NE, 5-HT, 3,4-dihydroxybenzylamine (IS)	Homogenization in 80% ethanol; centrifugation; Amberlite CG-50 chromatography	Lichrosorb RP-18	0.05 M NaH_2PO_4–0.02 mM EDTA–1 mM octane sulfonate (→pH 3)/methanol (19:1)	EC, +0.8 V	Chiu et al. (1981)

Table 4. (*Cont'd*)

Compound	Sample preparation	Column	Mobile phase	Detection	Reference
DOPAC, HVA, MHPG, 5-HIAA, DA, NE, 5-HT	Homogenization in formic acid-acetone; centrifugation; heptane–chloroform washing	µBondapak C₁₈	0.1 M Citrate–phosphate buffer, pH 3.5–0.431 mM sodium octyl sulfate/methanol (91:9)	EC, +0.9 V	Co *et al.* (1982)
DOPAC, HVA, 5-HIAA, 5-HT	Homogenization in HClO₄; centrifugation	µBondapak C₁₈	0.1 M Acetate buffer, pH 4.5– 1 mM EDTA/methanol (19:1)	EC, +0.55 V or +0.8 V	Sperk (1982)
DOPAC, HVA, 5-HIAA, DA, NE, 5-HT, DOPA, 3-methoxytyrosine, 5-HTP, isoproterenol (IS)	Homogenization in HClO₄; centrifugation	Lichrosorb RP-18	0.1 M NaH₂PO₄/methanol (21:4) + 2.6 mM octne sulfonic acid + 0.10 mM EDTA + 0.25 mM triethylamine; → pH 3.35	EC, +0.9 V	Wagner *et al.* (1982)
DOPAC, HVA, DA	Homogenization in HClO₄; centrifugation	Polygosil C₈ coated with tri-n-butylphosphate	0.05 M Citrate buffer, pH 4.9 + 0.20 M ClO₄⁻	EC, +0.80 V	Valkenburg *et al.* (1982)
DOPAC, HVA, 5-HIAA	Homogenization in HClO₄ or citrate-phosphate buffer, pH 3.3; centrifugation	µBondapak C₁₈	0.1 M Citrate-phosphate buffer, pH 3.3/methanol (7:1) + 0.172 mM sodium octyl sulfate + 20 µM EDTA	EC, +0.95 V	Lyness (1982)
DOPAC, HVA, 5-HIAA, DA, 5-HT	Homogenization in mobile phase; centrifugation	µBondapak C₁₈	0.1 M Citrate-phosphate buffer, pH 2.6/methanol (7:1) + 1.38 mM sodium octyl sulfate	EC, +0.75 V	Nielsen and Johnston (1982)
5-HIAA, 5-HT, 5-HTP	Homogenization in mobile phase; centrifugation	RCM-100 C₁₈	0.5 M Acetate buffer, pH 5.35/ methanol (17:3)	EC, +0.65 V	Narasimhachari *et al.* (1982)
HVA, VMA, 5-HIAA, DA, NE, 5-HT, octopamine	Homogenization in HCl; *n*-butanol extraction; alkaline solution back-extraction	µBondapak/phenyl	Methanol/water (7:93) + PIC-B7	UV, 280 nm	Shibuya *et al.* (1982)

Table 4. (*Cont'd*)

Compound	Sample preparation	Column	Mobile phase	Detection	Reference
DOPAC, HVA, 3-MT, DA	Homogenization in $HClO_4$; centrifugation	μBondapak C_{18}	Methanol/acetic acid/water (9:1:90) + 1 mM EDTA−2Na + 2 mM sodium 1-heptane sulfonate	EC, +0.76 V	Cheng and Wooten (1982)
5-HIAA, 5-HT, TRP, N-acetylserotonin, melatonin, 5-hydroxytryptophol	Homogenization in $HClO_4$; centrifugation	μBondapak C_{18}	0.01 M Acetate buffer, pH 4.25/ methanol (13:7)	EC, +0.7−0.8 V; FL of 345 nm excited at 285 nm	Anderson et al. (1982)
5-HIAA, 5-HT, 5-HTP, TRP	Homogenization in $HClO_4$; centrifugation	Hitachi Gel No.3011-C	0.5 M Citrate buffer, pH 4.0/ methanol (4:1)	FL of 340 nm excited at 280 nm	Hori et al. (1982)
5-HIAA, 5-HT, 5-HTP	Homogenization in $HClO_4$; centrifugation	MC-18	0.1 M Monochloroacetic acid/ methanol (17:3)	EC, +0.65 V	Di Paolo et al. (1983)
DOPAC, HVA, 5-HIAA, DA, 5-HT	Homogenization in $HClO_4$; centrifugation	Bio-Sil ODS	0.4 M Acetate buffer, pH 4.0−1 mM EDTA/methanol (100:1.5)	EC, +0.7 V	Wilson et al. (1983)
DOPAC, DA, NE, E, 3,4-dihydroxybenzylamine (IS)	Homogenization in $HClO_4$; centrifugation	Bio-Sil ODS	0.1 M Citrate−acetate buffer, pH 4.35−0.513 mM EDTA−0.647 mM sodium octyl sulfate/ methanol (96.5:3.5); flow-gradient, linear from 0.5 to 2.0 ml/min in 15 min	EC, +0.70 V	Rossetti et al. (1983)

Table 4. (*Cont'd*)

Compound	Sample preparation	Column	Mobile phase	Detection	Reference
DOPAC, HVA, 5-HIAA, DA	Homogenization in HClO$_4$; centrifugation	Biophase ODS	Acetonitrile/stock solution/THF (3.5:96.5:1.8); stock solution: 14.2 g monochloroacetic acid, 4.68 g NaOH, 200 mg sodium octyl sulfate, 25 g EDTA−2Na per liter	EC, see article	Mayer and Shoup (1983)
5-HIAA, 5-HT, 5-HTP TRP, N-acetylserotonin (IS)	Homogenization in HClO$_4$; centrifugation	Spherisorb ODS	0.01 M Sodium acetate/methanol (17:3); →pH 4.85	FL of 345 nm excited at 285 nm	Wolf and Kuhn (1983)
5-HIAA[1], 5-HT[1], 5-HTP[1], tryptamine[2], TRP[2]	Homogenization in HClO$_4$; centrifugation	Polygosil C$_{18}$ coated with tri-*n*-butyl phosphate	0.05 M Phosphate buffer, pH 1.95/(4:1) + 0.14 M ClO$_4^-$	EC, +0.80 V; [1] FL of 333 nm excited at 296 nm [2] FL of 354 nm excited at 282 nm	De Jong *et al.* (1983)
DHPG[1], MHPG[2], NMN[3]	Homogenization in HClO$_4$; centrifugation; Sephadex G-10 chromatography	Nucleosil C$_{18}$	[1] 0.34 M Phosphate−0.17 M citrate buffer, pH 4.8 [2] Phosphate−acetate buffer, pH 4.8 [3] 0.1 M TCA, pH 3.4/methanol (47:3)	EC, [1]0.550 V [2]+0.750 V [3]+0.850V	Westerink (1984)
DOPAC, HVA, 5-HIAA, DA, NE, 5-HT, TRP, tyrosine, epinine (IS)	Homogenization in formic acid−acetone; centrifugation; heptane−chloroform washing	Biophase ODS	0.15 M Monochloroacetic acid−0.1 mM EDTA−2Na (→pH 3.00) + 0.862 mM sodium octyl sulfate + 7.3% acetonitrile	EC, +0.95 V	Lasley *et al.* (1984)

Table 4. (*Cont'd*)

Compound	Sample preparation	Column	Mobile phase	Detection	Reference
MHPG-SO$_4$	Homogenization in H$_2$SO$_4$; centrifugation; DEAE-Sephadex A-25 chromatography; hydrolysis with sulfatase; deproteinization with HClO$_4$	Lichrosorb RP-18	0.15 M NaH$_2$PO$_4$/methanol (89:11) + 0.13 mM EDTA; →pH 3.15	EC, +0.85 V	Hornsperger et al. (1984)
DOPAC, HVA, 5-HIAA, 3-MT, DA, NE, 5-HT, TRP, isoproterenol (IS)	Homogenization in phosphate buffer, pH 4.0; centrifugation	Ultrasphere ODS	0.1 M NaH$_2$PO$_4$ – 1 mM EDTA–2Na – 1 mM sodium octane sulfonate (→pH 4.0)/ acetonitrile (100:10.5)	EC, +0.8 V	Saller and Salama (1984)
MHPG	Homogenization in HClO$_4$; hydrolysis with sulphatase; PH Bond-Elut column extraction	μBondapak C$_{18}$	0.05 M Citrate–phosphate buffer, pH 3.5/methanol (10:1)	EC, +0.85 V	Karege (1984)
DOPAC, HVA, 5-HIAA, DA, NE, 5-HT	Homogenization in HClO$_4$; centrifugation	μBondapak C$_{18}$	0.1 M NaH$_2$PO$_4$/methanol (21:4) + 2.6 mM sodium octyl sulfate + 0.1 mM EDTA–2Na + 0.25 mM TEA; →pH 4.25	EC, +0.8 V	Seegal et al. (1986)

Anderson, G. M. and Purdy, W. C. (1979). Liquid chromatographic–fluorometric system for the determination of indoles in physiological samples. *Anal. Chem. 51*, 283–286.

Anderson, G. M., Young, J. G., and Cohen, D. J. (1979). Rapid liquid chromatographic determination of tryptophan, tyrosine, 5-hydroxyindoleacetic acid and homovenillic acid in cerebrospinal fluid. *J. Chromatogr. 164*, 501–505.

Anderson, G. M., Young, J. G., Cohen, D.J., Shaywitz, B. A., and Batter, D. K. (1981a). Amperometric determination of 3-methoxy-4-hydroxyphenylethyleneglycol in human cerebrospinal fluid. *J. Chromatogr. 222*, 112–115.

Anderson, G. M., Young, J. G., Batter, D. K., Young, S. N., Cohen, D. J., and Shaywitz, B. A. (1981b). Determination of indoles and catechols in rat brain and pineal using liquid chromatography with fluorometric and amperometric detection. *J. Chromatogr. 223*, 315–320.

Anderson, G. M., Young, J. G., Cohen, D. J., and Young, S. N. (1982). Determination of indoles in human and pineal. *J. Chromatogr. 228*, 155–163.

Anderson, G. M., Feibel, F. C., and Cohen, D. J. (1985a). Liquid-chromatographic determination of vanillylmandelic acid in urine. *Clin. Chem. 31*, 819–821.

Anderson, G. M., Schlicht, K. R., and Cohen, D. J. (1985b). Two-dimensional high-performance liquid chromatographic determination of 5-hydroxyindoleacetic acid and homovanillic acid in urine. *Anal. Biochem. 144*, 27–31.

Änggård, E., Sjöquist, B., and Sjöström, R. (1970). Chromatography of biogenic amine metabolites and related compounds on lipophilic Sephadex. I. The phenolic acids and alcohols. *J. Chromatogr. 50*, 251–259.

Annino, J. S., Lipson, M., and Williams, L. A. (1965). Determination of 3-methoxy-4-hydroxymandelic acid (VMA) in urine by thin-layer chromatography. *Clin. Chem. 11*, 905–913.

Anton, A. H. and Sayre, D. F. (1967). Improving the specificity of the nitrosonaphthol procedure for urinary 5-hydroxyindoleacetic acid. *Clin. Chem. 13*, 1017–1020.

Bacopoulos, N. G., Hattox, S. E., and Roth, R. H. (1979). 3,4-Dihydroxyphenylacetic acid and homovanillic acid in rat plasma: possible indicators of central dopaminergic activity. *Eur. J. Pharmacol. 56*, 225–236.

Bauersfeld, W., Diener, U., Knoll, E., Ratge, D., and Wisser, H. (1982). Determination of urinary vanilmandelic acid and homovanillic acid by high performance liquid chromatography with amperometric detection. *J. Clin. Chem. Clin. Biochem. 20*, 217–220.

Beck, O., Palmskog, G., and Hultman, E. (1977). Quantitative determination of 5-hydroxyindole-3-acetic acid in body fluids by high-performance liquid chromatography. *Clin. Chim. Acta 79*, 149–154.

Bell, M. (1968). Newer chemical diagnostic tests. *J. Am. Med. Assoc. 205*, 105–106.

Bertani-Dziedzic, L. M., Krstulovic, A. M., Ciriello, S., and Gitlow, S. E. (1979). Routine reversed-phase high-performance liquid chromatographic measurement of urinary vanillylmandelic acid in patients with neural crest tumors. *J. Chromatogr. 164*, 345–353.

Bertani-Dziedzic, L. M., Krstulovic, A. M., Dziedzic, S. W., Gitlow, S. E., and Cerqueira, S. (1981). Analysis of urinary metanephrines by reversed-phase high-performance liquid chromatography and electrochemical detection. *Clin. Chim. Acta 110*, 1–8.

Betto, P., Lucarelli, C., and Ricciarello, G. (1983). Separation and determination of homovanillic and vanilmandelic acid by high-performance liquid chromatography. *Anal. Chem. Sympos. Ser. 14*, 371–378.

Binder, S. R. and Sivorinovsky, G. (1984). Measurement of urinary vanilmandelic acid and homovanillic acid by high-performance liquid chromatography with electrochemical detection following extraction by ion-exchange and ion-moderated partition. *J. Chromatogr. 336*, 173–188.

Bockstaele, M. V., Delien, L., Claeys, M., and DePotter, W. P. (1983). Simultaneous determination of the three major monoamine metabolites in cerebrospinal fluid by high-performance liquid chromatography with electrochemical detection. *J. Chromatogr. 275*, 11–20.

Bottiglieri, T., Lim, C. K., and Peters, T. J. (1984). Isocratic analysis of 3-methoxy-4-hydroxyphenyl glycol, 5-hydroxyindole-3-acetic acid and 4-hydroxy-3-methoxyphenylacetic acid in cerebrospinal fluid by high-performance liquid chromatography with amperometric detection. *J. Chromatogr. 311*, 354−360.

Brown, R. T., Oliver, J., Kirk, K. L., and Kopin, I. J. (1984). Determination of urinary 4-hydroxy-3-methoxyphenylethylene glycol in man by high performance liquid chromatography with electrochemical detection. *Life Sci. 34*, 2313−2318.

Buchanan, D. N., Fucek, F. R., and Domino, E. F. (1979). Analysis of urinary 4-hydroxy-3-methoxyphenyleneglycol as vanillyl alcohol by high-performance liquid chromatography with amperometric detection. *J. Chromatogr. 162*, 394−400.

Calne, D. B., Karoum, F., Ruthven, C. R. J., and Sandler, M. (1969). The metabolism of orally administered L-DOPA in Parkinsonism. *Br. J. Pharmacol. 37*, 57−68.

Chang, W.-H. Scheinin, M., Burns, R. S., and Linnoila, M. (1983). Rapid and simple determination of homovanillic acid in plasma using high performance liquid chromatography with electrochemical detection. *Acta Pharmacol. Toxicol. 53*, 275−279.

Chauhan, J. and Darbre, A. (1980). Determination of homovanillic, isohomovanillic and vanillylmandelic acids in human urine by means of glass capillary gas-liquid chromatography with temperature-programmed electron-capture detection. *J. Chromatogr. 183*, 391−401.

Cheng, C. H. and Wooten, G. F. (1982). Dopamine turnover estimated by simultaneous LCEC assay of dopamine and dopamine metabolites. *J. Pharmacol. Methods 8*, 123−133.

Chilcote, D. D. (1972). Column-chromatographic analysis of naturally fluorescing compounds: II. Rapid analysis of indoleacetic acid and 5-hydroxyindoleacetic acid in biological samples. *Clin. Chem. 18*, 1376−1378.

Chilcote, D. D. and Mrochek, J. E. (1972). Chromatographic analysis of naturally fluorescing compounds: I. Rapid analysis of nanogram amounts of indoles in physiologic fluids. *Clin. Chem. 18*, 778−782.

Chiu, A. S., Godse, D. D., and Warsh, J. J. (1981). Determination of brain regional nor-metanephrine levels by liquid chromatography with electrochemical detection (LC-EC). *Proc. Neuro-Psychopharmacol. 5*, 559−663.

Chou, P. P. and Jaynes, P. K. (1985). Determination of urinary 5-hydroxyindole-3-acetic acid using solid-phase extraction and reversed-phase high-performance liquid chromatography with electrochemical detection. *J. Chromatogr. 341*, 167−171.

Co, C., Smith, J. E., and Lane, J. D. (1982). Use of a single compartment LCEC cell in the determinations of biogenic amine content and turnover. *Pharmacol. Biochem. Behav. 16*, 641−646.

Cross, A. J., and Joseph, M. H. (1981). The concurrent estimation of the major monoamine metabolites in human and non-human primate brain by HPLC with fluorescence and electro-chemical detection. *Life Sci. 28*, 499−505.

Davis, T. D., Gehrke, C. W., Gehrke, C. W. Jr, Cunningham, T. D., Kuo, K. C., Gerhardt, K. O., Johnson, H. D., and Williams, C. H. (1978). High-performance liquid-chromato-graphic separation and fluorescence measurement of biogenic amines in plasma, urine, and tissue. *Clin. Chem. 24*, 1317−1324.

De Jong, J., Tjaden, U. R., Van't Hoff, W., and Van Valkenburg, C. F. M. (1983). Analysis of serotonin and derivatives by reversed-phase ion-pair partition chromatography with fluoro-metric and electrochemical detection. *J. Chromatogr. 282*, 443−456.

De Met, E. M., Halaris, A. E., Gwirtsman, H. E., Reno, R. M., and Becker, P. I. (1982). Effects of desipramine on diurnal rhythms of plasma 3-methoxy-4-hydroxyphenylglycol (MHPG) in depressed patients. *Psychopharmacol. Bull. 18*, 221−223.

Di Paolo, T., Dupont, A., Savard, P., and Daigle, M. (1983). Determination of 5-hydroxy-tryptophan, 5-hydroxytryptamine, and 5-hydroxyindoleacetic acid in 20 rat brain nuclei using liquid chromatography with electrochemical detection. *Can J. Physiol. Pharmacol. 61*, 530−534.

Draganac, R. S., Steindel, S. J., and Trawick, W. G. (1980). Liquid-chromatographic separa-tion of urinary 5-hydroxy-3-indoleacetic acid, with measurement at 254 nm. *Clin. Chem. 26*, 910−912.

Dreux, C., Bousquet, B., and Halter, D. (1973). L'exploration biochimique des tumeurs carcinoïdes. *Ann. Biol. Clin. (Paris) 31*, 283−294.

Dutrieu, J. and Delmotte, Y. A. (1984). Simultaneous determination of vanilmandelic acid (VMA), homovanillic acid (HVA) and 5-hydroxy-3-indoleacetic acid (5-HIAA) in urine by high-performance liquid chromatography with coulometric detection. *Fresenius Z. Anal. Chem. 317*, 124–128.

Dziedzic, S. W., Dziedzic, L. B., and Gitlow, S. E. (1973). Separation and determination of urinary homovanillic acid and iso-homovanillic acid by gas-chromatography and electron capture detection. *J. Lab. Clin. Med. 82*, 829–835.

Falkowski, A. J. and Wei, R. (1981). Optimized isocratic conditions for the simultaneous determination of serotonin precursors and metabolites by reversed-phase high-performance liquid chromatography with electrochemical detection. *Anal. Biochem. 115*, 311–317.

Felice, L. J. and Kissinger, P. T. (1976). Determination of homovanillic acid in urine by liquid chromatography with electrochemical detection. *Anal. Chem. 48*, 794–796.

Felice, L. J. and Kissinger, P. T. (1977). A modification of the Pisano method for vanilmandelic acid using high pressure liquid chromatography. *Clin. Chim. Acta 76*, 317–320.

Felice, L. J., Bruntlett, C. S., and Kissinger, P. T. (1977). Liquid chromatography assay for 3,4-dihydroxyphenylacetic acid in urine. *J. Chromatogr. 143*, 407–410.

Flood, J. G., Granger, M., and McComb, R. B. (1979). Urinary 3-methoxy-4-hydroxymandelic acid as measured by liquid chromatography, with on-line post-column reaction. *Clin. Chem. 25*, 1234–1238.

Fornstedt, N. (1978). Determination of 5-hydroxyindole-3-acetic acid in urine by high performance liquid chromatography. *Anal. Chem. 50*, 1342–1346.

Frattini, P., Santagostino, G., Schinelli, S., Cusshi, M. L., and Corona, G. L. (1983). Assay of urinary vanilmandelic, homovanillic, and 5-hydroxyindole acetic acids by liquid chromatography with electrochemical detection. *J. Pharmacol. Methods 10*, 193–198.

Fujita, K., Maruta, K., Ito, S., and Nagatsu, T. (1983). Urinary 4-hydroxy-3-methoxymandelic (vanilmandelic) acid, 4-hydroxy-3-methoxyphenylacetic (homovanillic) acid, and 5-hydroxy-3-indoleacetic acid determined by liquid chromatography with electrochemical detection. *Clin. Chem. 29*, 876–878.

Gagnieu, M.-C., Menouni-Foray, V., Guardiola, P., Quincy, C., and Renaud, B. (1984). Liquid chromatographic determination of homovanillic acid, 5-hydroxyindoleacetic acid and probenecid levels in human cerebrospinal fluid during probenecid test. *Clin. Chim. Acta 139*, 1–12.

Garnier, J. P., Bousquet, B., and Dreux, C. (1981). Determination of 5-hydroxyindoleacetic acid in urine by high-performance liquid chromatography: A fully automated method. *J. Chromatogr. 204*, 225–230.

Garelis, E., Young, S. N., Lal, S., and Sourkes, T. L. (1974). Monoamine metabolites in lumbar CSF: the question of their origin in relation to clinical studies. *Brain Res. 79*, 1–8.

Geeraerts, F., Schimpfessel, L., and Crokaert, R. (1980). The stability of tryptophan metabolites prior to urine analysis. *Clin. Chim. Acta 102*, 247–251.

Genefke, I. K. (1972). A simple method for the fluorimetric determination of 5-hydroxyindoleacetic acid in human urine. *Acta Pharmacol. Toxicol. 31*, 554–558.

Ghebregzabher, M., Rufini, S., Castellucci, M. G., and Lato, M. (1981). Analysis of some tryptophan and phenylalanine metabolites in urine by a straight-phase high-performance liquid chromatographic technique. *J. Chromatogr. 222*, 191–201.

Gordon, E. K., Oliver, J., Black, K., and Kopin, I. J. (1974). Simultaneous assay by mass fragmentography of vanillyl mandelic acid, homovanillic acid, and 3-methoxy-4-hydroxyphenethylene glycol in cerebrospinal fluid and urine. *Biochem. Med. 11*, 32–40.

Graffeo, A. P. and Karger, B. L. (1976). Analysis for indole compounds in urine by high-performance liquid chromatography with fluorometric detection. *Clin. Chem. 22*, 184–187.

Greenberg, R. E. and Gardner, L. I. (1960). Catecholamine metabolism in a functional neural tumor. *J. Clin. Invest. 39*, 1729–1736.

Hallman, H., Farnebo, L-O., Hamberger, B., and Jonsson, G. (1978). A sensitive method for the determination of plasma catecholamines using liquid chromatography with electrochemical detection. *Life Sci. 23*, 1046–1052.

Harris, P. Q. and Bacopoulos, N. G. (1984). Measurement of homovanillic acid in human plasma by high-performance liquid chromatography with electrochemical detection. *J. Chromatogr. 309*, 379–384.

Hefti, F. (1979). A simple, sensitive method for measuring 3,4-dihydroxyphenylacetic acid and homovanillic acid in rat brain tissue using high-performance liquid chromatography with electrochemical detection. *Life Sci. 25*, 775–782.

Hjemdahl, P., Daleskog, M., and Kahan, T. (1979). Determination of plasma catecholamines by high performance liquid chromatography with electrochemical detection: comparison with a radioenzymatic method. *Life Sci. 25*, 131–138.

Higa, S., Suzuki, T., Hayashi, A., Tsuge, I., and Yamamura, Y. (1977). Isolation of catecholamines in biological fluids by boric acid gel. *Anal. Biochem. 77*, 18–24.

Hoeldtke, R. and Stetson, P. L. (1980). Separation of urinary catecholamines and catecholamine metabolites by high-pressure liquid chromatography. *Anal. Biochem. 105*, 207–217.

Hori, S., Ohtani, K., Ohtani, S., Kayanuma, K., and Ito, T. (1982). Simultaneous determination of tryptophan, serotonin and 5-hydroxyindoleacetic acid in rat brain by high-performance liquid chromatography using a week acidic cation-exchange resin. *J. Chromatogr. 231*, 161–165.

Hornsperger, J.-M., Wagner, J., Hinkel, J.-P., and Jung, M. J. (1984). Measurement of 3-methoxy-4-hydroxyphenylglycol sulfate ester in brain using reversed-phase liquid chromatography and electrochemical detection. *J. Chromatogr. 306*, 364–370.

Howes, L. G., Miller, S., and Reid, J. L. (1985). Simultaneous assay of 3,4-dihydroxyphenylethylene glycol and norepinephrine in human plasma by high-performance liquid chromatography with electrochemical detection. *J. Chromatogr. 338*, 401–403.

Huber-Smith, M. J., Nesse, R., Mazhar, M., and McCann, D. S. (1986). Evaluation of plasma 3-methoxy-4-hydroxyphenylglycol. *J. Chromatogr. 377*, 91–99.

Iinuma, F., Mawatari, K., Tabara, M., and Watanabe, M. (1984). Fluorometric determination of 5-hydroxyindole derivatives by high-performance liquid chromatography with cobalt(II) chloride, sodium carbonate, and sodium hydroxide. *Bunseki Kagaku (Jap. Anal.) 33*, E323–E330.

Ishikawa, K. and McGaugh, J. L. (1982). Simultaneous determination of monoamine transmitters, precursors and metabolites in a single mouse brain. *J. Chromatogr. 229*, 35–46.

Ishimitsu, T. and Hirose, S. (1985). Simultaneous assay of 3,4-dihydroxyphenylalanine, catecholamines and *o*-methylated metabolites in human plasma using high-performance liquid chromatography. *J. Chromatogr. 337*, 239–248.

Iwatani, A., and Nakamura, H. (1984). Determination of urinary tryptophan, 5-hydroxytryptamine and 5-hydroxyindoleacetic acid in neonatal hyperbilirubinaemic infants using reversed-phase high-performance liquid chromatography with fluorescence detection. *J. Chromatogr. 309*, 145–150.

Jackman, G., Snell, J., Skews, H., and Bobik, A. (1982). Effects of noradrenergic neuronal activity on 3,4-dihydroxyphenylethylene glycol (DHPG) levels. Quantitation by high performance liquid chromatography. *Life Sci. 31*, 923–929.

Javaid, J. I., Liu, T. S., Maas, J. W., and Davis, J. M. (1983). Measurement of 3-methoxy-4-hydroxyphenylacetic acid (HVA) in plasma by high-performance liquid chromatography with electrochemical detector (HPLC-EC). *Anal. Biochem. 135*, 326–331.

Javors, M. A., Bowden, C. L., and Maas, J. W. (1984). 3-Methoxy-4-hydroxyphenylglycol, 5-hydroxyindoleacetic acid, and homovanillic acid in human cerebrospinal fluid; storage and measurement by reversed-phase high-performance liquid chromatography and coulometric detection using 3-methoxy-4-hydroxyphenyllactic acid as an internal standard. *J. Chromatogr. 336*, 259–269.

Jimerson, D. C., Insel, T. R., Reus, V. I., and Kopin, I. J. (1983). Increased plasma MHPG in dexamethasone-resistant depressed patients. *Arch. Gen. Psychiatry 40*, 173–176.

Joseph, M. H., Baker, H. F., Johnstone, E. C., and Crow, T. J. (1976). Determination of 3-methoxy-4-hydroxyphenylglycol conjugates in urine. Application to the study of central noradrenaline metabolism in unmedicated chronic schizophrenic patients. *Psychopharmacology, 51*, 47–51.

Joseph, M. H., Kadam, B. V., and Risby, D. (1981). Simple high-performance liquid chromatographic method for the concurrent determination of the amine metabolites vanillylmandelic acid, 3-methoxy-4-hydroxyphenylglycol, 5-hydroxyindoleacetic acid, dihydroxyphenylacetic acid and homovanillic acid in urine using electrochemical detection. *J. Chromatogr.* 226, 361–368.

Jouve, J., Mariotte, N., Sureau, C., and Much, J. P. (1983). High-performance liquid chromatography with electrochemical detection for the simultaneous determination of the methylated amines, normetanephrine, metanephrine and 3-methoxytyramine in urine. *J. Chromatogr.* 274, 53–62.

Karege, F. (1984). Method for total 3-methoxy-4-hydroxyphenylglycol extraction from urine, plasma and brain tissue using bonded-phase materials: comparison with the ethyl acetate extraction method. *J. Chromatogr.* 311, 361–368.

Karoum, F., Anah, C. O., Ruthven, C. R. J., and Sandler, M. (1960). Further observations on the gas-chromatographic measurement of urinary phenolic and indolic metabolites. *Clin. Chim. Acta* 24, 341–348.

Kempf, E. and Mandel, P. (1981). Reverse-phase high-performance liquid chromatographic separation and electrochemical detection of norepinephrine, dopamine, serotonin, and related major metabolites. *Anal. Biochem.* 112, 223–231.

Kendler, K. S., Heninger, G. R., and Roth, R. H. (1981). Brain contribution to the Haloperidol-induced increase in plasma homovanillic acid. *Eur. J. Pharmacol.* 71, 321–326.

Kilts, C. D., Breese, G. R., and Mailman, R. B. (1981). Simultaneous quantification of dopamine, 5-hydroxytryptamine and four metabolically related compounds by means of reversed-phase high-performance liquid chromatography with electrochemical detection. *J. Chromatogr.* 225, 347–357.

Kissinger, P. T. (1983). Determination of biogenic amines and their metabolites by liquid chromatography/electrochemistry. In: *Methods in Biogenic Amine Research*, S. Parvez, T. Nagatsu, I. Nagatsu and H. Parvez (Eds). Elsevier, Amsterdam, pp. 75–99.

Kissinger, P. T., Bratin, K., Davis, G. C., and Pachla, L. A. (1979). The potential utility of pre- and post-column chemical reactions with electrochemical detection in liquid chromatography. *J. Chromatogr. Sci.* 17, 137–146.

Ko, G. N., Elsworth, J. D., Roth, R. H., Rifkin, B. G., Leigh, H., and Redmond, D. E. (1983). Panic-induced elevation of plasma MHPG levels in phobic–anxious patients. *Arch. Gen. Psychiatry.* 40, 425–430.

Koch, D. D. and Kissinger, P. T. (1979). Determination of tryptophan and several of its metabolites in physiological samples by reversed-phase liquid chromatography with electrochemical detection. *J. Chromatogr.* 164, 441–445.

Koch, D. D. and Kissinger, P. T. (1980). Liquid chromatographyw ith pre-column sample enrichment and electrochemical detection. Regional determination of serotonin and 5-hydroxyindoleacetic acid in brain tissue. *Life Sci.* 26, 1099–1107.

Kodama, K., Nakata, T., and Aoyama, M. (1984). Direct determination of urinary vanillylmandelic acid and homovanillic acid by high-performance liquid chromatography on an anion-exchange column. *J. Chromatogr.* 311, 369–374.

Krstulovic, A. M., Zakaria, M., Lohse, K., and Bertani-Dziedzic, I. (1979). Diagnosis of neural crest tumors by reversed-phase high-performance liquid chromatographic determination of urinary catecholamine metabolites. *J. Chromatogr.* 186, 733–748.

Krstulovic, A. M., Matzura, C. T., Bertani-Dziedzic, L., Cerqueira, S., and Gitlow, S. E. (1980). Endogenous levels of free and conjugated urinary 3-methoxy-4-hydroxyphenylethyleneglycol in control subjects and patients with pheochromocytoma determined by reversed-phase liquid chromatography with electrochemical detection. *Clin. Chim. Acta 103*, 109–116.

Krstulovic, A. M., Bertani-Dziedzic, L., Dziedzic, S. W., and Gitlow, S. E. (1981). Quantitative determination of 3-methoxy-4-hydroxyphenylethyleneglycol and its sulfate conjugate in human lumbar cerebrospinal fluid using liquid chromatography with amperometric detection. *J. Chromatogr.* 223, 305–314.

Krstulovic, A. M., Bertani-Dziedzic, L., Bautista-Cerqueira, S., and Gitlow, S. E. (1982). Simultaneous determination of 4-hydroxy-3-methoxyphenylacetic (homovanillic) acid and other monoamine metabolites in human lumbar cerebrospinal fluid. An improved high-performance liquid chromatographic study with electrochemical detection. *J. Chromatogr.* *227*, 379–389.

LaBrosse, E. H., Com-Nouqué, C., Zucker, J.-M., Comoy, E., Bohuon, C., Lemerle, J., and Schweisguth, O. (1980). Urinary excretion of 3-methoxy-4-hydroxymandelic acid and 3-methoxy-4-hydroxyphenylacetic acid by 288 patients with neuroblastoma and related crest tumors. *Cancer Res. 40*, 1995–2001.

Lackovic, Z., Parenti, M., and Neff, N. H. (1981). Simultaneous determination of fentomole quantities of 5-hydroxytryptophan, serotonin and 5-hydroxyindoleacetic acid in brain using HPLC with electrochemical detection. *Eur. J. Pharmacol. 69*, 347–352.

Laganà, A. and Rotatori, M. (1983). High-performance liquid chromatographic procedure for the analysis of urinary 3-methoxy-4-hydroxymandelic acid. *J. Chromatogr. 275*, 168–173.

Lange, H.-W., Mannl, H. F. K., and Hempel, K. (1970). Quantitative and rapid fractionation of acidic and neutral catabolites from catecholamines by ion-exchange chromatography. *Anal. Biochem. 38*, 98–104.

Langlais, P. J., McEntee, W. J., and Bird, E. D. (1980). Rapid liquid-chromatographic measurement of 3-methoxy-4-hydroxyphenylglycol and other monoamine metabolites in human cerebrospinal fluid. *Clin. Chem. 26*, 786–788.

Lasley, S. M., Michaelson, I. A., Greenland, R. D., and McGinnis, P. M. (1984). Simultaneous measurement of tyrosine, tryptophan and related monoamines for determination of neurotransmitter turnover in discrete rat brain regions by liquid chromatography with electrochemical detection. *J. Chromatogr. 305*, 27–42.

Le Quan-Bui, K. H., Elghozi, J.-L., Devynck, M.-A., and Meyer, P. (1982). Rapid liquid chromatographic determination of 5-hydroxyindoles and dihydroxyphenylacetic acid in cerebrospinal fluid of the rat. *Eur. J. Pharmacol. 81*, 315–320.

Lyness, W. H., Friedle, N. M., and Moore, K. E. (1980). Measurement of 5-hydroxytryptamine and 5-hydroxyindoleacetic acid in discrete brain nuclei using reverse phase liquid chromatography with electrochemical detection. *Life Sci. 26*, 1109–1114.

Lyness, W. H. (1982). Simultaneous measurement of dopamine and its metabolites, 5-hydroxytryptamine, 5-hydroxyindoleacetic acid and tryptophan in brain tissue using liquid chromatography and electrochemical detection. *Life Sci. 31*, 1435–1443.

Magnusson, O., Nilsson, L. B., and Westerlund, D. (1980). Simultaneous determination of dopamine, DOPAC and homovanillic acid. Direct injection of supernatants from brain tissue homogenates in a liquid chromatography-electrochemical detection. *J. Chromatogr. 221*, 237–247.

Mais, D. E., Lahr, P. D., and Bosin, T. R. (1981). Determination of serotonin, its precursors, metabolites and [^{3}H]serotonin in lung by high-performance liquid chromatography with fluorescence detection. *J. Chromatogr. 225*, 27–35.

Martinez, E., Artigas, F., Sunol, C., Tusell, J. M., and Gelpi, E. (1983). Liquid-chromatographic determination of indole-3-acetic acid and 5-hydroxyindole-3-acetic acid in human plasma. *Clin. Chem. 29*, 1354–1357.

Mayer, G. S. and Shoup, R. E. (1983). Simultaneous multiple electrode liquid chromatographic–electrochemical assay for catecholamines, indoleamines and metabolites in brain tissue. *J. Chromatogr. 255*, 533–544.

Mefford, I. N. and Barchas, J. D. (1980). Determination of tryptophan and metabolites in rat brain and pineal tissue by reversed-phase high-performance liquid chromatography with electrochemical detection. *J. Chromatogr. 181*, 187–193.

Mefford, I. N., Gilberg, M., and Barchas, J. D. (1980). Simultaneous determination of catecholamines and unconjugated 3,4-dihydroxyphenylacetic acid in brain tissue by ion-pairing reverse-phase high-performance liquid chromatography with electrochemical detection. *Anal. Biochem. 104*, 469–472.

Mefford, I. N., Ward, M. M., Miles, L., Taylor, B., Chesney, M. A., Keegan, D. L., and Barchas, J. D. (1981). III. Determination of plasma catecholamines and free 3,4-dihydroxyphenylacetic acid in continuously collected human plasma by high performance liquid chromatography with electrochemical detection. *Life Sci. 28*, 477–483.

Minegishi, A. and Ishizaki, T. (1984). Rapid and simple method for the simultaneous determination of 3,4-dihydroxyphenylacetic acid, 5-hydroxyindole-3-acetic acid and 4-hydroxy-3-methoxyphenylacetic acid in human plasma by high-performance liquid chromatography with electrochemical detection. *J. Chromatogr. 308*, 55–63.

Mitchell, J. and Coscia, C. J. (1978). Application of paired-ion high-pressure liquid column chromatography to the analysis of L-3,4-dihydroxyphenylalanine metabolites. *J. Chromatogr. 145*, 295–301.

Moleman, P. and Borstrok, J. J. M. (1982). Analysis of urinary 3-methoxy-4-hydroxyphenyl-glycol by high-performance liquid chromatography with electrochemical detection. *J. Chromatogr. 227*, 391–405.

Moleman, P. and Borstrok, J. J. M. (1983). Determination of urinary vanilmandelic acid by liquid chromatography with electrochemical detection. *Clin. Chem. 29*, 878–881.

Molnár, I. and Horváth, C. (1976). Reverse-phase chromatography of polar biological substances: separation of catechol compounds by high-performance liquid chromatography. *Clin. Chem. 22*, 1497–1502.

Molnár, I. and Horváth, C. (1977). Rapid separation of urinary acids by high-performance liquid chromatography. *J. Chromatogr. 143*, 391–400.

Molyneux, S. G. and Franklin, M. (1985). Routine determination of unconjugated 3-methoxy-4-hydroxyphenylglycol in plasma using high-performance liquid chromatography with electrochemical detection. *J. Chromatogr. 341*, 160–166.

Momose, A., Uchikura, K., and Kabasawa, Y. (1983). New post-column detection system using electrochemical reaction for high performance liquid chromatography. *Bunseki Kagaku (Jap. Anal.) 32*, 142–143.

Morita, I., Masujima, T., Yoshida, H., and Imai, H. (1981). Enrichment and high-performance liquid chromatography analysis of trylptophan metabolites in plasma. *Anal. Biochem. 118*, 142–146.

Morrisey, J. L. and Shihabi, Z. K. (1979a). Assay of urinary 4-hydroxy-3-methoxymandelic (vanillylmandelic) acid by liquid chromatography with electrochemical detection. *Clin. Chem. 25*, 2043–2045.

Morrisey, J. L. and Shihabi, Z. K. (1979b). Assay of 4-hydroxy-3-methoxyphenylacetic (homovanillic) acid by liquid chromatography with electrochemical detection. *Clin. Chem. 25*, 2045–2047.

Mrochek, J. E., Butts, W. C., Rainey, W. T., and Burtis, C. A. (1971). Separation and identification of urinary constituents by use of multiple-analytical techniques. *Clin. Chem. 17*, 72–77.

Mrochek, J. E., Dinsmore, S. R., and Ohrt, D. W. (1973). Monitoring phenylalanine-tyrosine metabolism by high-resolution liquid chromatography of urine. *Clin. Chem. 19*, 927–936.

Muskiet, F. A. J., Fremouw-Ottevangers, D. C., Wolthers, B. G., and de Vries, J. A. (1977). Gas-chromatographic profiling of urinary acidic and alcoholic catecholamine metabolites. *Clin. Chem. 23*, 863–867.

Muskiet, F. A. J., Nagel, G. T., and Wolthers, B. G. (1980). Simultaneous determination of unconjugated homovanillic acid, vanilmandelic acid, and 3-methoxy-4-hydroxyphenylethylene glycol, in serum by mass fragmentography and deuterated internal standards. *Anal. Biochem. 109*, 130–136.

Nabeshima, T., Hirata, M., Noma, S., Uki, M., Amano, M., and Kameyama, T. (1982). Determination of methionine-enkephalin, norepinephrine, dopamine, 3,4-dihydroxyphenylacetic acid (DOPAC) and 3-methoxy-4-hydroxyphenylacetic acid (HVA) in brain by high-pressure liquid chromatography with electrochemical detector. *Res. Commun. Chem. Pathol. Pharmacol. 35*, 421–442.

Narasimhachari, N., Boadle-Biber, M. C., and Friedel, R. O. (1982). A critical evaluation of LC–EC for the simultaneous determination of 5-hydroxytryptophan, 5-HT, 5-HIAA and HVA in biological samples using GC–MS for validation. *Res. Commun. Chemical. Pathol. Pharmacol. 37*, 413–430.

Nielsen, J. A. and Johnston, C. A. (1982). Rapid, concurrent analysis of dopamine, 5-hydroxytryptamine, their precursors and metabolites utilizing high performance liquid

268   *A. Yoshida et al.*

chromatography with electrochemical detection: Analysis of brain tissue and cerebrospianl fluid. *Life Sci. 31*, 2847−2856.

Ong, H., Capet-Antonini, F., Yamaguchi, N., and Lamontagne, D. (1982). Simultaneous determination of free 3-methoxy-4-hydroxymandelic acid and free 3-methoxy-4-hydroxyphenylethyleneglycol in plasma by liquid chromatography with electrochemical detection. *J. Chromatogr. 233*, 97−105.

Orsulak, P. J., Kizuka, P., Grab, E., and Schidkraut, J. J. (1983). Determination of urinary normetanephrine and metanephrine by radial compression liquid chromatography with electrochemical detection. *Clin. Chem. 29*, 305−309.

Persson, B.-A. and Karger, B. L. (1974). High performance ion pair partition chromatography: the separation of biogenic amines and their metabolites. *J. Chromatogr. Sci. 12*, 521−528.

Petruccelli, B., Bakris, G., Miller, T., Korpi, E. R., nd Linnoila, M. (1982). A liquid chromatographic assay for 5-hydroxytryptophan, serotonin and 5-hydroxyindoleacetic acid in human body fluids. *Acta Pharmacol. Toxicol. 51*, 421−427.

Pisano, J. J., Crout, J. R., and Abraham, D. (1962). Determination of 3-methoxy-4-hydroxymandelic acid in urine. *Clin. Chim. Acta 7*, 285−291.

Reinhard, J. F. Jr, Moskowitz, M. A., Sved, A. F., and Fernstrom, J. D. (1980). A simple, sensitive and reliable assay for serotonin and 5-HIAA in brain tissue using liquid chromatography with electrochemical detection. *Life Sci. 27*, 905−911.

Rich, W., Johnson, E., Lois, L., Karla, P., Stafford, B., and Marton, L. (1980). Determination of organic acids in biological fluids by ion chromatography: plasma lactate and pyruvate and urinary vanilmandelic acid. *Clin. Chem. 26*, 1492−1498.

Rosano, T. G. and Brown, H. H. (1979). Liquid-chromatographic assay for urinary 3-methoxy-4-hydroxymandelic acid, with use of a periodate oxidative monitor. *Clin. Chem. 25*, 550−554.

Rosano, T. G., Brown, H. H., and Meola, J. M. (1981). Liquid chromatographic assay for urinary homovanillic acid, with fluorescent detection. *Clin. Chem. 27*, 228−231.

Rossetti, Z. L., Mercuro, G., and Piavano, C. A. (1983). A study of the parameters affecting flow gradient analysis of catecholamines, DOPA and DOPAC by ion pair liquid chromatography with electrochemical detection. *Life Sci. 33*, 2387−2397.

Saller, C. F. and Salama, A. I. (1984). Rapid automated analysis of biogenic amines and their metabolites using reversed-phase high-performance liquid chromatography with electrochemical detection. *J. Chromatogr. 309*, 287−298.

Sandler, M., Johnson, R. D., Ruthven, C. R. L., Reid, J. L., and Calne, D. B. (1974). Transamination is a major pathway of L-Dopa metabolism following peripheral decarboxylase inhibition. *Nature (London) 247*, 364−366.

Sankoff, I. and Sourkes, T. L. (1963). Determination by thin-layer chromatography of urinary homovanillic acid in normal and disease states. *Canad. J. Biochem. Physiol. 41*, 1381−1388.

Saraswat, L. D., Holdiness, M. R., Justice, J. B., Salamone, J. D., and Neill, D. B. (1981). Determination of dopamine, homovanillic acid and 3,4-dihydroxyphenylacetic acid in rat brain striatum by high-performance liquid chromatography with electrochemical detection. *J. Chromatogr. 222*, 353−362.

Sato, T., Wada, Y., and Maebashi, M. (1963). Simultaneous determination of urinary vanillylmandelic acid and 5-hydroxyindoleacetic acid. *Tohoku J. Exper. Med. 80*, 1−8.

Sawada, J. (1981). Neuroblastoma. *Shoni Igaku 14*, 41−70.

Scheinin, M., Chang, W.-H., Kirk, K. L., and Linnoila, M. (1983). Simultaneous determination of 3-methoxy-4-hydroxyphenylglycol, 5-hydroxyindoleacetic acid and homovanillic acid in cerebrospinal fluid with high-performance liquid chromatography using electrochemical detection. *Anal. Biochem. 131*, 246−253.

Schildkraut, J. J., Orsulak, P. J., LaBrie, R. D., Schatzberg, A. F., Gudeman, J. E., Cole, J. O., and Rohde, W. A. (1978). Toward a biochemical classification of depressive disorders. *Arch. Gen. Psychiatry 35*, 1436−1439.

Schinelli, S., Santagostino, G., Frattini, P., Cucchi, M. L., and Corona, G. L. (1985). Assay of

3-methoxy-4-hydroxyphenylglycol in human using high-performance liquid chromatography with amperometric detection. *J. Chromatogr. 338*, 396–400.

Scott, C. D. (1968). Analysis of urine its ultraviolet-absorbing constituents by high-pressure anion-exchange chromatography. *Clin. Chem. 14*, 521–528.

Scott, C. D., Attrill, J. E., and Anderson, N. G. (1967). Automatic, high-resolution analysis of urine for its ultraviolet-absoring constituents. *Proc. Soc. Exp. Biol. Med. 125*, 181–184.

Seegal, R. F., Brosch, K. O., and Bush, B. (1983). Direct determination of 4-hydroxy-3-methoxyphenylacetic (homovanillic) acid in urine by high-performance liquid chromatography with amperometric detection. *J. Chromatogr. 273*, 253–261.

Seegal, R. F., Brosch, K. O., and Bush, B. (1986). High-performance liquid chromatography of biogenic amines and metabolites in brain, cerebrosplinal fluid, urine and plasma. *J. Chromatogr. 377*, 131–144.

Semerdjian-Rouquier, L., Bossi, L., and Scatton, B. (1981). Determination of 5-hydroxytryptophan, serotonin and 5-hydroxyindoleacetic acid in rat and human brain and biological fluids by reversed-phase high-performance liquid chromatography with electrochemical detection. *J. Chromatogr. 218*, 663–670.

Sharpless, N. S., Halbreich, U., and Feldfogel, H. (1966). Determination of total 3-methoxy-4-hydroxyphenylglycol in plasma using reversed-phase liquid chromatography with electrochemical detection. *J. Chromatogr. 377*, 101–109.

Shea, P. A., and Howell, J. B. (1984). High-performance liquid chromatographic method for determining plasma and urine 3-methoxy-4-hydroxyphenylglycol by amperometric detection. *J. Chromatogr. 306*, 358–363.

Shibuya, T., Sato, K., and Salafsky, B. (1982). Simultaneous measurement of biogenic amines and related compounds by high performance liquid chromatography (HPLC). *Int. J. Clin. Pharmacol. Ther. Toxicol. 20*, 297–301.

Shihabi, Z. K. and Scaro, J. (1980). Liquid-chromatographic assay of urinary 5-hydroxy-3-indoleacetic acid, with electrochemical detection. *Clin. Chem. 26*, 907–909.

Shoup, R. E. and Kissinger, P. T. (1977). Determination of urinary normetanephrine, metanephrine, and 3-methoxytyramineby liquid chromatography, with amperometric detection. *Clin. Chem. 23*, 1268–1274.

Sjoerdsma, A., Weissbach, H., and Udenfriend, S. (1955). Simple test for diagnosis of metastatic carcinoid (argentaffinoma). *J. Am. Med. Assoc. 159*, 397.

Skrinska, V. and Hahn, S. (1984). High-performance liquid chromatography of 5-hydroxyindole-3-acetic acid in urine with direct sample injection. *J. Chromatogr. 311*, 380–384.

Soldin, S. J. and Hill, J. G. (1980). Simultaneous liquid-chromatographic analysis for 4-hydroxy-3-methoxymandelic acid and 4-hydroxy-3-methoxyphenylacetic acid in urine. *Clin. Chem. 26*, 291–294.

Sperk, G. (1982). Simultaneous determination of serotonin, 5-hydroxyindoleacetic acid, 3,4-dihydroxyphenylacetic acid and homovanillic acid by high performance liquid chromatography with electrochemical detection. *J. Neurochem. 38*, 840–843.

Stott, A. W., Smith, J. R. L., Hanson, P., and Robinson, R. (1975). A simple chromatographic procedure for the concurrent estimation of urinary 4-hydroxy-3-methoxymandelic acid (HMMA) and homovanillic acid (HVA) using a scanning technique. *Clin. Chim. Acta 63*, 7–12.

Stout, R. W., Michelot, R. J., Molnár, I., Horváth, C. H., and Coward, J. K. (1976). An analytical method for the separation of dopamine metabolites in cellular extracts by high-pressure liquid chromatography. *Anal. Biochem. 76*, 330–341.

Sunderman, F. W. Jr, Cleveland, P. D., Law, N. C., and Sunderman, F. W. (1960). A method for the determination of 3-methoxy-4-hydroxymandelic acid ('vanilandelic acid') for the diagnosis of pheochromocytoma. *Am. J. Clin. Pathol. 34*, 293–312.

Takahashi, S., Yoshioka, M., Yoshiue, S., and Tamura, Z. (1978). Mass fragmentographic determination of vanilmandelic acid, homovanillic acid and isohomovanillic acid in human body fluids. *J. Chromatogr. 145*, 1–9.

Taylor, J. T., Freeman, S., and Brewer, P. (1981). Liquid chromatography of 3-methoxy-4-hydroxyphenylethylene glycol in urine with fluorescence detection. *Clin. Chem.* 27, 173–175.

Tokuda, T., Yoshioka, M., Tamura, Z., and Yokomori, K. (1984). Automatic analysis of homovanillic acid and vanilmandelic acid by anion exchanged high performance liquid chromatograph*y*. *Bunseki Kagaku (Jap. Anal.)* 33, E331–E334.

Towell, J. F. and Erwin, V. G. (1981). Determination of the primary metabolite of central nervous system norepinephrine, 3-methoxy-4-hydroxyphenethyleneglycol, in mouse brain and brain perfusate by high-performance liquid chromatography with electrochemical detection. *J. Chromatogr.* 223, 295–303.

Udenfriend, S., Titus, E., and Weissbach, H. (1955). The identification of 5-hydroxy-3-indoleacetic acid in normal urine and a method for its assay. *J. Biol. Chem.* 216, 499–505.

Valkenburg, C. V., Tjaden, U., Van der Krogt, J., and Van der Leden, B. (1982). Determination of dopamine and its acidic metabolites in brain tissue by HPLC with electrochemical detection in a single run after minimal sample pretreatment. *J. Neurochem.* 39, 990–997.

von Studnitz, W. (1960). Methodische und klinische Untersuchungen über die Ausscheidung der 3-Methoxy-4-hydroxymandelsäure im Urin. *Scand. J. Clin. Lab. Invest. Suppl.* 48, 1–73.

von Studnitz, W. (1963). Über die Ausscheidung der 3-Methoxy-4-hydroxtphenylessigsäure (Homovanillinsäure) beim Neuroblastom und anderen neuralen Tumoren. *Klin. Wochenschr.* 40, 163–167.

Voorhess, M. L. (1974). Neuroblastoma–pheochromocytoma: products and pathogenesis. *Ann. N.Y. Acad. Sci.* 230, 187–194.

Wagner, J., Vitali, P., Palfreyman, M. G., Zraika, M., and Huot, S. (1982). Simultaneous determination of 3,4-dihydroxyphenylalanine, 5-hydroxytryptophan, dopamine, 4-hydroxy-3-methoxyphenylalanine, norepinephrine, 3,4-dihydroxyphenylacetic acid, homovanillic acid, serotonin, and 5-hydroxyindoleacetic acid in rat cerebrospinal fluid and brain by high-performance liquid chromatography with electrochemical detection. *J. Neurochem.* 38, 1241–1254.

Wahlund, K.-G. and Edlen, B. (1981a). Simple and rapid determination of 5-hydroxyindole-3-acetic acid in urine by direct injection on a liquid chromatographic column. *Clin. Chim. Acta* 110, 71–76.

Wahlund, K.-G. and Edlén, B. (1981b). Separation of 5-hydroxyindole-3-acetic acid and indole-3-acetic acid in urine by direct injection on a reversed-phase column containing a hydrogen-accepting stationary phase. *J. Chromatogr.* 204, 269–274.

Weise, V. K., McDonald, R. K., and Labrosse, E. H. (1961). Determination of urinary 3-methoxy-4-hydroxymandelic acid in man. *Clin. Chim. Acta* 6, 79–86.

Westerink, B. H. C. and Mulder, T. B. A. (1981). Determination of picomole amounts of dopamine, noradrenaline, 3,4-dihydroxyphenylalanine, 3,4-dihydroxyphenylacetic acid, homovanillic acid, and 5-hydroxyindoleacetic acid in nervous tissue after one-step purification on Sephadex G-10, using high-performance liquid chromatography with a Novel type of electrochemical detection. *J. Neurochem.* 36, 1449–1462.

Westerink, B. H. C., Bosker, F. J., and O'Hanlon, J. F. (1982). Use of alumina, Sephadex G-10, and ion-exchange columns to purify samples for determination of epinephrine, norepinephrine, dopamine, homovanillic acid and 5-hydroxyindoleacetic acid in urine. *Clin. Chem.* 28, 1745–1748.

Westerink, B. H. C. (1984). Determination of normetanephrine, 3,4-dihydroxyphenylethyleneglycol (free and total), and 3-methoxy-4-hydroxyphenylethyleneglycol (free and total) in rat brain by high-performance liquid chromatography with electrochemical detection and effects of drugs on regional concentrations. *J. Neurochem.* 42, 934–942.

Wielders, J. P. M. and Mink, C. J. K. (1984). Analysis of vanillylmandelic acid, homovanillic acid and 5-hydroxyindoleacetic acid in human urine by high-performance liquid chromatography and fluorometry. *J. Chromatogr.* 310, 379–385.

Wightman, R. M., Plotsky, P. M., Strope, E., Delcore, R. Jr, and Adams, R. N. (1977). Liquid chromatographic monitoring of CSF metabolites. *Brain Res.* 131, 345–349.

Williams, C. M. and Greer, M. (1963). Homovanillic acid and vanilmandelic acid in diagnosis of neuroblastoma. *J. Am. Med. Assoc.* 183, 836–840.

Wilson, W. E., Mietling, S. W., and Hong, J.-S. (1983). Automated HPLC analysis of tissue levels of dopamine, serotonin, and several prominent amine metabolites in extracts from various brian regions. *J. Liq. Chromatogr. 6*, 871–886.

Wolf, W. A. and Kuhn, D. M. (1983). Simultaneous determination of 5-hydroxytryptamine, its amino acid precursors and acid metabolite in discrete brain regions by high-performance liquid chromatography with fluorescence detection. *J. Chromatogr. 275*, 1–9.

Yamada, K., Kayama, E., Aizawa, Y., Oka, K., and Hara, S. (1981). Determination of vanillylmandelic acid in urine by pre-column dansylation using micro high-performance liquid chromatography with fluorescence detection. *J. Chromatogr. 223*, 176–178.

Yamaguchi, T., Yokota, K., and Uematsu, F. (1982). Separation of indole metabolites from urine with an ODS type resin by high-performance liquid chromatography. *J. Chromatogr. 231*, 166–172.

Yoshida, A., Yoshioka, M., Tanimura, T., and Tamura, Z. (1976a). Determination of vanilmandelic acid and homovanillic acid in urine by high-speed liquid chromatography. *J. Chromatogr. 116*, 240–243.

Yoshida, A., Yoshioka, M., Yamazaki, T., Sakai, T., and Tamura, Z. (1976b). Urinary levels of vanilmandelic acid and homovanillic acid determined by high-speed liquid chromatography. *Clin. Chim. Acta 73*, 315–320.

Yoshida, A., Yamazaki, T., and Sakai, T. (1977). Determination of urinary 5-hydroxyindole-3-acetic acid by high-speed liquid chromatography. *Clin. Chim. Acta 77*, 95–97.

Yoshida, A., Yoshioka, M., Sakai, T., and Tamura, Z. (1978). Determination of vanilpyruvic acid in urine by high-speed liquid chromatography. *Chem. Pharm. Bull. (Tokyo) 26*, 1177–1181.

Yoshida, A., Yoshioka, M., Sakai, T., and Tamura, Z. (1982). Simple method for the determination of homovanillic acid and vanillylmandelic acid in urine by high-performance liquid chromatography. *J. Chromatogr. 227*, 162–167.

Yoshida, A., Yamaguchi, Y., Yoshioka, M., and Tamura, Z. (1984). Simultaneous determination of homovanillic acid, vanillylmandelic acid and 5-hydroxyindole-3-acetic acid in urine by high performance liquid chromatography. *Bunseki Kagaku (Jap. Anal.) 33*, E257–E262.

Yoshida, A., Ichihashi, Y., and Yoshioka, M. (1985a). Identification of oxidation products from homovanillic acid and vanillylmandelic acid at electrochemical detector for high-performance liquid chromatography. *Biogenic Amines 2*, 119–124.

Yoshida, A., Ichihashi, Y., and Yoshioka, M. (1985b). Determination of homovanillic acid in human plasma by high-performance liquid chromatography with electrochemical detection. *J. Chromatogr. 343*, 155–159.

Yoshida, A., Yoshioka, M., and Parvez, H. (in submission). High performance liquid chromatography of metabolites of catecholamines and serotonin in urine, plasma, cerebrospinal fluid and brain tissue. II. Clinical applications. *Biogenic Amines.*

Yoshioka, M., Yoshida, A., Ichihashi, Y., and Saito, H. (1985). Homovanillic acid, vanillylmandelic acid and 5-hydroxyindole-3-acetic acid are undetectable in urine of the muskrat. *Chem. Pharm. Bull. (Tokyo) 33*, 2145–2148.

Progress in HPLC, Vol. 4, pp. 273—296.
Yoshioka *et al.* (Eds)
© 1989 VSP.

Solvent elimination approach for microcolumn liquid chromatography—infrared spectroscopy: buffer-memory technique

KIYOKATSU JINNO and CHUZO FUJIMOTO

School of Materials Science, Toyohashi University of Technology,
Toyohashi 440, Japan

INTRODUCTION

High-performance liquid chromatography (LC) and supercritical fluid chromatography (SFC) have become extremely powerful tools, sufficient to separate complex mixtures into their components. With LC and SFC, compounds of poor thermal stability and low volatility which cannot be separated by gas chromatography (GC), can be readily separated for either quantitative or qualitative analysis. As a general rule the amount of information that can be obtained from any chromatographic separation depends on the detector. As the field of application for LC and SFC has increased, the limitations of conventional detectors such as ultraviolet/visible (UV/VIS) and refractive index (RI) in LC, and UV/VIS and flame ionization (FID) in SFC have become increasingly restrictive to the growth of those techniques.

One of the ways in which to improve LC and SFC, to make them more informative, and more powerful analytical techniques, is a combination of those chromatographic techniques with spectroscopic techniques such as infrared spectrometry (IR) and mass spectrometry (MS). Using IR, one can obtain much information regarding the chemical structures of components separated by chromatography, because the large number of absorption bands present in an IR spectrum offers the possibility of having both universal and chemically selective or specific detection capabilities. In this instance the well-developed and analytically powerful application of various GC/IR interfaces (Griffiths *et al.*, 1983) has led many analysts to believe that a similarly effective and simple hybrid system may be developed for interfacing LC and SFC to IR spectroscopy.

Fundamental problems of compatibility have been encountered, however, and the development of the interfacing between LC and SFC, and IR has proven to be quite difficult. Unfortunately, most practical LC solvents, such as n-hexane, methanol, acetonitrile and water, have strong absorption bands in the mid-IR region, so that some spectral regions are opaque. Typical SFC mobile phase, carbon dioxide added with organic modifiers have the spectroscopic properties of increasing their absorptions in the IR region with increasing pressure and concentration of modifiers (Shafer and Griffiths, 1983; Olesik et al., 1984; Johnson et al., 1985). In order to maintain a sufficiently high transmittance to allow solute absorption bands to be observed at most wavelengths across the spectrum, the path length of the cell must be kept as short as possible. If the path length is increased the detection limit may be increased in the spectral windows, but the proportion of the spectrum lost through absorption by the mobile phase is increased. Therefore a compromise should be sought, to allow maximum cell thickness consistent with solvent transmission over all the bands of interest.

This intrinsic disadvantage of the flow cell configuration in LC—IR and SFC—IR can be overcome by eliminating the mobile phase solvent prior to the measurements of the IR spectrum of the solute. Kuehl and Griffiths (1979, 1980) used diffuse reflectance FTIR (DRIFT) with conventional LC columns; the effluent was concentrated by differential evaporation and the solutes were deposited on powdered KCl, which was held in small sample cups on a carousel. A drop monitor with a photocell and controlling electronics move the carousel after each drop is deposited. This method appears satisfactory, but it requires a relatively complicated interfacing device and also perfect solvent elimination is difficult, where typically 1 ml/min should be evaporated. From this point of view, if one should use a solvent elimination approach, the merit of microcolumn LC and SFC (micro-LC and SFC) which is the low flow rate of mobile phase can solve problems associated with LC—IR and SFC—IR interfacing. An approach in this direction has been proposed by various authors since 1980 (Jinno, 1981; Jinno and Fujimoto, 1981; Jinno et al., 1980, 1981, 1982a, b, 1985a; Fujimoto et al., 1983, 1985a, b), where micro-LC was employed for the separation, and a KBr crystal plate was used as the transport medium to accept the total column effluent. This is called the 'buffer-memory' technique.

In this contribution we will describe the analytical potential of the 'buffer-memory' technique to combine micro-LC and SFC with spectroscopic methods such as IR (including Fourier transform IR) and microprobe laser raman spectroscopy.

INTERFACING DEVICE

An interfacing device, which has been developed for the combination of microcolumn chromatography and infrared spectroscopic techniques, is shown in Fig. 1. The interface was made by modifying a commercially available microfeeder (MF-2, Azuma Electric, Tokyo, Japan). A collecting medium such as a KBr crystal plate is set in the holder attached to the end of the moving rod of the interface. The eluent from the microcolumn is deposited on the medium as a continuous narrow band. The medium is transferred past the exit of the microcolumn for sample collection. The speed of

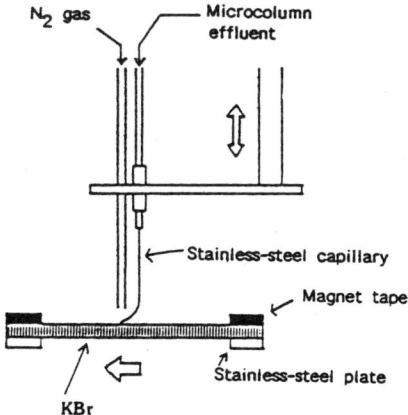

Figure 1. Sampling arrangement for the buffer-memory technique.

collection can be controlled according to the separation conditions used. The buffer-memory medium is automatically brought into the compartment of a spectrometer as the eluent is being collected. A spectrometer monitors the eluent as a single detector.

Figure 2 shows the buffer-memory crystal plate after deposition of sample components eluted from a microcolumn. The size of the plate is commonly 3 mm thickness, 8 mm width and 30 mm length. One can find traces on the plate which maintain the concentration distribution the same as the measured chromatograms.

The most important requirements in this kind of interfacing are signal, reproducibility and linearity. The experimental results relating to these requirements are shown in Figs 3 and 4. Figure 3 demonstrates signal linearity, where the sample probe is di-*n*-propylketone DNPH (dinitrophenylhydrazone). The various amounts of DNPH (between 1 µg and 6 µg) were deposited on the crystal plate with the buffer-memory interfacing device and IR chromatograms were measured. Measurements at fixed wave-number and at a wide wavenumber window both offer good linearity in their signal intensities. Figure 4 shows signal reproducibility for 2 µg of the same sample probe. Less than ±5% reproducibility was obtained. It therefore appears that the buffer-memory technique satisfies two minimum requirements for analytical techniques: good reproducibility and high linearity.

One of the demonstrations of the basic KBr buffer-memory technique is shown in Fig. 5, where size exclusion microcolumn chromatography (SEC) has been performed for the separation of polystyrene standards. The column used was a 1.0 mm i.d. × 220 mm length PTFE tube packed with TSK Gel 3000H (ToyoSoda, Tokyo, Japan) and the mobile phase was tetrahydrofuran (THF) at 8 µl/min. The spectrometer used was a JEOL (Tokyo, Japan) JIR-40X FTIR, in which a TGS detector was used and 64 times signal accumulation was performed for each data file. The samples were polystyrene standards with molecular weight ranging between 37000 and 500. In Fig. 5, UV and FTIR chromatograms are compared to clarify the

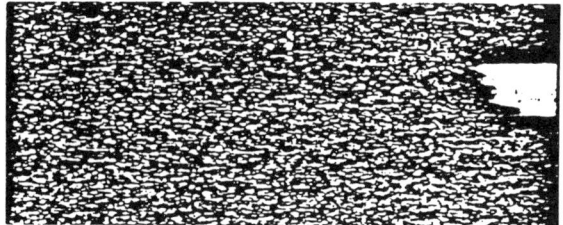

Figure 2. Photograph of the buffer-memory plate.

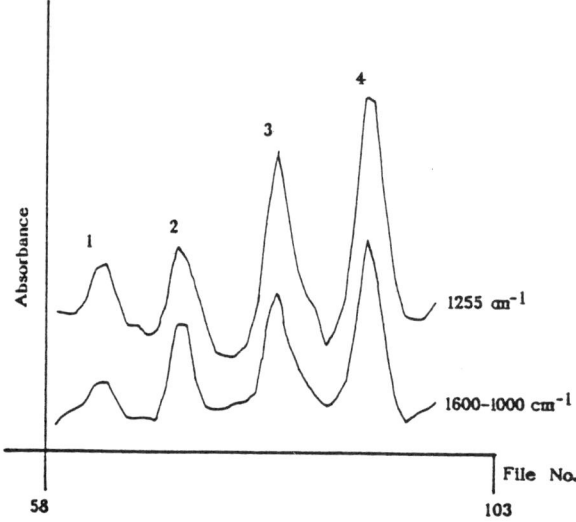

Figure 3. Peak height of the IR chromatogram dependent on the weight of sample injected. Solutions containing measured amounts of di-*n*-propyl DNPH were successively injected into a micro-LC column: 1; 1 µg, 2; 3 µg, 3; 4 µg, 4; 6 µg. (Copyright Elsevier, 1983.)

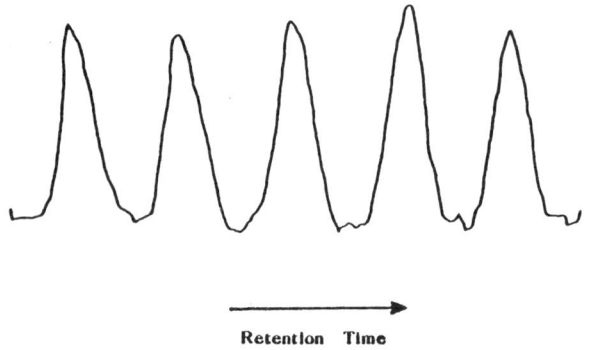

Figure 4. IR transmittance chromatogram illustrating the reproducibility of the buffer-memory technique. A solution containing di-*n*-propyl DNPH of 3 µg was successively injected. (Copyright Elsevier, 1983.)

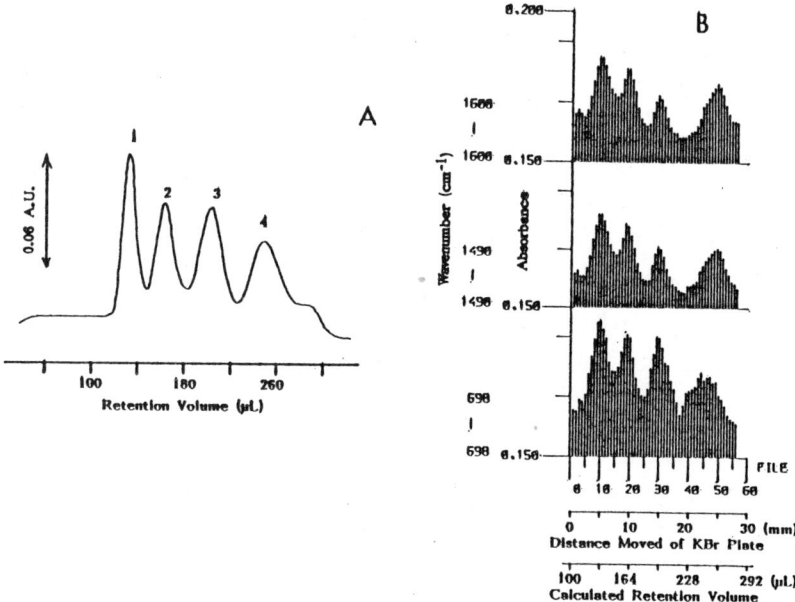

Figure 5. A: SEC chromatogram of polystyrene standards measured with a UV detector at 254 nm PTFE tube; packing, TSK Gel 3000H; mobile phase, THF 8 μl/min; peaks: 1 = MW, 37000, 2 = MW 10200, 3 = MW 2800, 4 = MW 500. B: FTIR chromatograms at various wavenumbers. Spectrometer: JEOL JIR 40X, detector: TGS, accumulation: 64 times, resolution: 8 cm^{-1}. (Copyright Elsevier, 1982.)

performance of the buffer-memory technique. Some minor resolution degradation was observed in FTIR chromatograms but the performance of the buffer-memory technique is clearly demonstrated.

REVERSED-PHASE LC—FTIR INTERFACING

The first application of 'buffer-memory' techniques to LC—spectroscopy combination is the interfacing between reversed-phase LC and FTIR. Until now, there have been few reports on reversed-phase LC—FTIR interfacing. Those approaches are commonly divided into direct flow-cell and solvent elimination. Because typical mobile phase solvents in reversed-phase LC, such as water, methanol or acetonitrile, have a number of big absorption bands in the mid-IR region (water is a big problem in common IR measurements), it is considered that a direct flow-cell approach is not feasible. Therefore, more recently, some modifications of the solvent elimination approach, using DRIFT, have been studied in an attempt to solve the water interference problem.

Conroy *et al.* (1984) have developed a new solvent elimination interface which includes extraction of solutes into dichloromethane; subsequently, the interface is operated in a manner similar to that in normal-phase separations (Kuehl and Griffiths, 1979, 1980). Kalasinsky and co-workers (1983) have utilized a post-column dehydrating agent (dimethoxypropane) which reacts with water in the effluent to form volatile products before being deposited on a moving sample 'train' which contains a KCl powder substrate. The spectra of the solutes were measured by DRIFT. However, both methods are still under investigation.

The authors have proposed a 'buffer-memory' technique for this interfacing problem, where stainless-steel wire nets were used as the collecting medium instead of KBr crystal plates in normal phase mode (Fujimoto et al., 1985b). Stainless-steel wire nets have many of the properties required for a substrate in IR sampling, in that they are completely insoluble in all solvents including water, not very fragile, conveniently handled, and easily regenerated. In addition, they possess enough open area to transmit the incident radiation to the detector, allowing the transmittance spectrum of a sample to be measured. Therefore the water problem in IR measurements will be circumvented.

Commercially available stainless-steel wire nets were formed into round disks (7 mm) or strips (10 mm × 50 mm) and, prior to use, were cleaned by rinsing in various solvents, e.g. methanol and carbon tetrachloride, and air-drying.

In this approach the performance of several stainless-steel wire nets having different opening window sizes was tested first. At first sight, none of these stainless-steel wire nets looks to be tight enough to trap the sample. However, all the sample seems to be retained on the surface because no breakthrough of solution was observed. Initially spectroscopic performance has been investigated. The spectra obtained by depositing 2 µg of carbaryl in methanol/water (1:1, v/v) onto each stainless-steel wire net surface are shown in Fig. 6. It can be seen that the intensities of the absorption bands increase substantially as the opening is reduced, and the 846-mesh stainless-steel wire net gives the most intense spectrum. Only a slight increase in background noise level was observed on the spectrum from the 846-mesh net. It is interesting to find that the band intensity is independent of the energy impinging on the detector. The percentage of open area (listed in Table 1) reflects the efficiency of incident radiation transmitted through the wire net without a beam condenser. Based on these results, the 846-mesh stainless-steel wire net was chosen for all further studies.

The most serious point to be considered in the buffer-memory technique is the possible loss of sample deposited on exposure to the atmosphere and IR beam, as well as losses during solvent evaporation. It should be realized that in all the LC–FTIR techniques involving a solvent elimination step, fairly low volatility of the sample is a prerequisite to their success.

In order to establish the suitability of buffer-memory technique with stainless-steel wire nets for various samples, some compounds of relatively low volatility were tested. One main advantage of the buffer-memory technique over other solvent elimination techniques is that the solutes are retained on a single substrate as a permanent record of the chromatogram, so that the solutes can be characterized by other analytical techniques if

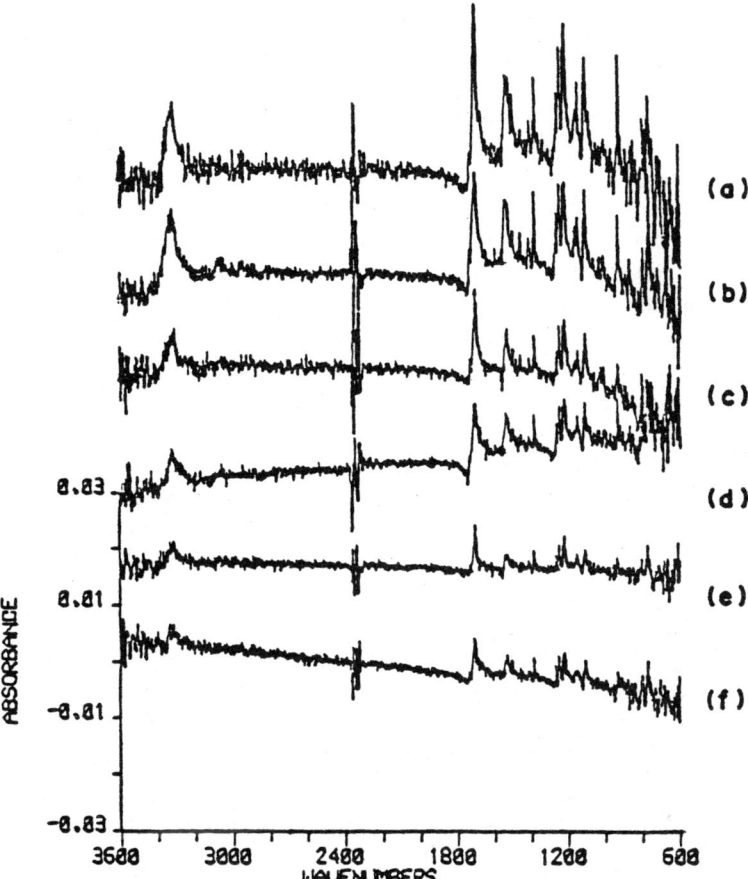

Figure 6. FTIR spectra of carbaryl deposited on stainless-steel wire nets with different screen-size opening (μm): (a) 15, (b) 20, (c) 37, (d) 53, (e) 74, (f) 105. Conditions: carbaryl solution; flow rate, 1 μl/min; collection time, 12 s. (Copyright Elsevier, 1985.)

Table 1.
The stainless-steel wine nets tested

Screen size opening (μm)	Taylor equivalent designation (mesh)	Nominal wire diam. (mm)	Open area (%)
15	846	0.012	30.9
20	635	0.015	32.7
37[a]	400	0.026	34.5
46	300	0.035	32.3
53[a]	270	0.038	33.9
74[a]	200	0.053	34.0
105[a]	145	0.070	36.0

[a] JIS standard

required (Fujimoto *et al.*, 1983). The selected compounds were *n*-octade-
cane, carbaryl, *n*-butyl stearate, and *p,p'*-DDT, of which vapor pressures at
20°C are of the order of 10^{-4}, 10^{-5}, 10^{-6} and 10^{-7} torr, respectively.
Variations of absorbance of each compound with exposure time for IR beam
are shown in Fig. 7. Obviously, the relative evaporation rate of the sample
decreases with decreasing vapor pressure. The evaporation loss is very rapid
with *n*-octadecane, which is liquid at room temperature. In these measurements the stainless-steel wire nets with sample deposited were placed at a
small focus between the IR source and beam splitter in the FTIR spectrometer and absorbances were obtained from 200 times co-added, 4 cm^{-1}
resolution spectra. Therefore, the heating effect of the source radiation on
the sample also creates sample losses. In fact, at higher data collection rates
in 8 cm^{-1} resolution, the absorbances decreased much more slowly. Such
losses can be greatly reduced by placing the stainless-steel wire nets just in
front of the detector. Compounds having a vapor pressure of not more than
10^{-5} torr or so could be handled by the buffer-memory technique.

In order to demonstrate the performance of the buffer-memory technique, a mixture of phenacetin, caffeine, and aspirin (8.4 µg of each) was
separated on a 10 µm C_{18} column with methanol/water (3:2) at 4 µl/min; a
UV detector was placed before the deposition point. In this instance, efficient evaporation required heating of the nitrogen stream to roughly 80°C.
The UV and FTIR chromatograms are shown in Fig. 8. Even with the high
water content of the mobile phase, the FTIR chromatogram maintains peak
shape and resolution. The response obviously depends on the properties of
the sample molecules, e.g. the sensitivity for phenacetin taken from a file

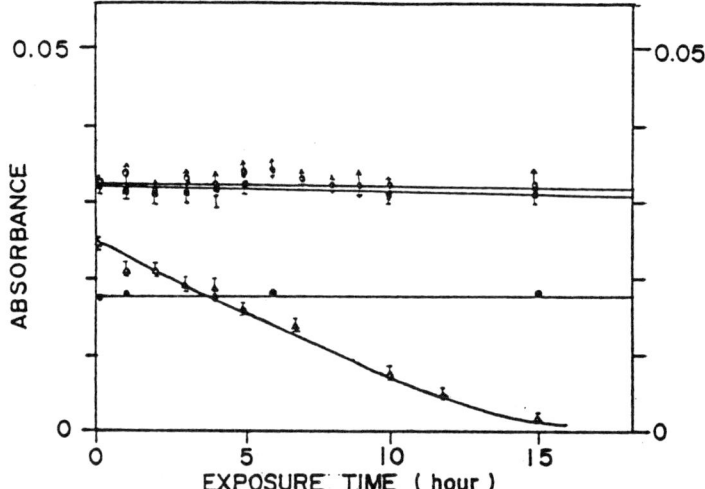

Figure 7. Variation of band intensity of sample with exposure time. Each sample (5 µg) was
deposited on the 846-mesh stainless-steel wire net. (a) *p,p'*-DDT at 1100 cm^{-1}; (b) *n*-butyl-
searate at 2920 cm^{-1}; (c) carbaryl at 1712 cm^{-1}; (d) *n*-octadecane at 2930 cm. Conditions: 5 µg
of sample in methanol or carbon tetrachloride was deposited; temperature, *ca.* 27°C, relative
humidity 40%. (Copyright Elsevier, 1985.)

used in Fig. 8 was compared with a spectrum obtained by depositing a phenacetin solution onto a KBr crystal directly without the column, as shown in Fig. 9. There were some differences in the intensities of certain bands but changes of bandwidth were not markedly observed. Figure 10 shows a three-dimensional chromatogram obtained by FTIR measurement, and no interferences are apparent in the chromatogram from mobile phase solvents, water and methanol.

In conclusion, the incompatibility between reversed-phase LC and FTIR seems to be largely overcome by the buffer-memory procedure with a stainless-steel wire net substrate. Even with the substrate, the concept of the buffer-memory technique remains the same; continuous infrared chromatograms can be detected without interference from the mobile phase. The substrate serves as a sample collector, concentrator, infrared cell, and storage device.

NORMAL-PHASE LC–FTIR-MICROPROBE LASER RAMAN SPECTROSCOPY

Although there are different molecular spectroscopic techniques, the most versatile ones for identification are IR and Raman spectroscopies. Even though these spectroscopies are powerful enough for the identification process in analysis, it is very hard to identify components in complex mixtures

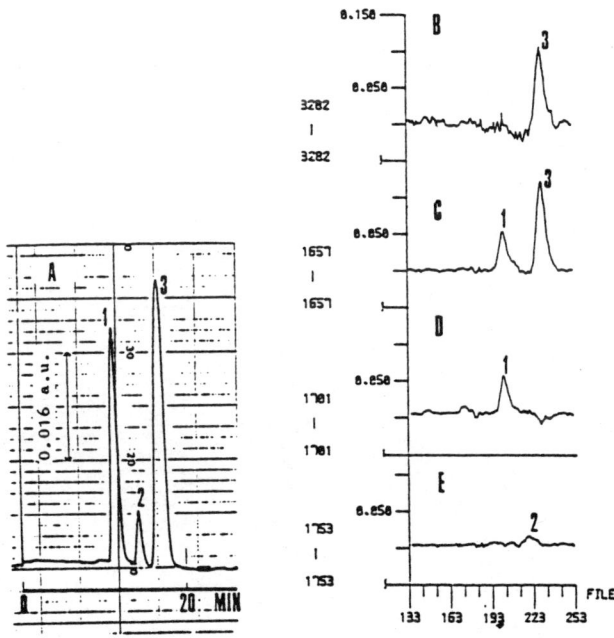

Figure 8. UV and FTIR chromatograms by the buffer-memory technique in reversed-phase micro-LC-FTIR. A: UV at 254 nm, B: FTIR at 3282 cm^{-1}, C: FTIR at 1657 cm^{-1}, D: FTIR at 1701 cm^{-1}, E: FTIR at 1753 cm^{-1}.

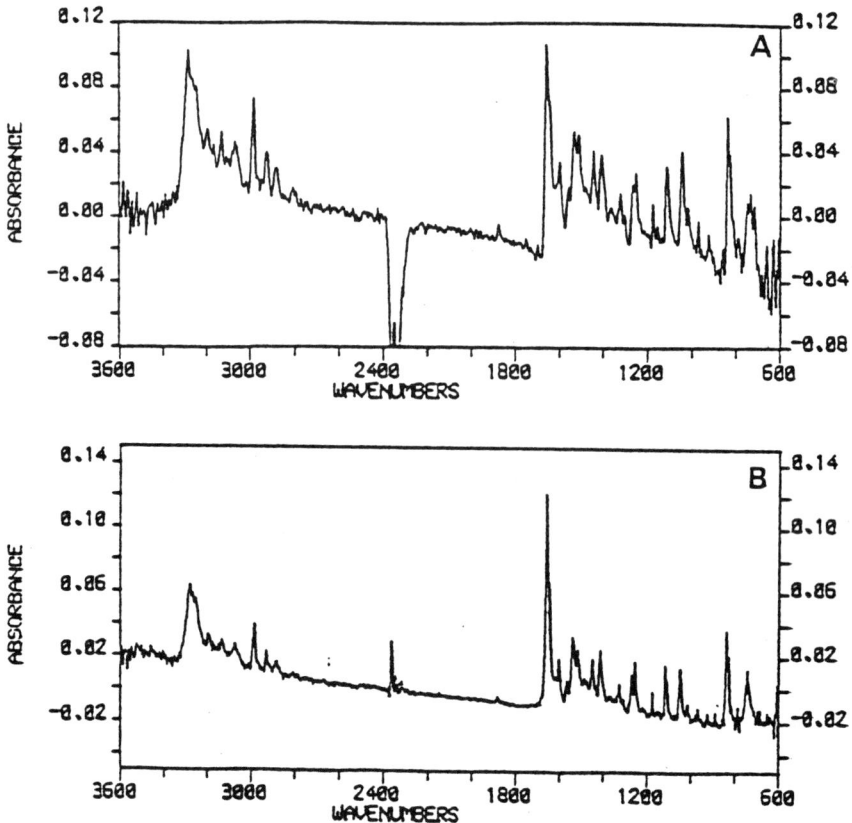

Figure 9. IR spectra of pehnacetin: A: spectrum obtained by the buffer-memory technique; B: standard spectrum.

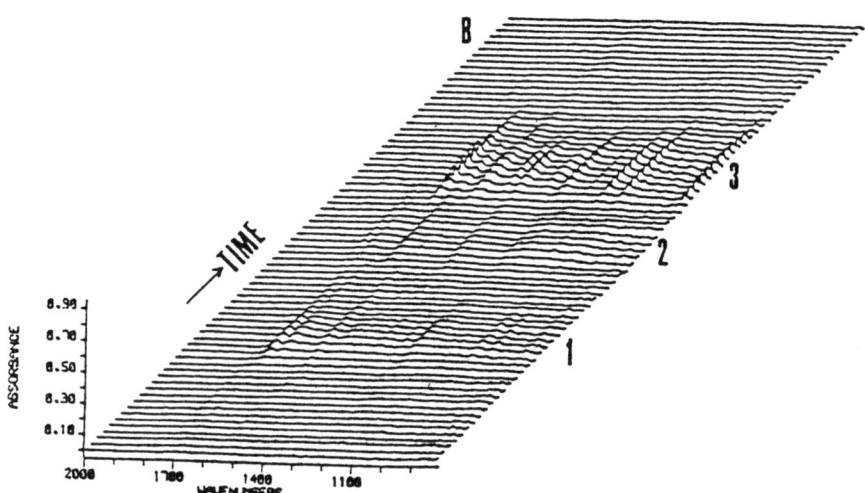

Figure 10. Three-dimensional chromatogram of the three components separation. (Copyright Elsevier, 1985.)

as they are without any pretreatments such as separation. Thus chromatography, as a separation tool, is regarded as the most important analytical technique in the past few decades. Techniques, which are generally a combination of chromatography and spectroscopy, attract much attention in practical analysis.

The merit of the buffer-memory technique, as previously mentioned, is that subsequent other spectroscopic analyses can be performed conveniently without any tedious procedures, and one of the examples shown has been for MS analysis (Fujimoto *et al.*, 1983). As a second demonstration to show the analytical potential of the buffer-memory technique, the combination of normal-phase LC, FTIR and microprobe laser Raman spectroscopy is described. This combination should be more informative for identification of components in mixtures than simple combined techniques such as LC—IR or LC—Raman.

The micro-LC consisted of a microfeeder MF-2, a model 7419 injector (Rheodyne, CA, USA) and a Jasco Uvidec 100-III UV detector (Jasco, Tokyo, Japan). Spectrometers used were a JIR-40X FTIR and a Jasco NR-1000 microprobe laser Raman spectrometer. A test mixture of di-*n*-propylketone DNPH and di-*n*-butylaldehyde DNPH was separated on a fused-silica capillary column of 0.53 mm i.d. × 250 mm long, packed with 5 μm silica (Develosil, Nomura Chemicals, Seto, Japan) with *n*-hexane/chloroform (7/3) at 2 μl/min. The collecting medium, KBr crystal plate, was transferred past the exit of the column for sample collection with the interfacing device shown in Fig. 1. Then the buffer-memory is brought into the FTIR spectrometer and afterwards into the Raman spectrometer.

As both DNPHs are easily detected by UV, IR and Raman spectroscopic measurements, they offer more precise identification of the components. Because, in conventional LC analysis with a UV detector, the only method for identification is by matching the retention time to the standard substance, except when one can use a UV multichannel detector where the UV spectra will be utilized for identification (Hoshino *et al.*, 1985, Jinno *et al.*, 1985b). This is the main reason why analysts want to combine chromatography to spectroscopy, to enhance the identification power in analysis.

The chromatograms monitored by UV at 254 nm and IR at 1621 cm^{-1} and 1260 cm^{-1} of two DNPHs are compared in Fig. 11. It is apparent that the trace of UV detection is almost consistent with those of IR detection, although close examination of Fig.11B indicates that the peak intensities of peak 1 and peak 2 are different at 1621 and 1260 cm^{-1}. To assign two peaks in the chromatogram, IR spectra of two peaks are usable. Therefore IR spectra of peaks 1 and 2 are called from the data file in the computer equipped with the FTIR spectrometer, as shown in Fig. 12. Using the IR data-interpretation system called Sigma-IR (Ishida and Sasaki, 1984), partially modified to make it compatible with the microcomputer in our laboratory, the partial structures listed in Table 2 should be present in the components of peak 1 and peak 2. From this information it is apparent that both peaks in Fig. 11 are DNPH derivatives. Raman information could

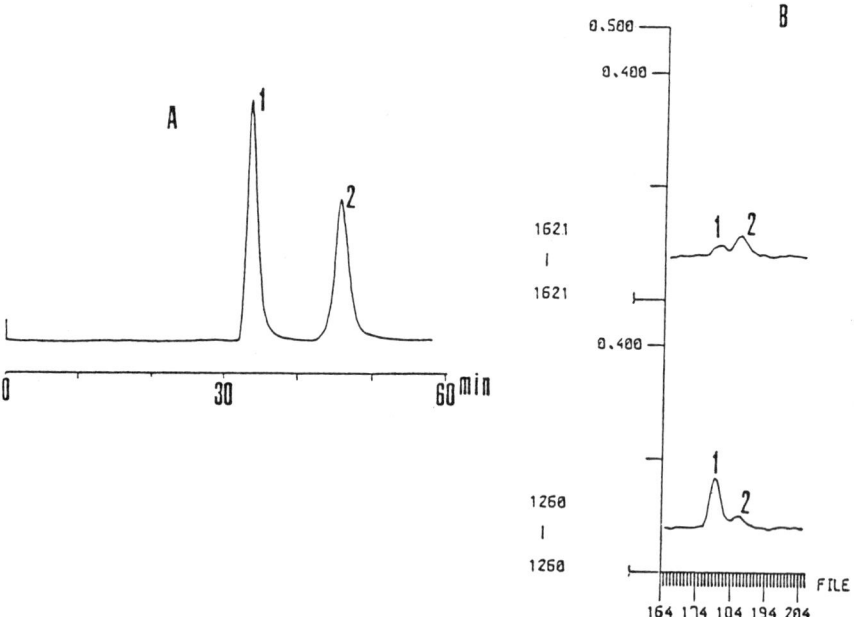

Figure 11. UV and FTIR chromatograms of di-*n*-propylketone DNPH and *n*-butylaldehyde DNPH: A: UV at 254 nm; B: FTIR chromatograms at 1621 and 1260 cm^{-1}.

support these identification processes in addition to IR information. Therefore, raman spectroscopic measurements were subsequently performed.

After the FTIR measurements the KBr crystal plate was transferred to the Raman spectrometer and the laser Raman spectra and raman chromatograms were measured. Figure 13 shows the typical three-dimensional laser Raman chromatogram of DNPHs in the wavenumber region between 1411 cm^{-1} and 1237 cm^{-1} obtained by moving the plate at a constant speed. Some differences in the Raman spectra are apparent, but it is still very difficult to assign these peaks. More interesting results to be noted are shown in Fig. 14, in which the contour plots by laser raman spectroscopy are demonstrated. These data clearly indicate the distribution of the separated components deposited on the crystal plate. After finding the most condensed part, i.e. high concentrations of the deposited components, microscopic observations can be performed to evaluate the crystal shape of the solutes. Figure 15 shows the microscopic picture of the peak 1 component. The crystal shape of this component is needle-like, while that of the peak 2 component cannot be defined well by this observation.

With complete solvent elimination in the buffer-memory technique, IR and raman spectra of submicrogram amounts of components separated by micro-LC can be measured. The obtained spectra, and other observations such as microscopy, offer important information for identification of the components.

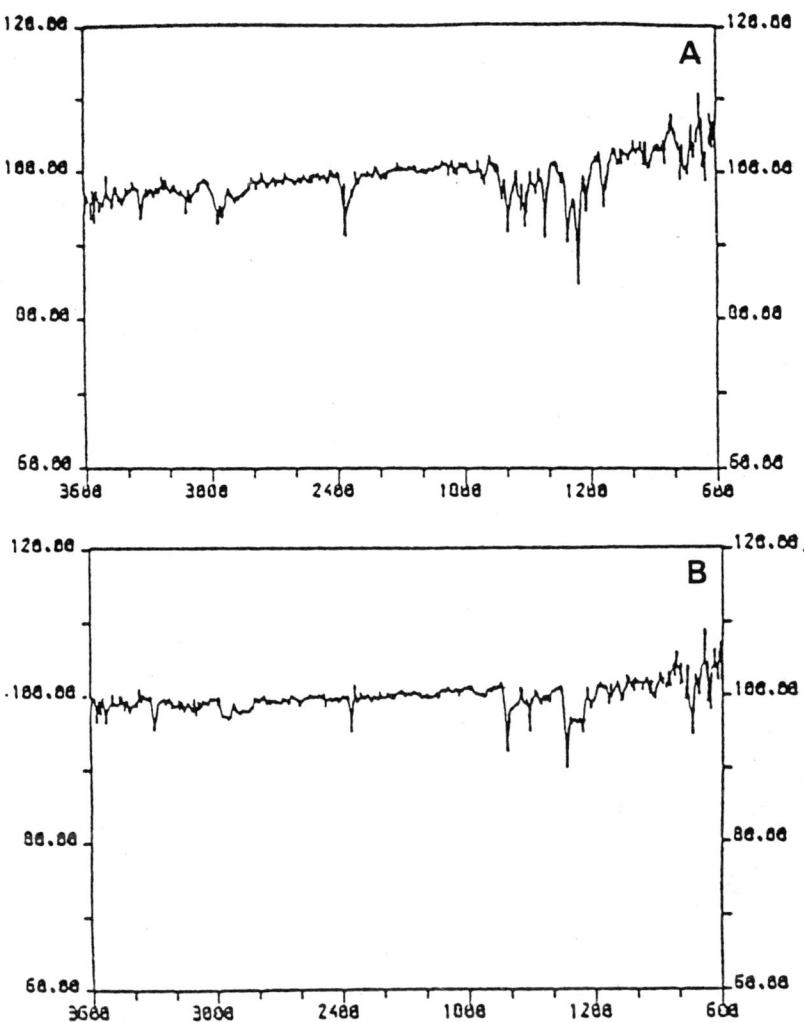

Figure 12. IR spectra of DNPHs: A: peak No. 1 in Fig. 11 (di-*n*-propylketone DNPH); B: peak No. 2 in Fig. 11 (di-*n*-butylaldehyde DNPH).

SFC−FTIR INTERFACING

The next demonstration is to show the feasibility of using the buffer-memory technique for SFC−IR.

Although SFC was described as early as 1962, by Klesper *et al.*, it has experienced rather slow growth and restricted acceptance as an analytical tool because of technological difficulties in handling supercritical fluids in a chromatographic system. However, the previous technical limitations have been overcome through the availability of high-pressure instrumentation developed for LC, encouraging results obtained by the use of open tubular

Table 2.
Assignments of partial structures in peaks 1 and 2 with sigma IR interpretation program

Assigned structure	Possibility
For peak 1	
**group: Double bond or aromatic ring group	
1 sub-st.:−CH=CH2	0.522
1 0 1629 0 913	
**group: Aromatic secondary amine group	
1 sub-st.:−NH−C6H4−	0.553
1 0 1597 1308 1254 0	
**group: Aromatic nitro compound	
1 sub-st.:−C6H4−NO2	0.668
1 1597 1308	
**group: Aromatic amide or anilide group	
2 sub-st.:−CO−N(H)− & −C6H5	0.716
2 3332 1629 1412 734	
3 sub-st.:−CO−N(H)−C6H4−	0.844
3 3332 1629 1529 1308 734	
4 sub-st.:−CO−N−C6H4−	0.805
4 1629 1496 1308 1064 734	
**group: Hydrazine group	
1 sub-st.:−NH−NH2 or >N−NH2	0.733
1 3332 1597 1412 0	
For peak 2	
**group: Hydrazine group	
1 sub-st.:−NH−NH2 or >−NH2	0.791
1 3302 1589 1417 824	
**group: Aromatic amide or anilide group	
2 sub-st.:−CO−N(H)− & −C6H5	0.715
2 3302 1614 1417 734	
3 sub-st.:−CO−N(H)−C6H4−	0.726
3 3302 0 1511 1327 716	
4 sub-st.:−CO−N−C6H4−	0.744
4 1614 1511 1327 1073 716	
**group: Primary amine group	
1 sub-st.:−NH2	0.553
1 0 3302 1589 1073 824	
**group: Aromatic nitro compound	
1 sub-st.: −C6H4−NO2	0.693
1 1614 1327	

columns and microcapillary columns packed with microparticles (Novotny and Ishii, 1985).

SFC has several advantages compared to GC and LC. As is well known, the properties of a supercritical fluid are intermediate between those of gases and liquids. Solute diffusivities are about 100 times higher in a super-critical fluid than in the corresponding liquid phase, and viscosities are

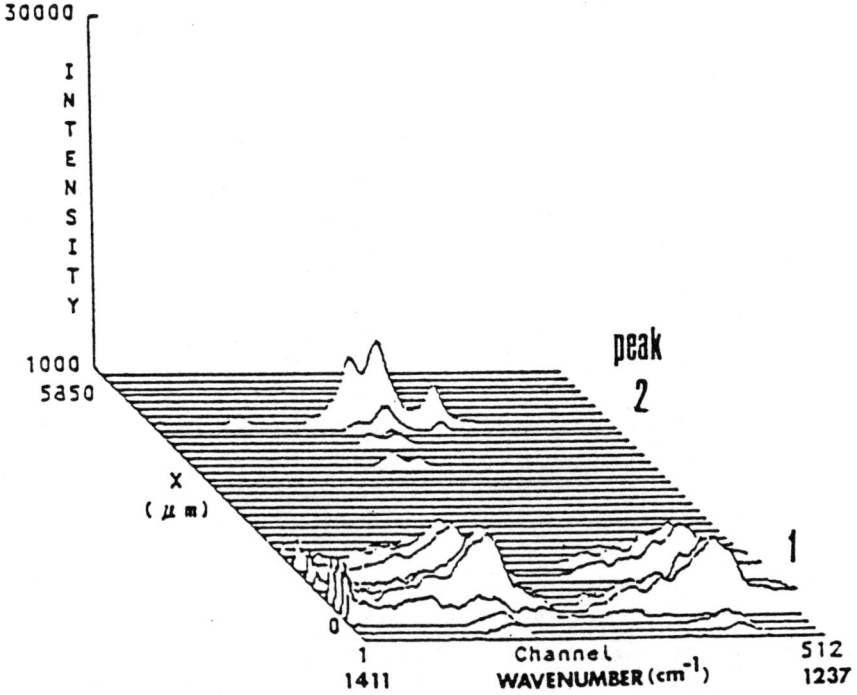

Figure 13. Three-dimensional laser Raman chromatogram of DNPHs: laser; Ar+ 5145 A, 40 mW (beam dia. 5–10 μm); peaks; 1 = di-*n*-propylketone DNPH, 2 = di-*n*-butylaldehyde DNPH; stacked condition; every 150 μm on the KBr plate.

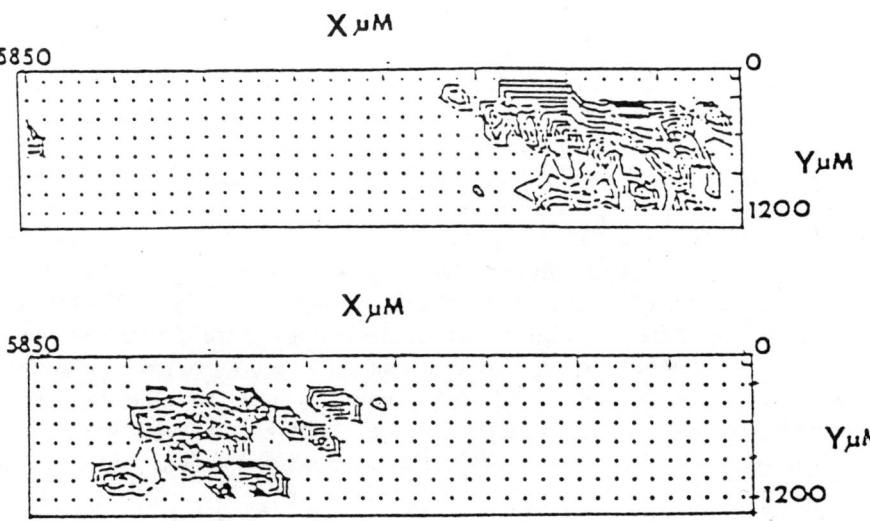

Figure 14. Contour plots of DNPHs by laser Raman spectroscopy. Peak channels 424 and 255 correspond to 1270 and 1324 cm^{-1}, respectively.

Figure 15. Microscopic picture of peak 1 deposited on the KBr buffer-memory plate.

similar to those in the gas phase. Furthermore, the greater density of supercritical fluids compared with gases imbues the mobile phase with solvating powers, which can readily be controlled by application of pressure. As a result, these properties should enable greatly enhanced chromatographic efficiency compared to LC, shorter analysis times than in LC, and the possibility of separating high-molecular weight and thermally labile compounds that cannot be separated by GC. However, as in the case of LC, detection is currently the weak point of SFC. Most GC and LC detectors have been used with SFC, including FID, thermal conductivity, UV, fluorescence and RI, but none of them have been used for identifying the eluted components. The most similar detector for SFC is a mass spectrometer, because of higher sensitivity compared with other detectors and its specificity in identifying unknown compounds, or for confirming the presence of suspected compounds. Therefore, a number of works have been published in recent years by Smith *et al.* (1982, 1983, 1984, 1985).

An alternative possibility is IR spectroscopy. This enables analysis of certain functional groups in the compounds' although the sensitivity obtained by IR spectroscopy is poorer than that achieved by MS. IR spectroscopy is more often complementary to MS than competitive. Shafer and Griffiths (1983) have reported the first results of FTIR detection in SFC using a high-pressure resistant flow cell and a mobile phase of carbon dioxide. More recently, a detailed discussion of this approach was provided by many groups (Johnson *et al.*, 1985, and Olesik *et al.*, 1984). However, as far as the flow-cell technique is concerned, the situation resembles that in LC. Measurements made by use of a flow cell show severe spectroscopic interference from the intense absorption bands of the mobile phase. Moreover, when pressure programming (which is equivalent to temperature programming in GC or gradient elution in LC) is performed in order to vary the solvent strength, solvent compensation is essentially impossible, since the absorptivities of the bands in the spectrum of a supercritical fluid increase with pressure.

A solution to this problem is to eliminate the mobile phase leaving the separated components for IR analysis. For LC–FTIR, at least two solvent-elimination techniques have been proposed, involving heating of an effluent containing the separated components and deposition of the concentrated sample onto a KBr plate for absorption spectroscopy (our buffer-memory technique), or onto a KCl powder for measurements by DRIFT. Excellent results using SFC—DRIFT have been published (Shafer *et al.*, 1986).

This section describes the first practical demonstration of SFC combined with IR spectroscopy with the buffer-memory technique for the analysis of oligomers. In this example a ratio-recording IR spectrometer is used, and the components are separated on a packed microcapillary column with a *n*-hexane-based solvent. An FTIR spectrometer is desirable because it has several distinct advantages over conventional dispersive instruments, but its use is often prohibited in many laboratories because of the high cost. The performance of the spectrometer used is essentially comparable to that FTIR spectrometers except for the longer scan time required. The SFC—IR system used is shown schematically in Fig. 16. It includes two pumping systems, a 7410 loop injector (1.0 µl, Rheodyne), a GC oven (Shimadzu, Kyoto, Japan) and a Jasco Uvidec 100-III detector. The mobile phase, *n*-hexane modified by ethanol, was delivered at room temperature into a fused-silica capillary column (0.5 mm i.d. × 100 cm long) packed with C-18 (Develosil, ODS-10) which was placed in the oven. Constant pressure operation was carried out solely by using pump A, which contained a stainless-steel coil partially filled with the mobile phase. The pressure programming was achieved by delivering the second mobile phase from pump B to pump A at a constant flow rate (about 1 ml/min), after setting the pressure of pump A and closing the valve located between a nitrogen cylinder and pump A, so that the pressure at the column inlet was allowed to increase with time. The flow rate of the mobile phase fed into the column was varied between 7 and 10 µl/min. The programming rate can be determined by the preset pressure of pump B or the amount of mobile phase stored in pump

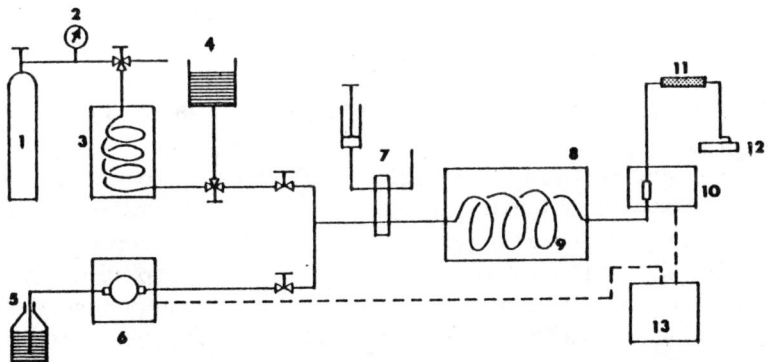

Figure 16. Schematic diagram of the buffer-memory technique for SFC-IR: 1 = nitrogen cylinder, 2 = pressure gauge, 3 = pump A, 4 and 5 = solvent reservoirs, 6 = pump B, 7 = sample injector, 8 = GC oven, 9 = packed capillary column (Develosil ODS-10, 100 cm × 0.53 mm i.d.), 10 = UV detector, 11 = restrictor (75 mm × 0.2 mm i.d. packed with Develosil ODS-5), 12 = KBr plate mounted on the interface, 13 = recorder. (Copyright Elsevier, 1985.)

A. The effluent from the column was transferred through a fused-silica capillary column, packed with Develosil ODS-5 (0.2 mm i.d. × 7.5 mm long), onto a KBr plate. This short column served as a pressure restrictor.

With the procedures described previously, the species eluted from the SFC column were deposited onto a continuously moving KBr plate. The mobile phase was evaporated on contact with the plate, leaving a permanent record of the solutes on the plate. When the chromatogram was complete the plate was simply transfered to an IR spectrometer and the IR chromatogram recorded by monitoring the absorption at a preset wavenumber. The chromatograms and spectra were obtained on a Jasco Model 810, double-beam ratio-recording IR spectrometer. It was fitted with a 3X beam condenser and a 4 × 1.7 mm aperture. The spectrum of each chromatography peak memorized on the plate was obtained as follows: a reference spectrum was obtained from ten scans of a virgin KBr crystal and then stored in the data system. At each peak maximum on the IR chromatogram the plate was stopped and the IR spectrum was scanned ten times. The spectra of both sample and reference were measured with spectral resolution: 5.4 cm^{-1} at 1000 cm^{-1}, and corrected for the absorption bands due to moisture on the beam condenser.

Carbon dioxide is the most convenient fluid for SFC, mainly because of its low critical temperature ($T = 31.3°C$). Throughout this work, n-hexane modified by 10% ethanol was used as the mobile phase. The critical temperature is 241.5°C, as obtained on linear interpolation with $T_c(n$-hexane$) = 234.2°C$ and $T = 243.4°C$. This mobile phase was chosen for several reasons. First, it is able to dissolve many oligomers even at room temperature, which enables sample injection, as well as UV flow cell detection, to be performed at that temperature. Consequently, the chromatographic instrumentation is simplified. Secondly, because it is a liquid at room temperature, the deposition of the effluent onto a KBr plate is easily accomplished. Thirdly, because it has a high vapor pressure under atmospheric conditions, the evaporation of the mobile phase is very easy. Finally, the utilization of the mobile phase prevents the generation of bubbles at the column exit; this is not the case when n-pentane modified by 10% ethanol is used. There may be a difference in selectivity between CO_2 and n-hexane, but there are no data to support this.

The first demonstration of SFC−IR is the analysis of styrene oligomers. Styrene A-500 was chromatographed under the conditions shown in the caption of Fig. 17. Then the effluent was continuously deposited onto a single KBr plate until eight peaks had been eluted. A typical UV and IR chromatogram are compared in Fig. 17. An IR chromatogram measured at 698 cm^{-1} which was generated from the buffer-memory looks suitable. It is seen that the resolution is somewhat diminished between almost all neighbouring peaks, but the 'chromatogram' is clearly memorized on the plate. The IR spectrum of the $n = 4$ peak (marked with an asterisk in Fig. 17B) is shown in Fig. 18. Despite the fact that the amount corresponding to this chromatographic peak is very small, as expected from the response of the

UV and IR chromatograms, this technique produces a high-quality spectrum, which is ideal for identification purposes.

The analytical potential of SFC is perhaps best realized when considering its applicability to a mixture of oligomers which span a wide range of molecular weights. Therefore the buffer-memory technique has been applied for the separation of DC-710 (Dow Corning, USA), which is one of the widely used GC stationary phases. DC-710 can be prepared by an equilibrium reaction involving cleavage and reformation of the siloxane bonds as shown at the top of Fig. 19, where *m* and *n* are equal to the number of difunctional units in the oligomer. Obviously, in addition to a linear polysiloxane, some cyclic oligomers result from residual starting material. Figure 19 shows the UV chromatogram of the sample where the separation was performed using essentially the same chromatographic conditions as those used for the separation shown in Fig. 15, except for the pressure applied.

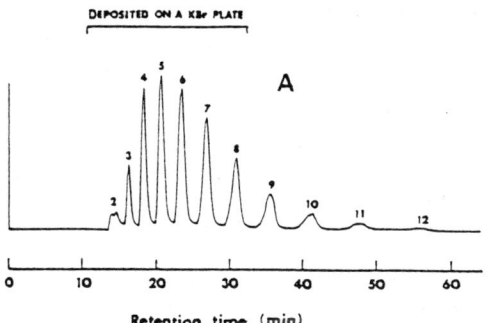

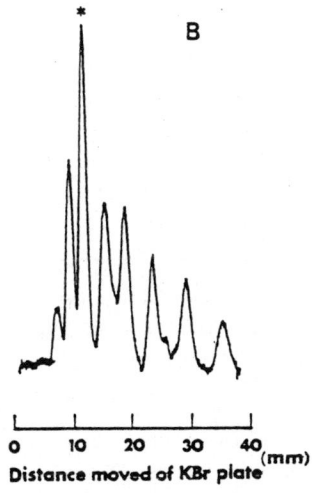

Figure 17. UV and IR chromatograms of polystyrene A-500: A: UV at 225 nm; B: IR at 698 cm⁻1; mobile phase: 10% ethanol in *n*-hexane; column temperature: 255°C. (Copyright Elsevier, 1985.)

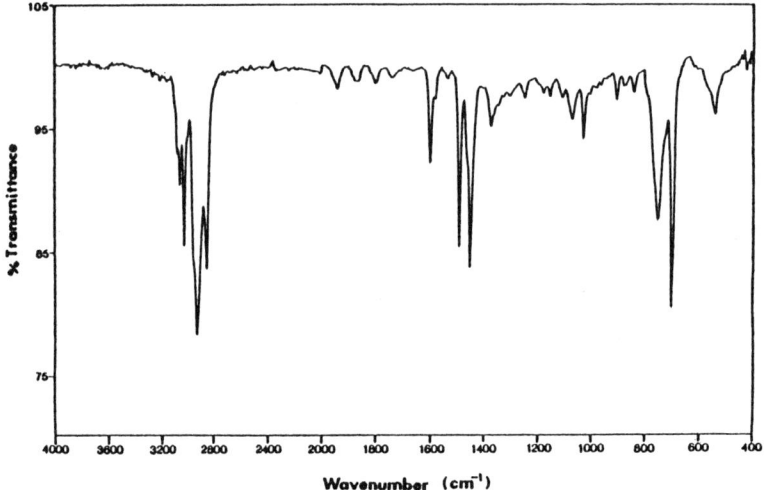

Figure 18. IR spectrum of the peak marked by the asterisk in Fig. 17B. (Copyright Elsevier, 1985.)

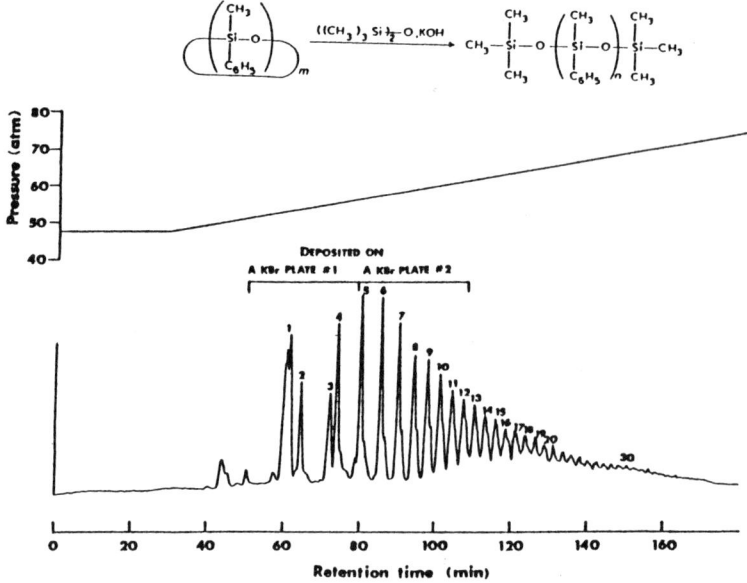

Figure 19. UV chromatogram of methylphenylpolysiloxane DC-710: column and mobile phase as in Fig. 17; column temperature: 260°C; detection: 254 nm. (Copyright Elsevier, 1985.)

The pressure was maintained isobarically at 47.5 atm for 30 min, after which it was programmed at 0.17 atm/min to 73.5 atm. It is seen that oligomers up to peak 33 are resolved, after which the trace returns to the baseline. On two sections of the KBr plate a limited number of oligomers, from peak 1 to 11, were deposited. The resulting IR chromatogram is shown in Fig. 20; the spectrometer was set at 1126 cm^{-1}. The chromatogram is similar to that in Fig. 19, except that the unresolved portion of the material decreased. The IR spectra corresponding to each peak deposited on the plate were scanned, and are shown in Fig. 21 for peaks 1—5. All of the peaks following peak 5 showed identical spectra to that of peak 5. It is interesting that for peaks 1 and 3 the intensity of the band at 1050 cm^{-1} relative to that of both the bands at 1126 and 1026 cm^{-1} is greater than in the case of peaks 2, 4 and 5. Also, the band at 842 cm^{-1} found for peaks 2, 4 and 5 is not present in the spectra of peaks 1 and 3. Therefore, peaks 2, 4 and 5 represent homologous species which are different from peaks 1 and 3.

Many reports have dealt with the spectra of polysiloxanes, but the vibrational assignments are by no means definite, even for simple oligomers such as dimethylpolysiloxanes. The absorption characteristic of the phenyl group in combination with a silicon atom can be seen at 1429 and 1124 cm^{-1}. The strong bands lying in the 1087—1025 cm^{-1} region are assigned to the anti-symmetric Si—O—Si stretching mode. Less certain are the methyl rocking and Si—C stretching modes, found between 842 and 698 cm^{-1}, since there is a great possibility of mixing of the vibrations. The absorption at 1260 cm^{-1} may be due to the methyl symmetrical deformation. Taking into account the fact that the bulk of the material is a linear polysiloxane, it can be concluded that peaks 1 and 3 are due to cyclic polysiloxanes, whereas peaks 2, 4 and 5 are linear ones. Also, neither peak 1 nor 3 seems to be a six-membered methylphenylsiloxane, in that an intense band at 1020—1010 cm^{-1}, expected for the cyclic trimers of siloxane oligomers, is not found in their spectra.

In summary, the present demonstration indicates that the buffer-memory technique is a powerful one for the SFC—IR analysis of oligomers. Unlike

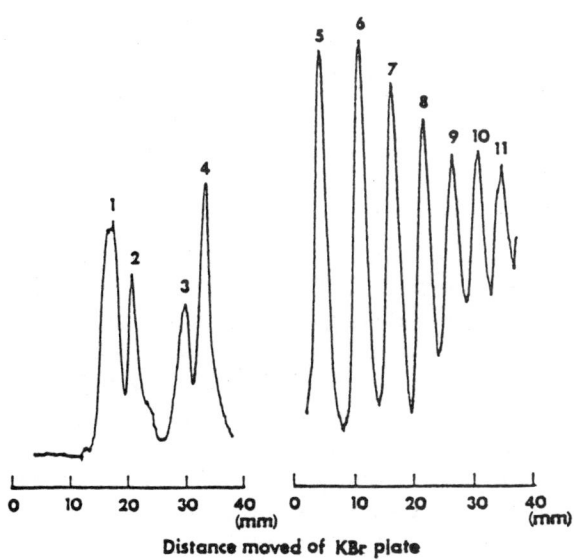

Figure 20. IR chromatogram of DC-710 measured at 1126 cm^{-1}. (Copyright Elsevier, 1985.)

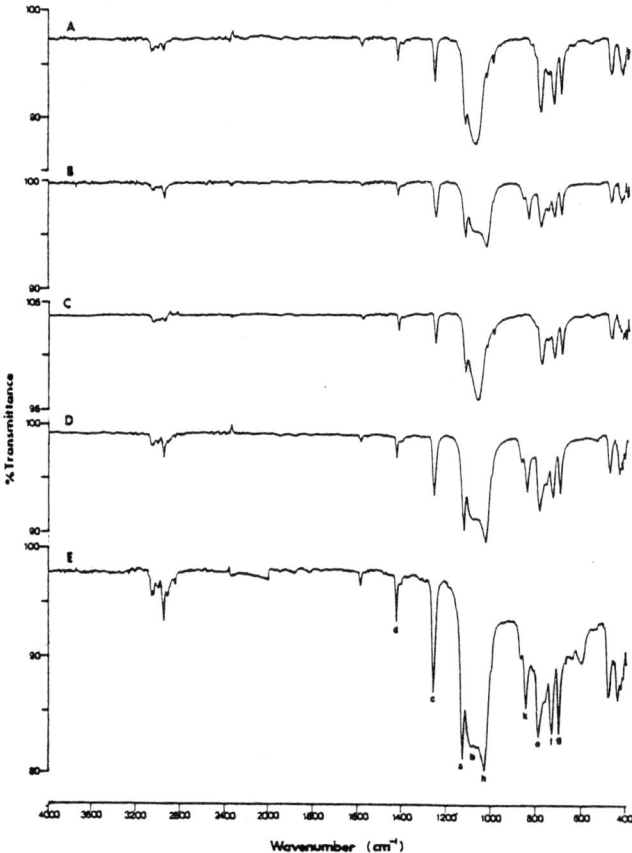

Figure 21. IR spectra of the components of DC-710: peak 1(A), 2(B), 3(C), 4(D) and 5(E); see Fig. 20 for peak numbers. (Copyright Elsevier, 1985.)

the flow-cell technique, it is compatible with a mobile phase containing a polar modifier.

CONCLUSION

The use of microcolumns does, of course, still present several disadvantages for LC–IR interfacings, although they offer various merits. The principal problem is the capacity of the microcolumns, which is often equal to, or even less than, the detection limits of the common IR spectrometer. It is well known that column capacity decreases in proportion to cross-sectional area; thus the capacity of a 0.5 mm i.d. column is about 80 times less than that of a 4.6 mm i.d. conventional column. To avoid this big problem, analysts should find the compromise which they consider more important, whether detection limit or much information. It is also necessary that commercial highly sensitive IR detectors should be available (much more sensitive than the present MCT detectors), in order to realize the merits of

LC—IR interfacings in practical applications. The authors believe, however, that realization will come in the near future.

ACKNOWLEDGEMENTS

The authors would like to express their sincere thanks to Y. Hirata and T. Oosuka of Toyohashi University of Technology for their collaboration in SFC—IR and reversed-phase LC—FTIR, respectively. We also acknowledge the assistances of T. Ikeda of Jasco, in microprobe laser raman spectroscopic measurements.

REFERENCES

Conroy, C. M., Griffiths, P. R., Duff, P. J., and Azarrage, L. V. (1984). Interfacing of a reverse-phase high performance liquid chromatography with a diffuse reflectance fourier transform infrared spectrometer. *Anal. Chem. 56*, 2636—2642.

Fujimoto, C., Jinno, K., and Hirata, Y. (1983). Liquid chromatography—spectrometry with the buffer-memory technique. *J. Chromatogr. 258*, 81—92.

Fujimoto, C., Hirata, Y., and Jinno, K. (1985a). Supercritical fluid chromatography—infrared spectroscopy of oligomers: use of buffer-memory technique. *J. Chromatogr. 332*, 47—56.

Fujimoto, C., Oosuka, T., and Jinno, K. (1985b). A new sampling technique for reversed-phase liquid chromatography/fourier transform infrared spectrometry, *Anal. Chim. Acta 178*, 159—168.

Griffiths, P. R., deHaseth, J. A., and Azarraga, L. V. (1983). Capillary GC/FT-IR. *Anal. Chem. 55*, 1361A—1387A.

Hoshino, T., Hondo, T., Senda, M., Saito, M., and Tohei, S. (1985). Quantitative deconvolution of heavily fused chromatographic peaks of biological compounds using a multiwavelength UV detector. *J. Chromatogr. 332*, 139—146.

Ishida, Y. and Sasaki, S. (1984). Application of membership function to automated structure analysis of infrared spectra of organic compounds, *Com. Enhan. Spectrosc. 1*, 173—184.

Jinno, K. (1981). The concept of non-dispersive infrared detector for micro high performance liquid chromatography using KBr buffer-memory technique. *Spectrosc. Lett. 14*, 659—663.

Jinno, K. and Fujimoto, C. (1981). Combination of micro high performance liquid chromatography and fourier transform infrared spectrometry using pottasium bromide buffer-memory technique, *J. High Res. Chromatogr. Chromatogr. Commun. 4*, 532—533.

Jinno, K., Fujimoto, C., Takeuchi, T., Ideriha, M., and Ishii, D. (1980). Investigation of buffer-memory of the chromatograms with KBr discs. *Bunseki Kagaku* (Japanese), *29*, 612—617.

Jinno, K., Fujimoto, C., and Hirata, Y. (1982a). An interface for the combination of micro high performance liquid chromatography and infrared spectrometry. *Appl. Spectrosc. 36*, 67—69.

Jinno, K., Fujimoto, C., and Ishii, D. (1982b). Buffer-memory technique for the combination of micro high performance liquid chromatography and infrared spectrometry, *J. Chromatogr. 239*, 625—632.

Jinno, K., Fujimoto, C., and Ishii, D. (1985a). The combination of micro-HPLC/IR spectroscopy. In: *Microcolumn Separations*, M. Novotny and D. Ishii (Eds). Elsevier, Amsterdam.

Jinno, K. Hondo, T., and Saito, M. (1985b). Retention prediction system coupled with an ultraviolet multichannel detector for identification of polycyclic aromatic hydrocarbons. *Chromatographia 20*, 351—356.

Johnson, C. C., Jordan, J. W., Taylor, L. T., and Vidrine, D. W. (1985). On-line supercritical fluid chromatography with fourier transform infrared spectrometric detection employing packed column and a high pressure lightpipe flow cell. *Chromatographia 20*, 717—723.

Kalasinsky, K. S., McDonald, J. T. Jr, and Kalasinsky, V. F. (1983). Development of an LC/FT−IR accessory for aqueous reverse-phase solvents. *FT−IR Spectral Lines 5*, 14−17.

Klesper, E., Corwin, A. H., and Turner, D. A. (1962). High pressure gas chromatography above critical temperatures. *J. Org. Chem. 27*, 700−701.

Kuehl, D. and Griffiths, P. R. (1979). Novel approach to interfacing a high performance liquid chromatograph with a fourier transform infrared spectrometer. *J. Chromatogr. Sci. 17*, 471−476.

Kuehl, D. and Griffiths, P. R. (1980). Microcomputer-controlled interface between a high performance liquid chromatograph and a diffuse reflectance infrared fourier transform spectrometer. *Anal. Chem. 52*, 1394−1399.

Novotny, M. and Ishii, D. (1985). *Microcolumn Separations*. Elsevier, Amsterdam.

Olesik, S. V., French, S. B., and Novotny, M. (1984). Development of capillary supercritical fluid chromatography/fourir transform infrared spectrometry. *Chromatographia 18*, 489−495.

Shafer, K. H. and Griffiths, P. R. (1983). On-line supercritical fluid chromatography/fourier transform infrared spectrometry. *Anal. Chem. 55*, 1939−1942.

Shafer, K. H., Pentoney, S. L., and Griffiths, P. R. (1986). Supercritical fluid chromatography/fourier transform infrared spectrometry with an automatic diffuse reflectance interface. *Anal. Chem. 58*, 58−64.

Smith, R. D., Wright, B. W., and Udseth, H. R. (1986). Capillary supercritical fluid chromatography and supercritical fluid chromatography/mass spectrometry. In: *Chromatography and Separation Chemistry*, S. Ahuja (Ed.). ACS Symposium Series No. 297. American Chemical Society, Washington, DC.

Author index

Abu-Zeid, M. M., 181

Doehl, J., 53

Fujimori, H., 181
Fujimoto, C., 273
Fuke, A., 181

Gemmel, B., 73
Goto, A., 181
Greibrokk, T., 53

Hirai, K., 181
Hondo, T., 87

Ishii, M., 181

Jinno, K., 157, 273

Kalinoski, H. T., 111
Klesper, E., 3
Kojima, K., 211

Leyendecker, D., 3

Leyendecker, D., 3
Lundanes, E., 53

Nagatsu, T., 211

Parvez, H., 181, 211, 229
Parvez, S., 211

Saito, M., 87
Saito, T., 25
Schmitz, F. P., 3, 73
Senda, M., 87
Sugimoto, T., 181
Sugiyama, K., 87
Smith, R. D., 111

Takeuchi, M., 25

Wright, B. W., 111

Yamada, K., 181
Yoshida, A., 229
Yoshioka, M., 181, 229

Subject index

Adenine, 186, 189—192, 201
Adenine nucleoside, 181
Adenine nucleotide, 181
Adenosine, 181, 192, 201, 203
Adenylate cyclase, 186, 188, 202
ADP, 181, 192, 193, 197—201
Alumina, 233
AMP, 191—193, 200, 201
ATP, 181, 188, 191—193, 197—201
Alachlor, 145
Alkaline phosphatase, 186, 188, 201
Anthracene, 48

Biphenyl, 48
Blood, 189, 197
Boric acid gel, 233
Brain, 229, 247, 255
Breadth—length ratio, 161
Bromoacetaldehyde, 185—187, 201, 205
Brucine, 146
Buffer—memory plate, 276, 288
Buffer—memory technique, 275, 281, 282, 285

Caffeine, 103
Capillary column, 118, 135
Carbamate pesticide, 144
Carbazole, 151
Cascade pump, 25, 30—35, 37, 38, 40, 42—44
Catecholamine, 211, 213, 229, 234, 236, 248, 252, 255
Cerebrospinal fluid, 229, 247, 252
Coal tar, 123
Coffee bean, 103—106
Collision-induced dissociation, 151
Column connection, 68
Column dimension, 67
Column packing, 56
Column switching, 60
Compressibility, 25, 29, 31—33, 36, 40, 50
Compression pump, 29, 34
Compression value, 39
Computer-assisted, 157, 168

Connectivity index, 161
Correlation coefficient, 163
Correlation factor, 161
Critical density, 113
Critical pressure, 5, 113, 114
Critical temperature, 5, 19, 113
Cyclic AMP (cAMP), 181, 187, 191—193, 202

Damping device, 27, 36, 37
Density, 32, 33, 40, 46, 47, 112, 113
Deoxy-ADP, 203
Deoxy-ATP, 203
Diesel engine particulate matter, 165, 169
Diesel fuel, 147, 148, 150, 151
Diffusion rate, 112
Dimethoxystrychnine, 146
Direct fluid injection, 140
Diuron, 145
Double-plunger pump, 36

Eluent, 3, 16, 17, 19, 32, 33, 75, 76, 78, 82
Epinephrine, 202
Extraction vessel, 89

Flame ionization detection, 144
Flame ionization detector, 56
Flow diagram, 34
Flow rate, 29, 37, 41—44, 50, 133
Fluoranthene, 48
Fluorene, 48
Fluorescence detector, 56, 188
Fluorescent reaction, 187
Fourier transformed infrared spectroscopy (FTIR), 277, 281, 285
Fused silica capillary, 125
Fused silica capillary tubing, 127
Fused silica column, 69, 70

Gradient, 48
Gradient elution, 73, 81, 83

HVA (homovanillic acid), 229, 230
5-Hydroxytryptamine (serotonin, 5-HT), 229, 234, 236, 248, 255

Inductive interaction, 161
Infrared spectroscopy (IR), 273, 276, 282, 286, 293, 294
Injector, 55
Interface, 138, 139, 285
Isobaric condition, 45, 46

Laser Raman chromatogram, 287
Lemon peel, 90, 93, 94, 95, 98
Linuron, 145

Mass spectrometry, 122, 136, 138
Measured chromatogram, 168, 174
Measured retention, 165
Metabolite, 234, 248, 252, 255
Metering pump, 29, 30, 34, 38, 39, 42, 49
Microbore, 211, 213, 217, 224, 225
Microcolumn, 273
Mobile phase velocity, 57
Modifier, 13, 62, 63, 66, 101, 121
Multiwavelength UV detector, 92, 106, 169

Na, K-ATPase, 188, 200
Naphthalene, 48
Neuroblastoma, 230, 235
Neuroblastoma cell, 190, 199
Non-polar-bonded phase, 59
Non-silica hydrophobic material, 59

Oligomer, 73, 78
Oligostyrene, 81
Ouabain, 200

Packed column, 53
Packed fused silica column, 67
Packing pressure, 195
Peptide, 211, 217, 221
Phenacetin, 282
Piston pump, 55
Plasma, 198, 229, 246, 248
Plate height, 57
Plunger pump, 26, 27, 50
Plunger volume, 28, 36, 37, 41—43, 44
Polar-bonded packing material, 58
Polarizability, 164
Polycyclic aromatic hydrocarbon (PAH), 60, 63, 64, 158, 163, 165, 167, 169, 171, 177
Polyethylene glycol, 142
Polymeric mixture, 141
Polystyrene, 57, 182, 235
Polystyrene oligomer, 43, 45, 46
Position sensor, 41
Predicted chromatogram, 168, 174

Predicted retention, 165
Pressure, 3, 9—11, 18, 20, 26, 29, 31, 36—40, 43—46, 48, 50, 77, 96, 97, 105, 114, 117, 118, 120, 125, 150, 163, 164
Pressure gradient, 49, 54, 63
Pressure ripple, 35
Pressure sensor, 41
Protein, 211, 217, 221
Pyrene, 48

Real-time display, 50
Resolution, 83, 194, 195
Restrictor, 56, 69, 125—127, 136
Retention time, 47, 48
Reversed-phase column, 63

Selectivity, 121
SFC—MS interface, 132, 133, 142—144
SFE/SFC, 101, 102, 104, 106, 107
Semi-microbore, 196
Silica, 58
Sodium chloride, 193, 196
Sodium fluoride, 202
Solubility, 114
Supercritical fluid chromatography, 99, 103
Supercritical fluid chromatography—mass spectrometry (SFC—MS), 111, 115, 123, 125, 128—131, 134—136, 140, 145, 146—148, 157
Supercritical fluid expansion, 124
Supercritical fluid extraction (SFE), 87, 88, 90—93, 95, 99, 103, 104, 106
Supercritical fluid extraction—mass spectrometry, 149
Supercritical fluid injection, 138
Styrene oligomer, 77, 80

Temperature, 3, 8—16, 18—20, 40, 77, 80, 93, 96, 97, 105, 114, 120, 125, 194
Temperature program, 81
Temperature sensor, 41
Thermal expansion, 49
Thermal expansion coefficient, 33
Thermionic detector, 56
Thiram, 146

Urine, 229, 236
UV multichannel detector, 92, 106, 169

Van der Waals volume, 161
Viscosity, 32, 33, 38, 112, 125
VMA (vanillylmandelic acid), 229, 230

Wheat, 148